Three-dimensional Analysis
of Human Locomotion

ISB International
Society
of Biomechanics

This is the second volume in the International Society Biomechanics Series, published in association with The International Society of Biomechanics, University of Calgary, Canada.

Three-dimensional Analysis of Human Locomotion

Edited by

PAUL ALLARD

Laboratoire d'Étude du Mouvement, Research Center, Sainte-Justine
Hospital, Montreal, and University of Montreal, Canada

AURELIO CAPPOZZO

Cattedra di Tecnologie Biomediche, Università degli Studi, Sassari,
Italy

ARNE LUNDBERG

Department of Orthopaedic Surgery, Huddinge University Hospital,
Sweden

and

CHRISTOPHER L. VAUGHAN

Department of Biomedical Engineering, University of Cape Town,
South Africa

JOHN WILEY & SONS
Chichester · New York · Weinheim · Brisbane · Singapore · Toronto

National   01243 779777
International (+44) 1243 779777
e-mail (for orders and customer service enquiries): cs-books@wiley.co.uk
Visit our Home Page on http://www.wiley.co.uk
or http://www.wiley.com

Reprinted August 1998

Published in association with The International Society of Biomechanics (ISB),
University of Calgary, Canada.

Other Wiley Editorial Offices

John Wiley & Sons, Inc., 605 Third Avenue,
New York, NY 10158-0012, USA

WILEY-VCH Verlag GmbH, Pappelallee 3,
D-69469 Weinheim, Germany

Jacaranda Wiley Ltd, 33 Park Road, Milton,
Queensland 4064, Australia

John Wiley & Sons (Asia) Pte Ltd, 2 Clementi Loop #02-01,
Jin Xing Distripark, Singapore 129809

John Wiley & Sons (Canada) Ltd, 22 Worcester Road,
Rexdale, Ontario M9W 1L1, Canada

Library of Congress Cataloging-in-Publication Data

Three-dimensional analysis of human locomotion/edited by Paul Allard ... [et al.].
    p. cm.
  Includes bibliographical references and index.
  ISBN 0-471-96949-4 (alk. paper)
  1. Human locomotion. 2. Three-dimensional imaging in biology.
I. Allard, Paul, 1952-.
  [DNLM: 1. Locomotion–physiology. 2. Gait–physiology.
  3. Computer Simulation. 4. Biomechanics. WE 103 T531 1997]
  QP303.T585 1997
  612.7'6–dc21
  DNLM/DLC
  for Library of Congress   97–11505
                                                             CIP

British Library Cataloguing in Publication Data

A catalogue record for this book is available from the British Library

ISBN 0 471 96949 4

Typeset in 10/12pt Times by Keytec Typesetting Ltd, Bridport, Dorset.
Printed and bound in Great Britain by Bookcraft (Bath) Ltd.
This book is printed on acid-free paper responsibly manufactured from sustainable forestry,
in which at least two trees are planted for each one used for paper production.

Contents

Contributors

MARK F. ABEL — Department of Orthopedic Surgery, University of Virginia, Charlottesville, VA, USA

RACHID AISSAOUI — Laboratoire d'Étude du Mouvement, Research Center, Sainte-Justine Hospital, 3175 Côte Ste-Catherine, Montreal, PQ, H3T 1C5, Canada; and Department of Physical Education, University of Montreal, PQ, Canada

PAUL ALLARD — Laboratoire d'Étude du Mouvement, Research Center, Sainte-Justine Hospital, 3175 Côte Ste-Catherine, Montreal, PQ, H3T 1C5, Canada; and Department of Physical Education, University of Montreal, PQ, Canada

SCOTT BARNES — Department of Biomedical Engineering, Ohio State University, 206 West 18th Avenue, Columbus, OH 43210-1107, USA

MARC BÉLANGER — Département de Kinanthropologie, Université du Québec à Montréal, Montréal, Canada

NECIP BERME — Department of Biomedical Engineering, Ohio State University, 206 West 18th Avenue, Columbus, OH 43210-1107, USA

F. L. BUCZEK — Motion Analysis Laboratory, Shriner's Hospital for Children, 1645 West 8th Street, Erie, PA, USA

M. J. BURGESS-MILLIRON — Converse Inc., One Fordham Road, North Reading, MA 01864-2680, USA

ANGELO CAPPELLO — Dipartimento di Elettronica, Informatica e Sistemistica, Università degli Studi di Bologna, Viale Risorgimento 2, 40136 Bologna, Italy

AURELIO CAPPOZZO — *Cattedra di Tecnologie Biomediche, Università degli Studi, Viale San Pietro 43c, 07100 Sassari, Italy*

P. R. CAVANAGH — *The Center for Locomotion Studies, Pennsylvania State University, University Park, PA 16802, USA*

DIANE L. DAMIANO — *Department of Orthopedic Surgery, University of Virginia, Charlottesville, VA, USA*

B. L. DAVIS — *Department of Biomedical Engineering, Lerner Research Institute, The Cleveland Clinic Foundation, 9500 Euclid Avenue, Cleveland, OH 44195-5254, USA*

ROY B. DAVIS — *Newington Children's Hospital, Gait Analysis Laboratory, Newington, CT 06111, USA*

UGO DELLA CROCE — *Cattedra di Tecnologie Biomediche, Università degli Studi, Viale San Pietro, 43c, 07100 Sassari, Italy*

PETER A. DELUCA — *Connecticut Children's Medical Center, Hartford, CT, USA*

JACQUES DUBOY — *UMR 6630 CNRS, Université de Poitiers, Poitiers, France*

MORRIS DUHAIME — *Laboratoire d'Étude du Mouvement, Research Center, Sainte-Justine Hospital, 3175 Côte Ste-Catherine, Montreal, PQ, H3T 1C5, Canada; Orthopedic Surgery, Shriner's Hospital, Montreal, PQ, Canada and McGill University, Montreal, PQ, Canada*

SANDRO FIORETTI — *Dipartimento di Eletronica ed Automatica, Università degli Studi di Ancona, Via Brecce Bianche, 60131 Ancona, Italy*

HANS FURNÉE — *Delft University of Technology, Motion Study Laboratory, PO Box 5046, GA Delft, The Netherlands*

MARK D. GRABINER — *Department of Biomedical Engineering, Lerner Research Institute, The Cleveland Clinic Foundation, 9500 Euclid Avenue, Cleveland, OH, 44195-5254, USA*

EWALD M. HENNIG — *Biomechanik Laboratorie, Sport FB-2, Universität Essen, PF 103-764, 43000 Essen, Germany*

ALAIN JUNQUA — *Laboratorie de Metallurgie Physique, Faculté des Sciences, Université de Poitiers, 40 Avenue du Recteur Pineau, 86022 Poitiers Cedex, France*

REGIS LACHANCE

Laboratoire d'Étude du Mouvement, Research Center, Sainte-Justine Hospital, 3175 Côte Ste-Catherine, Montreal, PQ, H3T 1C5, Canada; and Department of Physical Education, University of Montreal, Montreal, PQ, Canada

PATRICK LACOUTURE

UMR 6630 CNRS, Université de Poitiers, Poitiers, France

MARIO A. LAFORTUNE

Nike Sport Research Laboratory, 1 Bowerman Drive, Beaverton, OR 97005, USA

ALBERTO LEARDINI

Laboratorio di Analisi del Movimento, Istituti Ortopedici Rizzoli, Via di Barbrano 1/10, 40136 Bologna, Italy

FRANCK LEPLANQUAIS

UMR 6630 CNRS, Université de Poitiers, Poitiers, France

WEN LIU

Department of Mechanical Engineering and Institute of Biomedical Engineering, Drexel University, 32nd and Chestnut Streets, Philadelphia, PA 19104, USA

GLEN A. LIVESAY

Musculoskeletal Research Center, Department of Orthopedic Surgery, University of Pittsburgh, E1641 Biomedical Science Tower, 200 Lothrop Street, PO Box 71199, Pittsburgh, PA 15213, USA

LUIGI LUCCHETTI

Istituto di Fisiologia Umana, Università degli Studi 'La Sapienza', Piazzale Aldo Moro 5, 00185 Rome, Italy

ARNE LUNDBERG

Department of Orthopaedic Surgery, Karolinska Institute, Huddinge University Hospital, S-141 86 Huddinge, Sweden

BRADFORD J. MCFADYEN

Département de Physiothérapie, Faculté de Médecine, Université Laval, 525 Boulevard Wilfrid-Hamel, Québec, Canada

ELENA OGGERO

Department of Biomedical Engineering, Ohio State University, 206 West 18th Avenue, Columbus, OH 43210-1107, USA

SYLVIA ÕUNPUU

Connecticut Children's Medical Center, Hartford, CT, USA

GUIDO PAGNACCO

Department of Biomedical Engineering, Ohio State University, 206 West 18th Avenue, Columbus, OH 43210-1107, USA

JOHN P. PAUL *Bioengineering Unit, Wolfson Centre, University of Strathclyde, 106 Rottenrow, Glasgow G4 0NW, UK*

THEODORE W. RUDY *Musculoskeletal Research Center, Department of Orthopedic Surgery, University of Pittsburgh, E1641 Biomedical Science Tower, 200 Lothrop Street, PO Box 71199, Pittsburgh, PA 15213, USA*

THOMAS J. RUNCO *Musculoskeletal Research Center, Department of Orthopedic Surgery, University of Pittsburgh, E1641 Biomedical Science Tower, 200 Lothrop Street, PO Box 71199, Pittsburgh, PA 15213, USA*

HEYDAR SADEGHI *Department of Physical Education, University of Montreal, Montreal, PQ, Canada, and Laboratoire d'Étude du Mouvement, Research Center, Sainte-Justine Hospital, 3175 Côte Ste-Catherine, Montreal, PQ, H3T 1C5, Canada*

SORIN SIEGLER *Department of Mechanical Engineering and Institute of Biomedical Engineering, Drexel University, 32nd and Chestnut Streets, Philadelphia, PA 19104, USA*

W. E. THORNTON *701 Coward's Creek Road, Friendswood, TX 77546, USA*

CHRISTOPHER L. VAUGHAN *Department of Biomedical Engineering, University of Cape Town Medical School, Observatory, Cape 7925, South Africa*

SAVIO L-Y. WOO *Musculoskeletal Research Center, Department of Orthopedic Surgery, University of Pittsburgh, E1641 Biomedical Science Tower, 200 Lothrop Street, PO Box 71199, Pittsburgh, PA 15213, USA*

1

Instrumental Observation of Human Movement: Historical Development

AURELIO CAPPOZZO[1] AND JOHN P. PAUL[2]

[1]Cattedra di Tecnologie Biomediche, Università degli Studi, Sassari, Italy
[2]Bioengineering Unit, University of Strathclyde, Glasgow, UK

PRIOR TO HISTORY

Recent findings indicate that the first ape-like creatures began to walk upright at least 4 million years ago (*Nature,* August 1995). Although the most intriguing question, why they stood up, remains unanswered, it may be said with confidence that the locomotor function in humans as well as in all animal species has been shaped by the evolutionary history and, in particular, by the fact that the probability of survival of each individual species and that of its progeny depends upon the capacity to adapt its locomotion to the environment and relevant ecological scenery.

It therefore goes without saying that both the shape and the function of a living organism largely depend on its proper locomotor mechanisms and that the comprehension of the latter helps in understanding the former. For this reason, and perhaps because movement is immediately discernible, its intimate nature and the causes of it have puzzled the human mind since antiquity. This applies to biological movements as well as to movements of inanimate bodies.

Three-dimensional Analysis of Human Locomotion. Edited by P. Allard, A. Cappozzo, A. Lundberg and C. Vaughan
© 1997 John Wiley & Sons Ltd. ISBN 0 471 96949 4

MOVEMENT AS A CONCEPT

The Greek philosophers approached the nature of movement by thought. They held that the senses deceive and therefore no experimental method can lead to truth. This was the case with one far-seeing exception: the Kos Medical School, founded by Hippocrates, which claimed the opposite, i.e. experience is the way of reaching truth.

The great interest that Greeks had in movement is well shown in this sentence by Heraclitus (540–?475 BC) 'we cannot know the water but only its perpetual movement'.

Lorini, Bossi and Specchia (1992), in a learned dissertation on the concept of movement according to Greek philosophy, concluded that the latter may be summarized as follows:

- Movement is the characteristic that distinguishes what is alive from what is not alive.
- Movement is harmony (the heavenly spheres, music).
- Only the heavenly spheres autonomously possess perpetual movement.
- Space and time are not absolute concepts but only philosophically detectable bodies. Based on Greek philosophy, the Romans constructed their enlightened pragmatism.

Quoting again Lorini, Bossi and Specchia (1992), we may say that they did it 'with the simplicity of a scientist and the perfection of a poet'.

The great Christian thinkers faced the phenomenon of movement having well in mind, of course, the imperfection and limits of human intelligence: 'If no one asks me, I know what it is; if I wish to explain it to him who asks me, I do not know' (from St Augustine's Confessions, AD 400). The following sentences define the cultural path with respect to the concept of movement:

> The Creator . . . moves the slowest bodies and halts those which are too fast, brings back to the right path those which have strayed . . . (Severino Boethius, 480–524)

> The mover gives what he has to the one who is moved in that it causes him to be in motion. (St Thomas Acquinas, 1226–1274)

> Every body has an 'impetus' which permits it to continue to move. Being moved is passing from potential to the act . . . (William Ockam, 1290–1349)

Pierre Gassendi (1592–1655) goes as far as this: *ambulo ergo sum* (I walk, therefore I am).

Artists have always sensed imminent cultural revolutions and have often given the first impulse to them. Vasari writes about a depiction of an equestrian monument of John Hawkwood, painted by Paolo Uccello in 1436: 'This work was and still is held to be very beautiful for a painting of that kind, and if Paolo had not made that horse move its legs on one side only, which naturally horses do not do or they would fall ... this work would have been absolutely perfect' (Dagg, 1977). Looking at Michelangelo's Creation of Adam in the Sistine Chapel, or at Bernini's statues, is like catching a glimpse of a scene in motion, like looking at a frame of a film. It is not shape alone any more; it is the relationship between shape and time, between anatomy and function. Leonardo da Vinci, a contemporary of the two above-mentioned artists, interprets this relationship in a sublime manner. We may say about him what we have already said about the Roman thinkers: he looked at nature and described it with the simplicity of a scientist and the perfection of a poet.

In 1564 Michelangelo died and Galileo Galilei was born. The intimate nature of movement, of any movement, was disclosed, and the fact that mere thought cannot lead to truth but only experimental observation which can be repeated and tested was finally established.

THE SCIENCE OF BIO-MOVEMENTS: THE BEGINNING

One of Galileo's disciples, Giovanni Alfonso Borelli (1608–1680), postulated that every function in the living body, animal or vegetable, manifests itself through movement: macroscopic and apparent, as in locomotion, or microscopic, on an atomic dimension, as in the movement by which atoms come in contact to form living matter. He approaches 'the difficult matter of Physiology as part of the Physics, adorned and enriched with mathematical demonstrations'. Clearly stated in this phrase is Borelli's fundamental belief that the same laws which govern the inanimate world are applicable to living matter and those laws can be expressed through mathematical equations. Borelli accomplished his findings using a rudimentary microscope, in collaboration with Malpighi, a ruler and a balance. He analysed the complex anatomical reality using simplified physical models and described relevant phenomena using geometry and mathematics. One of his most interesting findings was that the skeletal segments are subjected to loads which may reach magnitudes that far exceed those of the external forces (Borelli (1680, 1681), translated into English and published in 1989 (Borelli, 1989)).

One hundred years later, a milestone in the cultural development of human-kind was produced by a relatively small number of enlightened men, *les illuminés: L'Encyclopedie*, the encyclopedia by Diderot and D'Alambert. This was published starting in 1751 and distributed throughout Europe. From beginning to end this work represents one unbroken process of exaltation of scientific knowledge on the one hand and peaceful industry on the other. As biomechanicians we are bound to be interested in the word *mouvement:* 'movement, says the encyclopedia, is the action of a living body which is necessary for the conservation of its health; the lack of movement as well as the excess of it are extremely prejudicial to the body'. This definition and the concept of adaptation to function which it implies was not at all obvious in the eighteenth century. In fact, this is still one of the most challenging issues in biomechanics.

In 1894, Wilhelm and Eduard Weber's book was published (Weber and Weber, 1894). This book has been translated by Maquet and Furlong (1992). This review of the work of the Weber brothers is conducted with the benefit of the translation. The Weber brothers based their book on the principles of physics and anatomical dissection studies. Their measuring equipment for gait analysis comprised a telescope with a glass scale, so that the movements of a test subject a fixed distance away could be measured in the telescope. Despite the lack of sophistication of this measurement system when compared to current optoelectronic measuring systems, which can also be supplemented by ground-to-foot force transducers, the conclusions which the Weber brothers drew still form the basis of much analysis of locomotion at the present day. Naturally, because of the lack of instrumentation they were mistaken in some of their concepts and findings, and their few errors will be mentioned during the course of this chapter. Work on cadavers was conducted at the Anatomy Institute of Leipzig and the physical measurements were made using instruments from Gottingen.

The brothers were ahead of modern thinking because they proposed the concept that the activities which they were measuring would be undertaken by their test subjects in such a way that the amount of muscular effort would be minimized. They were also concerned that in their day the spine and the pelvis were conceived to be positioned mainly in a straight line, and their own measurements of the inclination of the pelvis are frequently mentioned in the book. In their survey of the theory of walking and running, they start by dividing the body into supporting and supported parts, commenting on the very mobile connection between trunk and legs and erroneously stating that in locomotion the legs oscillate on the trunk like a pendulum. The useful feature of the legs is that they can lengthen and shorten by considerable amounts by extending or flexing the knee. It was inferred that in walking the legs support the trunk not only by the rigidity of the bones but also by the development of force in the muscles, and they discuss in depth the muscles which flex and extend the joints of the leg. Probably because their observations were mostly taken from a lateral position, their discussion of locomotion corresponds to sagittal plane movements only and they thus make no reference to the abductor muscles of the hip and their importance in these activities. In their studies of walking and running, they define stance and swing for each leg and identify periods of double support in walking and of double swing in running. They have the interesting concept that in movement the trunk sits on the femoral head with no applied moment (their words are different) but, as with an inverted pendulum, the inclination of the trunk is such as to balance any acceleration or external loads on it. It is correctly stated that gait velocity can be increased by making quicker and longer steps, and distinctions between the characteristics of slow and rapid walking are clearly made. They determined the amount of the vertical oscillations of the body and also measured the foot-to-ground clearance, which they state as being one-ninth of leg length. The vertical oscillation, they say, is about 32 mm and does not change markedly with walking speed. They are aware that anterior–posterior force between ground and foot in the single support phase of walking would introduce rotation of the trunk unless the support area of the foot is vertically below the centre of mass. The tendency to rotation of the trunk due to acceleration of the swinging leg, they state, is balanced by alternate swinging of the arms.

For running, the Webers convey many interesting pieces of information such as the following. The step length is twice as long as in walking and the cadence is 1.5 times that of walking. The rise and fall of the trunk in running is 20–30 mm and the stance duration is between 0.25 and 0.33 s. The double swing phase is of duration 0.1 s, and of this 0.067 s is the period during which the height of the centre of mass diminishes by about 22 mm. Additionally, in running the step duration can be modified less and the step length more than in walking. The Webers define springing as differing from quick running in that it enables one to carry out slower and longer steps than are possible in walking, and they conclude their examination of springing by indicating that it is to quick running what solemn stepping is in relation to rapid walking.

The second part of the Webers' book relates to the anatomical structures of the body and is not of direct relevance to the study of locomotion. It is interesting, however, to note that they are measuring the height of the vertebral bodies to a second decimal place in millimetres, and angles of flexion and extension to the first decimal place in degrees, and, having undertaken a test with a tilting board to determine the position of the centre of mass of a human volunteer, they quote the height of the centre of mass from the heel as 947.7 mm! It is difficult to imagine how such precision was attained.

The second half of the book relates to experiments and measurements concerning walking and running. They report that their measurements were undertaken on a horizontal path protected from wind in a large room allowing a length of more than 43 m to be used for the measurements. Some experiments, however, were undertaken by measurement in the open air. The timing of events was conducted by the test subject, who carried the chronometer in his hand; this was started by the test subject as he reached the beginning of the measurement section, and the experimenter also counted the number of steps taken along the measured length. This brings to mind some current assessments of the progress of rehabilitation in which the test subject is challenged by requiring to undertake mental calculations or to change locomotion on receipt of a signal, either audible or visual. They also attempted to make sure that the tests were conducted 'blind' by not calculating any of the derived results before the end of the whole series of tests.

In the description of walking, we feel that there may have been some difficulties in expression of terms either by the Webers or their translators in respect of the activity. For instance, in the measurements relating to the vertical movements of the body in walking, they distinguish between the mean of 31.7 mm in 'striking ground with the whole foot' and 21.0 mm in 'striking the ground with the tip of the foot'. There seems to be no reference to present concepts of normal gait, in which 'heel strike' is a specific instant. This is confirmed by Figure 1, illustrating successive positions of leg segments during walking, with the ground contact being exclusively on the forefoot. Interestingly, this corresponds to an initial knee flexion and hip flexion angle of approximately 40°, which is greatly in excess of what we consider to be 'normal' gait at the present time. This is achieved by having the plantar surface of the foot more or less parallel to the ground at the end of swing phase and again involving a large angle of knee flexion as a result. In their discussion of the swing phase of locomotion, the authors state categorically that the movement by which the leg overtakes the supporting leg during swing phase is 'not produced by muscular force but by gravity alone'. They support the contention by describing an

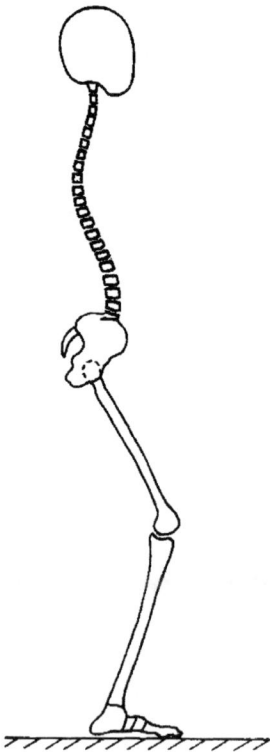

Figure 1. A diagram of the skeletal structure during walking estimated from limb length dimensions by Weber and Weber (1894). Reproduced from *Mechanics of the Human Walking Apparatus* by kind permission of the translators Maquet and Furlong, and of the publishers Springer Verlag, Berlin

experiment in which the leg was made free by the test subjects standing on a raised surface and bending forwards. The normal diminution of the range of angular movement was restored at intervals by externally applied force, and the authors claim that this did not affect the cycle time of the phase when it is applied. Neverless, electromyography (EMG) studies in the twentieth century show peaks of activity of the flexor muscles starting the swing and extensor muscles terminating the swing. The authors claim additionally that the maximum possible speed in walking depends on the length of the legs and the speed at which they swing under the action of their own weight. A theory of walking is presented based on three tenets:

1. That the vertical component of force in the leg is sufficient but no higher than is necessary to maintain the middle point of the body in the same horizontal plane.
2. That the line of action of the force on the legs always passes through the middle point of the body and the point of support of the stance-phase foot.
3. That the stance leg is vertical (interpreted as the line from ankle to hip) at the instant of toe-off of the other leg.

Tenet 1 is, of course, Newton's first law applied to the equilibrium of vertical components of forces. Tenet 2 relates to the moments corresponding to tilting the body forwards and backwards being zero. In normal locomotion the trunk does undertake cyclical forward and backward bending, and thus this is not exactly true. Tenet 3 is not supported by the film data in Eberhart (1947).

The Weber brothers offer no guidance on the movements of the body from side to side, although in their historical review they indicate that both Borelli (1680) and Gassendi (cited by Weber and Weber, 1894) undertook simple tests which illustrated the side-to-side movement of the head during walking. The Weber brothers report the work of Gerdy (1829) but do not take account of many of Gerdy's statements, such as those corresponding to the tilting of the pelvis on a cyclical basis in the frontal plane and the alternate rightwards and leftwards movements of the trunk during the walking cycle.

Thus the Weber brothers, with primitive equipment, inferred many of the aspects of the body's movement in walking, and the acquisition of high-technology equipment has not made a correspondingly large contribution to our knowledge of this complex phenomenon.

THE GRAPHIC METHOD

Dans une étude sur la locomotion, c'est le nom de M. le professeur Marey qui doit etre prononcé le premier. (In any study about locomotion, the first name mentioned must be that of Professor Marey) (Richer, 1901)

Etienne Jules Marey (1830–1904) writes:

> When a variation occurs which, because of its smallness or its rapidity, we cannot
> perceive it or when it develops in such a slow manner that it challenges our patience;
> when in any case, for its own nature, the variation escapes our appreciation, then it
> is needed, if possible, to apply to a recording device. ... When the eye stops seeing,
> the ear stops hearing, the sense of touch stops feeling, or when our senses give us
> erroneous appearances, then these devices are as new senses and have an astonishing
> precision. (Marey, 1878)

In these sentences, the seed of a renewed scientific revolution resides. As
mentioned by Marey himself, the first reported recording device appears to be
the 'anémographe' described by the Marquis of Ons-en-Bray in the *Mémoire de
l'Académie,* 1734, which was able to trace curves on a sheet of paper rolled
around a turning cylinder. Further developments of this device were achieved at
the end of the eighteenth century by Rutherford, MacLennan and Watt, with the
aim of recording the variation versus time of temperature, atmospheric pressure,
wind velocity, air humidity, mechanical work performed by a steam engine, etc.
However, the first application in physiological sciences is due to the German

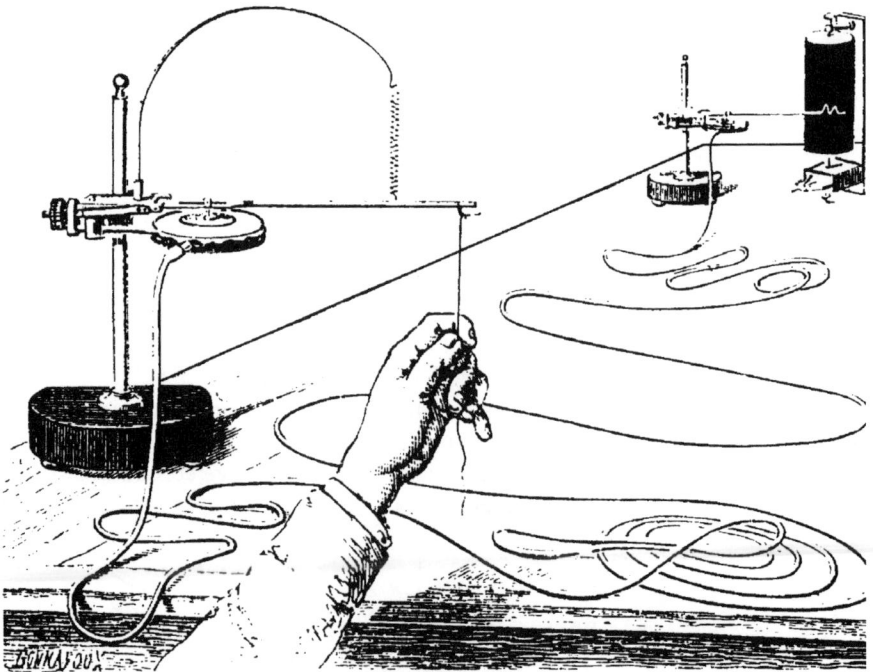

Figure 2. A kymograph with the device for transmission of movement ('exploring drum'
and the 'scribing drum'). From Richer (1901)

scientists Karl Ludwig, in 1847, with his 'kimographion', and Vierordt of Tubingen in 1857 with his 'sphygmographe', both for the recording of arterial pressure. The common problem exhibited by these devices was inertia and unwanted vibrations.

Marey, with the collaboration of his student Carlet (Carlet, 1872), made extensive use of these recording devices, improved their accuracy and associated them with either mechanical or electrical time-marking devices (*chronographes*). However, the most intriguing invention of this eclectic scientist is the *tambour á levier*, often referred to as Marey's drum. Two of these devices, the 'exploring drum' and the 'scribing drum', linked by a rubber tube filled with air, permit the transmission of a movement through a distance (Figure 2): the displacement of the 'exploring lever' is faithfully reproduced by the movement of the 'scribing lever', which in turn is recorded on a kymograph.

Using the above-described system, Marey could record the movement of points of the human body along a given direction during gait. It is evident that, in order to achieve this, a fixed base of support had to be provided to the exploring drum. When, for instance, the movement of a point defined on the pelvis had to be recorded during walking, the exploring drum was placed on an arm rotating on a plane at a given height. The arm was pushed by the subject while walking in a circle. An example of a trace thus obtained is shown in Figure 3.

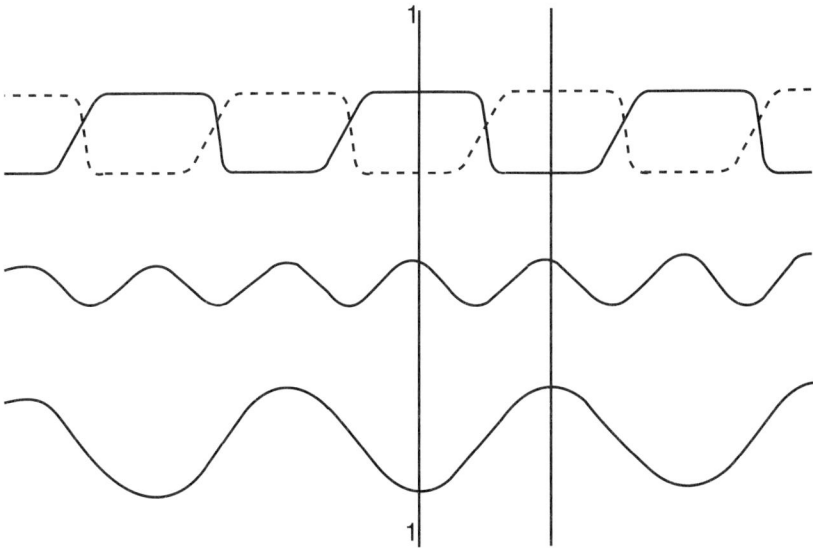

Figure 3. Upper curves represent stance (high) and swing (low) phases of the two limbs. The middle curve depicts the vertical oscillations of the pubis. The lower curve depicts the lateral oscillations of the pubis. Adapted from Marey (1873)

In Figure 4 the very famous *chaussure exploratrice* is depicted. With it the temporal factors of different locomotor acts were extensively investigated. A disputable attempt was also made to interpret the resulting traces as carrying information concerning the vertical component of the ground reaction (Marey, 1873, p. 120). Another device worthy of mention is also depicted in Figure 4 and enlarged in Figure 5. With it Marey recorded the vertical displacement of the head during walking without the need for a fixed reference point. He is somewhat vague in the illustration of this device:

> We shall not enter into the details of the experiments which have been used to verify the accuracy of the traces obtained with this device; these consisted in the

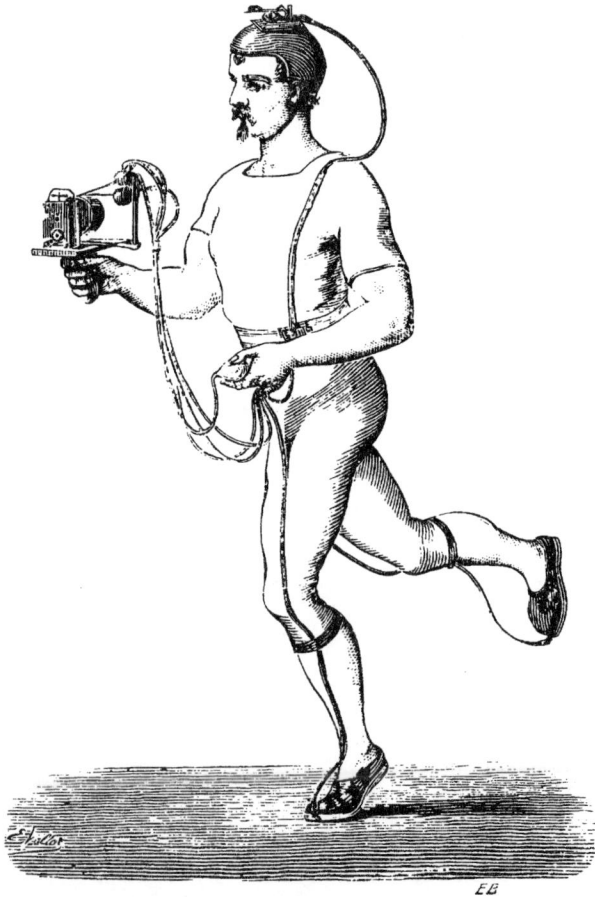

Figure 4. Experimental set-up for the recording of gait phases and the vertical movement of the vertex of the head. From Marey (1873)

Figure 5. Head vertical displacement transducer (see also Figure 4). From Marey (1873)

graduation of the weight of the leaded disk and of the elasticity of the drum membrane, until the movements impressed to the device faithfully represented those reported in the trace.

An analysis of this system shows that it may be modelled as a mass linked to the base of support through a spring and a dash-pot in parallel. The second-order equation which describes the movement of the mass (output) when the base of support undergoes oscillation (input) shows that if the frequency of this oscillation is lower than the natural frequency of the system, then the output of the transducer is proportional to the acceleration of the input movement. When, on the contrary, the device is made to oscillate at higher frequencies, then the output is proportional to the input displacement. Thus, Marey had to tune mass and elastic coefficient so that the oscillations associated with the locomotor act under analysis fell in the latter range. An example of results obtained with the above-mentioned experimental set-up is given in Figure 6.

PHOTOGRAPHY

P. J. C. JANSSEN

'*La plaque photographique est la véritable rétine du savant*' [The photographic plate is the real retina of the scientist]: this is the intuition of Pierre-Jules-César

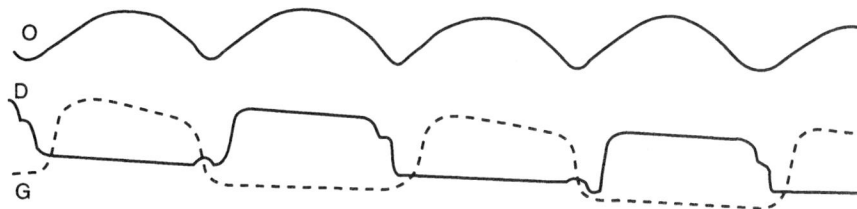

Figure 6. Curves obtained during running and using the experimental set-up shown in Figure 4. Adapted from Marey (1873)

Janssen (1824–1907), a French–Norwegian astronomer, the inventor of the photographic revolver. He started this project in 1873 and used the device for the first time in Japan in the following year to record the passage of Venus in front of the sun. A photographic sequence of 48 frames was taken at an approximate rate of a frame every second. The exposure time of each frame was 1.5 s. This is an anticipation of what is called today 'time-lapse cinematography', through which the timescale of slow phenomena is shrunk and the movement made visible.

EADWEARD MUYBRIDGE

Meanwhile, the USA started to play its provocative and stimulating role in the field of biomovements. On 19 October 1878, *Scientific American* published Muybridge's well-known photographic sequence of a horse in motion.

Eadweard Muybridge (1830–1904) was born in Kingston-upon-Thames, near London, a member of the Muggeridge family and given the name of Edward James. He lived a very adventurous life during which he changed his surname first to Muygridge and then to the one he is known by now. The name Eadweard derives from an archaic version of 'Edward', with reference to a Saxon king.

In 1872, Leland Stanford, former Governor of California and enlightened horse breeder, asked Muybridge to photograph his famous horse, The Californian Wonder. The first attempts to photograph this horse in motion took place at Sacramento. The results, though little better than silhouettes, were apparently sufficient to settle an old controversy by showing that a trotting horse did, at certain stages of its movement, have all four feet off the ground at once. The work continued in 1877 with another horse, Occident. However, no original negatives or photographs of these first attempts have been traced, though drawings exist which may be based on them. One of these was published in the 3 August 1877 issue of *High California*. This was accompanied by a caption saying that the original photograph, called an 'automatic electro-photograph', had been exposed for 1/2000 of a second, which was a remarkable achievement at that time; so much so that doubts were cast on the veracity of this statement. The other interesting aspect of this experiment is the electromagnetic mechanism which triggered the shutter. A more elaborate investigation, again financed by Stanford, took place at Palo Alto in 1878–1879. Twenty-four cameras were placed in line, at a distance apart of 0.53 m, in a long shed, and in front of each was a special shutter triggered electromagnetically by the horse or the wheels of a sulky as they made contact with wires stretched across the track. Photographs were taken at intervals of 40 ms (25 frames per second) and the exposure of each was about 1/2000 of a second. The method was successful in achieving the first photographs of sequences of movement, and results were first published, as already mentioned, in the summer of 1878 under the title of 'Horses in motion'. In a subsequent publication by Muybridge, 'Attitudes of Animals in Motion'

(1881), the photographic sequences were mostly of horses, but included some other animals and a few human athletes as well. The photographs were highly praised in the USA and Europe, and artists and scientists were swift to appreciate their potential significance (see Muybridge 1955, 1957, 1972, 1979).

Dr Marta Braun has recently examined Muybridge's plates most carefully and has shown how he renumbered his photographic sequences or pasted in other images, in order to give the illusion of continuity despite a faulty sequence, where the failure of one or more units in his battery of cameras had occurred. Marta Braun writes: 'the relationship of the images, in some 40 percent of the plates, is not what Muybridge states it to be' (Braun, 1993). However, this circumstance does not reduce the historical importance of Muybridge's work.

E. J. MAREY

Etienne Jules Marey (1830–1904) saw Muybridge's photographic sequences in the 14 December 1878, issue of *La Nature* and immediately understood the importance of photography for scientific purposes. It was thus in Paris, at the College de France, that physiologists first used photography as a major instrument for investigating biomovement. It was Marey, with the collaboration of Georges Demenÿ, who invented photochronography on a still plate (later called chronophotography). His first instrument was the 'photographic gun', with which he was able to take a sequence of 12 photographs on a single rotating disc at an exposure time of, first 1/720 (*La Nature*, 22 April 1882), and, later, of 1/1440 of a second. His favourite subjects were birds in flight, but he made pictures of other animals and humans during locomotion. However, the intermittent nature of the disc motion caused vibrations and, therefore, moved images. In addition, pictures were too small to allow proper analysis. Based on this experience, Marey developed a new chronophotographic device which exhibited a stationary and adequately large plate and a shutter made of a continuously rotating disc with one or more slits. By setting the number and dimensions of the slits and the rotation velocity of the shutter, both exposure time and frame rate could be regulated. Since the plate was exposed several times, the ambient light had to be obscured, the background black and the moving figure, or relevant details of it, white. A ruler and a clock were also photographed on the same plate, allowing for both space and time calibration. The limit of this device was, of course, the overlapping of the figure images, which imposed a limit on the picture rate.

In 1885 Marey published a second edition (Marey 1885) of *The Graphic Method,* augmented with a supplement on 'The Development of the Graphic Method Through the Use of Photography', in which he reported the above-mentioned experiences.

A further development of chronophotography was achieved by the use of a device with two side-by-side plates and two objectives but one rotating shutter.

Even images were thus obtained on one plate and odd images on the other, allowing for the doubling of the picture rate.

Another problem posed by these chronophotographic devices was solved. When the subject moved towards the camera, overlapping of the images made the recording impossible to analyse. Marey placed a rotating mirror within the camera so that images were projected onto the plate side by side, avoiding overlapping.

Marey also tackled the problem of the three-dimensional reconstruction of a figure in motion. In fact, he performed experiments in which three cameras with orthogonal optical axes were used, thus obtaining the projection of the figure onto three planes. However, due to technical difficulties, he could not use the cameras simultaneously and nor could he devise an analytical model for three-dimensional (3D) reconstruction. With this experimental set-up he collected sufficient information to mould sequential plastic models of a seagull in flight which anticipate the dynamic sculptures of the Italian futurists.

In 1887 the Eastman company made available on the market a photosensitive strip of paper which could be rolled on a spool, allowing a camera to be loaded once for a series of photographs. Marey took immediate advantage of this product and, it might be said, invented the cine camera. On 29 October 1888, Marey, at the Académie des Sciences of Paris, said: 'I have the honour of presenting today a strip of sensitive paper on which a series of images have been obtained in the measure of twenty per second,' and he added 'Film photochronography [*la chronophotographie à pellicule*] contains the solution of all problems of physiology, physics or of mechanics in which the position of bodies in different points of space and at equal intervals of time must be determined.' In the photochronograph film the strip of paper could reach, through an intermittent motion, a speed of 1.60 m/s, and its dimensions were 9 cm by approximately 1 m. The device was portable and had dimensions of only 18 × 24 cm.

An interesting iconography relating the work of both Muybridge and Marey is reported in Tosi (1992) and Bouisset (1992).

A. LONDE

At La Salpetrière, no longer a gunpowder factory but a mental hospital, at the end of the last century, Dr Charcot gave his lectures on various nervous and mental illnesses. They must have been fascinating lectures; it was, in fact, common for the upper classes to attend them. Very likely, the charm of these lectures also derived from the photographic material provided by Albert Londe, which was projected for the public. The photographer Londe, a most enthusiastic follower of Marey, was the initiator of what we would call today the audiovisual department of the above-mentioned hospital. He documented with a series of 12 photographs the movements of neurological patients. Pictures were taken on a still plate through 12 objectives in an interval of time equal to or greater than

1/10 of a second. Dr Charcot interpreted these photographic sequences and used them for clinical as well as for educational aims, a remarkable example of movement analysis for clinical purposes.

THE THREE-DIMENSIONAL POINT POSITION RECONSTRUCTION

BRAUNE AND FISCHER

The first investigators to perform a comprehensive analysis of the 3D movements of parts of the human body were Braune and Fischer from Leipzig. They started publishing their work in the *Proceedings of the Royal Saxon Society of Sciences* in 1895. Subsequently, five papers were published in 1899, 1900 and 1901, and in 1904 by Fischer alone, after Braune had died. These works were translated into English and published in 1987 (Braune and Fischer, 1987).

These authors performed three tests on one male young adult subject while walking at a natural speed of progression. In two experiments the subject was unloaded, and in the third one he was loaded with the 23.3 kg of field equipment of German infantrymen. The subject was dressed in a black jersey suit similar to the one used by Marey. Eleven Geissler tubes were strapped to the following body segments: the head, right and left thigh, shank, foot, upper arm and forearm. Black varnish was used to establish interruptions in the line of white in the pictures which represented marker points. These corresponded to the joint centres and to the segment centres of gravity. A point indicating the vertex of the head was also marked. Further markers were made available along the thigh tube to allow reconstruction of the hip marker in those frames where it was hidden by the forearm. The tubes were connected to wires which were gathered in the back of the subject and made to hang from the ceiling. It took from 6 to 8 h to dress the subject.

Two sets of two still cameras were used. Each pair was placed on one side of the subject. Calibration of the photogrammetric system was carried out by photographing on the same plate as the actual walking trial a network of 1-cm squares printed on glass. Experiments were carried out in July 1891. During the experiments the tubes were lit 26.09 times per second.

The image coordinates of the markers were measured on the photographic plates using a device specially built for this purpose. Using a stereophotogram-metric model and the measured image coordinates of the markers, the instanta-neous 3D positions of the markers were reconstructed. The positions of the markers on the Geissler tubes enabled Fischer to determine the positions of points which would better approximate the joint centres. Hip and shoulder joints were assumed to lie on the line connecting the respective markers. The other joints were assumed to lie on lines passing through the relevant markers and perpendicular to the plane of progression. Relevant distances were measured. Fischer acknowledges the error intrinsic in this procedure and alludes to the fact that, for a better estimation, tubes should have also been placed on the medial side of the segments. The instantaneous position of the segment centre of gravity markers, and of the vertex of the head marker, and the positions of the tips of the feet, were also reconstructed.

From the above information the following kinematic quantities were calculated for the three walking experiments: velocity and acceleration of the centres of gravity of individual body segments, trajectory, velocity and acceleration of the total body centre of gravity. Differentiation was carried out using a graphical method. In addition, joint kinematics were assessed by projecting a stick model of the skeletal structure onto the plane of progression and onto the quasi-frontal plane perpendicular to this latter plane.

The following dynamic quantities were also estimated: the three components of the resultant ground reaction force, inertia forces and inertia couples acting on the lower limb segments and relevant intersegmental loads. The ground reaction force and the intersegmental loads were limited to the sagittal plane and to the single support phase, since, as underlined by the authors, the values of ground reactions acting on individual limbs were not available. The inertial parameters of the body segments involved, which were needed for the above-mentioned dynamic analysis, were obtained by the same authors in previous cadaver studies (Braune and Fischer, 1889, 1892). The myodynamic indeterminate problem was also addressed, although no solution was proposed.

Knowledge of the intersegmental moments at the lower limb joints allowed Fischer to demonstrate that the Weber brothers' conclusion, according to which the swinging leg moves like a pendulum, was incorrect.

N. BERNSTEIN

Nikolaj Bernstein (1896–1964) was born and worked in Russia. His work in biomechanics has been alluded to as the microscopy of movement. From the methodological point of view, this definition refers to the fact that he gave great emphasis to both space and time resolution of his photographic recordings; he called this cyclogrammetry. Frame rate, for example, was brought up to 150 frames/s. Stroboscopic light was used and stereoscopy was obtained using a mirror, so that a single camera observed the field of view both directly and reflected from a perpendicular direction, and, in addition, synchronization problems were removed. Displacement data differentiation was also improved with respect to the graphical method used by Fischer using a five-point window approach.

Bernstein's belief that he had devised highly accurate experimental and analytical methods led him to state that 'cyclogrammetry puts into the clinician's hands an instrument which permits him to uncover, with uncommon precision and range of selection, all the tiniest peculiarities of movement which are invisible to the naked eye' (Bernstein, 1935). Whether this is true or not is still a matter of discussion. For further information about this well-reputed physiologist, see Bernstein (1967), Whiting (1984) and Jansons (1992).

THE THREE-DIMENSIONAL POSITION AND ORIENTATION RECONSTRUCTION

Herbert D. Eberhart was an associate Professor of Civil Engineering at the University of California, Berkeley, and Verne T. Inman an assistant Professor of Orthopedic Surgery at the San Francisco Medical Center. Over the period September 1945 to June 1947 they were the project leaders of a major research programme at the College of Engineering, Berkeley, funded by the United States National Research Council through their Committee on Artificial Limbs in association with the Veterans' Administration and the Office of the Surgeon General of the United States Army. They produced a final report entitled 'Fundamental Studies of Human Locomotion and other Information Relating to Design of Artificial Limbs' which occupies approximately 750 quarto-sized pages. Their 42 associates in this contract involved medical and engineering professionals, an anthropologist, a prosthetist, a mechanical manager and electronic and photographic specialists: but unfortunately the report was never published as a volume in its entirety, although various parts of it have appeared in the scientific literature and work derived from it continued for many years afterwards.

The work was specifically related to problems in locomotion experienced by amputees and included comparisons with the gait of normal individuals, since normal locomotion was seen as a target for the amputees' locomotion training. The kinematics of locomotion were measured from recordings on 35-mm film, generally exposed in three cameras viewing along orthogonal axes. The cameras were generally viewing the subject from the front, from one side and either from above or from below using a glass walkway and mirror. The images were synchronized, since each camera had the single time clock in its field of view. They produced diagrams generally on a time base showing the linear displacements of anatomical landmarks during the walking cycle, the angular positions of the limb segments and trunk relative to the system axes and also relative angular movements between adjacent segments. The cameras were fixed in position; in some cases they had calibration markers in the field of view, and in other cases records were taken of calibration frames in the fields of view of the individual cameras. Passive markers were fixed to anatomical locations at the joints by adhesive tape. Generally, for markers in the fields of view of two cameras, a parallax correction was made. At this time all calculations were done by hand with the assistance of mechanical calculators.

Bresler and Frankel (1950) performed a full 3D analysis of a locomotion cycle in this way using a force platform for external force measurement and indicated that the calculation of the derived data occupied between 250 and 500 person-hours for each test subject. The investigators were original in designing the first successful six quantity force platform and a six quantity force transducer (Figure 7) for incorporating in an artificial limb, together with a multichannel system for acquisition of EMG from relevant muscles in the leg and trunk during the 2-year investigation—a truly magnificent contribution to the science of gait analysis.

At the time of their investigations, users of artificial limbs complained of discomfort, particularly corresponding to rotational movements of the prosthesis relative to the body. The first investigation cited in the report is one into the rotations of the limb segments during locomotion. Normal test volunteers were used who had volunteered to receive bone pins into the pelvis, the medial condyle of the femur and the lateral proximal condyle of the tibia at the level of the tibial tubercle. Spheres attached to light wooden rods were joined to the stainless steel bone pins and projected 30–50 cm lateral to the body so that they could be viewed by an overhead camera. The pins were inserted in the leg and pelvis on the right side of the body and the pelvic marker interfered with the free swing of the right arm. The atypical locomotion pattern adopted was for the right arm to be held with the hand against the chest while the left arm was free. Since arm action is a major controller of body rotations in locomotion, it appears that the data may not be fundamentally error-free. Over 30 volunteers were used and just over half of the tests gave useful data. The remainder were rejected because of looseness of the pins or complaints of pain from the volunteers. The principal records of the segment rotation were obtained from the overhead camera, although frontal and lateral cameras were also used. Initially, corrections were made from the 3D angulation data to give the true angles in the horizontal plane corresponding to the segment rotations. However, since the major interest was in the rotations during the stance phase, it was found that the uncorrected data from the overhead camera gave values within 2° of the true values during the middle 60% of this phase, and further correction was not made. A typical diagram for one test subject is shown in Figure 8.

The investigators were aware that leg prostheses at that time were basically passive structures with very limited possibility of control by the amputee. Their ambition was to develop a series of controls or power sources for the amputee by utilizing the remaining muscles in the stump, and for this reason they put much effort into determining the magnitude of muscular contraction and the phases of the gait cycle at which these occurred using surface electrodes connected by trailing leads to the best thermionic valve amplifiers which they could construct. The amplified signals were recorded on photographic paper by galvanometer recorders. The test subjects were 10 young college males, and their gait parameters were acquired by heel and toe contact switches fixed to the footwear.

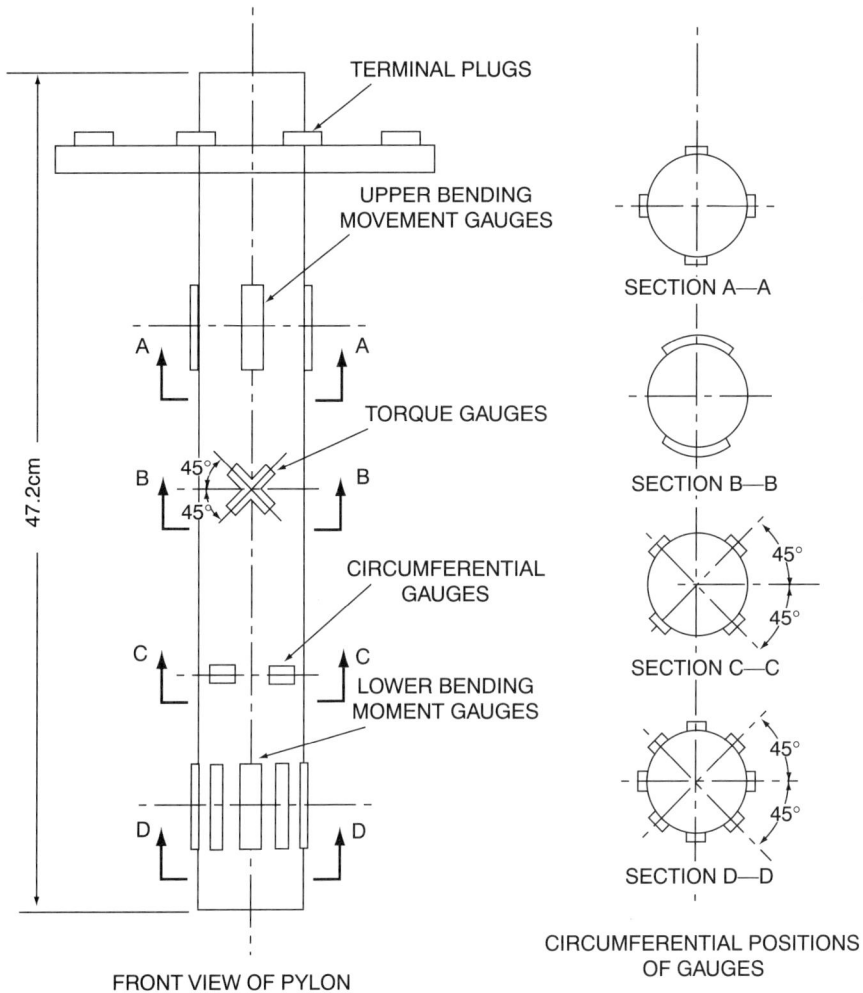

Figure 7. Pylon dynamometer for leg prostheses. Adapted from Eberhart (1947)

EMG signals were obtained from eight major muscle groups from the pretibial up to erector spinae. The investigators were unsuccessful in their ambitions to obtain control signals from the muscles investigated, but they contributed much to knowledge of the science of gait analysis. At a subsequent stage the contractile power of the pectoral muscles was utilized, acting on an arm prosthesis through a tunnel constructed in the skin of the upper chest which allowed active control of an artificial hand. This project, however, was not continued, since problems of hygiene and skin breakdown at the area loaded by the connection became too

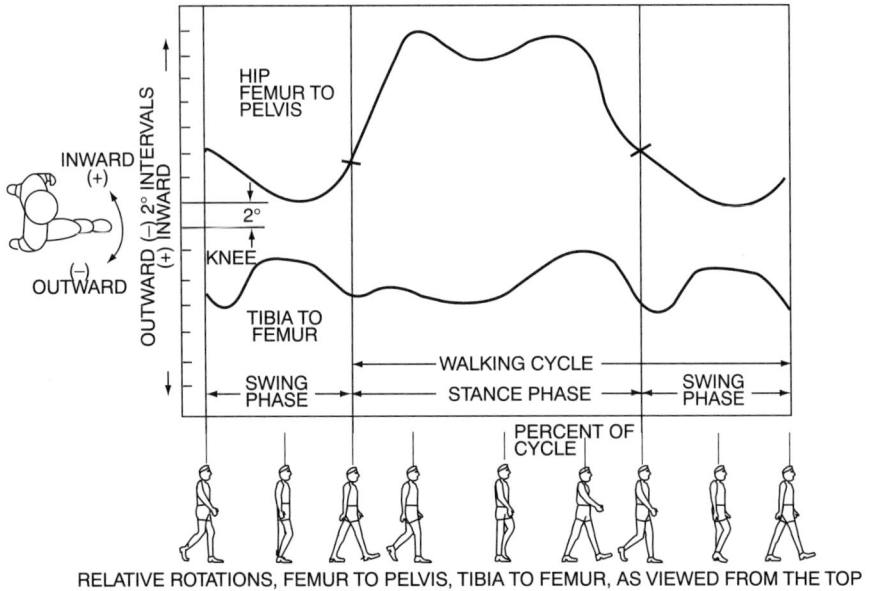

RELATIVE ROTATIONS, FEMUR TO PELVIS, TIBIA TO FEMUR, AS VIEWED FROM THE TOP

Figure 8. Segment rotations about their longitudinal axis in walking. Adapted from Eberhart (1947)

serious. The EMG work is reported in a subsequent publication (University of California, 1953).

Sagittal plane records of locomotion of normal subjects and amputees were made specifically to establish the kinematics of locomotion. Single-plate recording film was used, with multiple exposures produced by a rotating shutter mechanism. The test subjects wore several small light bulbs secured over the apparent positions of the joints of the leg, and the records on the single-plate film were analysed to obtain 2D coordinates of marker positions. Simple numerical differentiation formulas were used to obtain linear and angular velocities and accelerations. These data were acquired to give insight into the requirements for swing-phase controls for leg prostheses, and tests were conducted on negotiating stairs and ramps upwards and downwards.

The California group added much to the science and technology of gait analysis. Had they been able to use digital computers, the course of the study of gait analysis would have been greatly accelerated. Volume 2 of the group's Contractors Report was largely concerned with details of selection of materials for leg prostheses and the mechanical engineering of their function. These concepts were based on the gait analysis studies and have changed little over the years.

DYNAMOMETRY

Another crucial stimulus arrived in Europe from the USA. This was the work by Taylor concerning the organization of work and the optimization of production: the search for a good-quality product at a low cost. Jules Amar, at the Conservatoire des Arts et Metiers in Paris, acknowledges the importance of Taylor's work, but also understands that the human motor, *le moteur humain,* must be investigated using modern scientific methods and possibly with an approach independent of economic considerations. We have here a remarkable example of biomechanics research as applied to the man and woman at work.

The analysis of the force systems present during movement, already tackled by Marey and Carlet, as mentioned previously, and by Demenÿ (quoted in Bouisset, 1992) using pneumatic systems, was significantly advanced by Amar (1916) with the invention of a dynamometric table, i.e. an instrument to measure ground-to-foot forces. The measuring system was mechanical, with springs deflected under load operating an indicator. The instrument measured vertical, sideways and backward force components only, and gave no measure of the forward force component or of moment actions about any reference axis. Amar's force plate was used in investigations of the gait of subjects wearing external prostheses, and the findings were used to suggest improvements.

A further contribution in force plate design was made by Elftman (1938), although this device involved force measurement by deflections of springs and did not give a full six quantity specification of the ground-to-foot force system.

Kinetics was seen to be an essential feature of gait analysis, and a review was made of existing instruments for measurement of ground–foot load actions. The Northrop company attempted to modify the Elftman force platform using electrical resistance strain gauge technology, but this was not successful. A new design of force platform was produced by Cunningham and Brown as described in their 1952 paper. This involved two rectangular platforms joined to each other at the corners by cylindrical columns. The columns carried patterns of electrical resistance strain gauges connected in bridge circuits to measure the three components of force relative to the centre of the top plate of the force platform and the three moments about reference axes through the centre of the top platform. Problems arose with oscillations of the top plate in a horizontal plane relative to the base, and the magnitude of these in a typical gait record obscured some of the characteristics of the ground–foot force data. A special viscous fluid damping system was therefore set up and satisfactory results were obtained. Subsequent users of this type of force platform have found large cross-sensitivities between shear force and the corresponding moments. These, however, have generally been amenable to solution using a cross-sensitivity matrix, which is, of course, the effective way to use modern force platforms, taking account of the fact that the reference axis origin of the force platform is at a prescribed distance below the top surface (Hall et al., 1996). Associated with this was the first ever design of a six quantity transducer for incorporation into the load-bearing structure of a leg prosthesis. This was strain gauged in the same way as one of the columns of the force platform and could give signals corresponding to three components of force and three components of moment. The original California prosthesis dynamometer (pylon dynamometer) was too long to allow use in most patients requiring a transtibial prosthesis, and a revised design with a better measurement of shear forces was introduced by Berme et al. (1975). Such force transducers have the benefit that several consecutive load cycles on the prostheses can be acquired, and a goniometer for knee angle is usually incorporated in the data acquisition system, to allow extrapolation to obtain hip movements.

INTERNAL FORCE ACTIONS

Inman (1947) published a classical paper on the forces developed in the abductor muscles and at the hip joint in the human while standing stationary on one leg. Paul (1967) and Morrison (1968) took this work and that of Bresler and Frankel a step further. Their interest was the variation with time of the magnitude and direction of the force transmitted at the articulating surfaces of the hip and knee joints during level walking, and stair and ramp negotiation. A single force platform similar to that of Cunningham and Brown (1952) was used in the same way as in Bresler and Frankel (1950) with a motion analysis system. This system comprised two 16 mm cine cameras working at a frequency of 50 Hz which viewed the test subject from the front and the left side. After every test the film was rewound and re-exposed while a calibration grid was in the field of view, giving an intrinsic calibration of the photographic record regardless of lens aberration and film distortion in processing. The force platform and camera records were synchronized by the firing of a flash bulb in the field of view of both cameras, synchronizing them to within 0.02 s, and a corresponding pulse was made on one channel of the galvanometer recorder collecting the force platform data. Markers were placed on bony landmarks where skin movement was determined to be minimal: left and right ASIS (anterior superior iliac spines); a light rod projecting posteriorly from its fixation in the sacral region; the most lateral point of the tibial plateau, the tibial tubercle and the lateral malleolus; a marker on a light rod projecting in an anterior–lateral direction from the anterior medial plane of the tibia (this marker allowed the definition of the angular position of the tibia about its longitudinal axis); and a marker on the lateral surface of the footwear over the fifth metatarsal–phalangeal joint.

The positions of the centres of the hip joints were marked by examination and palpation and the position relative to the markers of the ASIS and the sacrum measured, so that the joint positions in space could subsequently be calculated from the measured marker coordinates. Some activities of the test subjects involved natural arm swing, and consequently the left ASIS marker was obscured by the wrist during several frames of film. To acquire data during this time, an auxiliary marker was placed at least a wrist breadth posterior to the ASIS and, although it was on more mobile skin, records of its position assisted in interpolation of the positions of the ASIS marker while it was obscured. The data were used to calculate the intersegment forces and moments, taking account of segment mass and acceleration in the same way as Bresler and Frankel. Paul and Morrison then associated the moments and forces with the tensions in muscles and ligaments at the knee and hip regions. Since these systems are statically indeterminate, the analysis was undertaken using a simplified system of equivalent muscles having approximately the same areas of origin and insertion as the anatomical muscles. This allowed the calculation of forces in the equivalent muscles and hence the joint forces. Since that time many other experimenters have tackled this problem using the complex functions available in high-powered computers, adopting minimization procedures, although justification of the quantities minimized is open to question.

CONCLUSIONS

Our review ends in the late 1960s. From then up to the present day, two major events have occurred, both involving digital computers: the introduction on the market of optoelectronic stereophotogrammetric systems, on the instrumental side, and, on the data-processing side, the optimal estimation of bone position and orientation of bones during movement starting from surface marker positions. Both of these achievements will be thoroughly dealt with in subsequent chapters.

REFERENCES

Amar, J. (1916) Trottoir dynamographique. *CR Acad. Sci.,* **163**, 130–132.

Berme, N., Lawes, P., Solomonidis, S. E. and Paul, J. P. (1975) A shorter pylon transducer for measurement of prosthetic forces and moments during amputee gait. *Eng. Med.,* **4**(4), 6–8.

Bernstein, N. A. (1935) *Untersuchngen ueber die Biodynamic der Lokomotion,* Verlag WIEM, Moscow/Leningrad, pp. 5–13.

Bernstein, N. (1967) *The Coordination and Regulation of Movement.* Pergamon Press, Oxford.

Borelli, G. A. (1680) *De Motu Animalium. Pars Prima.* A. Bernabò, Rome.

Borelli, G. A. (1681) *De Motu Animalium. Opus Posthumun Pars Altera.* A. Bernabò, Rome.

Borelli, G. A. (1989) *On the Movement of Animals* (translation). Springer, Berlin.

Bouisset, S. (1992) Marey: when motion biomechanics emerged as science. In: *Bioloco-motion: a Century of Research Using Moving Pictures* (ed. A. Cappozzo, M. Marchetti and V. Tosi), ISB Book Series, Vol. 1, Promograph Publishers, Rome, pp. 71–88.

Braun, M. (1993) *Picturing Time—The work of Etienne Jules Marey.* University of Chicago Press, Chicago.

Braune, W. and Fischer, O. (1889) *Uber den Schwerpunkt des menschlichen Körpers mit Rücksicht auf die Ausrüstung des Deutschen Infanteristen.* S. Hirzel.

Braune, W. and Fischer, O. (1892) Ueber den schwerpunkt des menschlichen körpers und seiner glieder. *Abhandl. Math. Phys. Kl. K. Sächs. Gesellsch. Wissensch.* **18**, 409–492.

Braune, W. and Fischer, O. (1987) *The Human Gait* (translation). Springer, Berlin.

Bresler, B. and Frankel, J. P. (1950) Forces and moments in the leg during walking. *Trans Am. Soc. Mech. Eng.*, **72**, 27–36.

Carlet, G. (1872) Essai expérimental sur la locomotion humaine. Étude de la marche. *Am. D. Sci. Nat. Zool.*, **16**, 1–92.

Cunningham, D. and Brown, G. W. (1952) Two devices for measuring the forces acting on the human body during walking. *Proc. Soc. Exp. Stress Anal.*, **1X**(2), 75–90.

Dagg, A. I. (1977) *Running, Walking and Jumping.* Wykeham Publishing, London.

Eberhart, H. (ed.) (1947) Fundamental Studies of Human Locomotion and other Information Relating to Design of Artificial Limbs. Report to US Veterans' Association, University of California, Berkeley.

Elftman, H. (1938) The measurement of the external force in walking. *Science,* **88**(2276), 152.

Gerdy, P. N. (1829) On the mechanism of the human gait. *J. Physiol.*, **1X**, 2–12.

Hall, M. G., Fleming, H. E., Dolan, M. J., Millbank, S. F. D. and Paul, J. P. (1996) Static in situ calibration of force platforms. *J. Biomech.*, **29**(5), 659–665.

Inman, V. T. (1947) Functional aspects of the abductor muscles of the hip. *J. Bone Joint Surg.*, **39**(3), 607–619.

Jansons, H. (1992) Bernstein: the microscopy of movement. In: *Biolocomotion: a Century of Research Using Moving Pictures* (ed. A. Cappozzo, M. Marchetti and V. Tosi), ISB Book Series, Vol. 1, Promograph Publishers, Rome, pp. 137–174.

Lorini, G., Bossi, D. and Specchia, N. (1992) The concept of movement prior to Giovanni Alfonso Borelli. In: *Biolocomotion: a Century of Research Using Moving Pictures* (ed. A. Cappozzo, M. Marchetti and V. Tosi), ISB Book Series, Vol. 1, Promograph Publishers, Rome, pp. 23–32.

Maquet, P. and Furlong, R. (1992) *Wilhelm Weber, Eduard Weber: Mechanics of the Human Walking Apparatus.* Springer-Verlag, Berlin.

Marey, E. J. (1873) *La Machine Animal.* Librairie Germer Baillière, Paris.

Marey, E. J. (1878) *La Méthode Graphique dans les Sciences Expérimentales et Principalement en Physiologie et en Médicine.* G. Masson Editeur, Paris.

Marey, E. J. (1885) *La Méthode Graphique dans les Sciences Expérimentales et Principalement en Physiologie et en Médicine, Deuxiéme tirage augmenté d'un supplément sur Le Développment de le Méthode Graphique.* G. Masson Editeur, Paris.

Morrison, J. B. (1968) Bioengineering analysis of force actions transmitted by the knee joint. *Biomed. Eng.*, **4**, 164–170.

Muybridge, E. (1955) *The Human Figure in Motion.* Dover, New York.

Muybridge, E. (1957) *Animals in Motion.* Dover, New York.

Muybridge, E. (1972) *The Stanford Years, 1872–1882 Catalogue of Expilation.* Museum of Modern Art, Stanford University, San Francisco.

Muybridge, E. (1979) *Complete Human and Animal Locomotion,* Dover, New York.

Paul, J. P. (1967) Forces transmitted by joints in the human body. *Proc. Inst. Mech. Eng.*, **181**(3J), 8–15.

Richer, P. (1901) Locomotion humaine. In: *Traité de Physique Biologique* (ed. D'Arsonval, Chauveau, Gariel and Marey), Masson Éditeur, Paris, pp. 137–228.

Tosi, V. (1992) Marey and Muybridge: how modern biolocomotion analysis started. In:

Biolocomotion: a Century of Research Using Moving Pictures (ed. A. Cappozzo, M. Marchetti and V. Tosi), ISB Book Series, Vol. 1, Promograph Publishers, Rome, pp. 51–69.

University of California (1953) *The Phasic Activity of the Muscles of the Lower Extremity.* Prosthetic Devices Research Report 3, University of California, Berkeley.

Weber, W. and Weber, E. (1894) *Mechanics of the Human Walking Apparatus.* Springer Verlag, Berlin.

Whiting, H. T. A. (ed) (1984) *Human Motor Actions–Bernstein Reassessed.* Advances in Psychology Series 17, North-Holland, Amsterdam.

2

Functional Anatomy

ARNE LUNDBERG

Department of Orthopaedic Surgery, Karolinska Institute, Huddinge University Hospital, Huddinge, Sweden

GENERAL ASPECTS

The anatomy of the musculoskeletal system has been extensively described in basic anatomical and physiological texts (Warwick and Williams, 1973). It is not the purpose of this chapter to cover the fundamentals of these issues, but rather to highlight some modern concepts and controversies relating to functional anatomy, especially as related to movement science. Thus, the basics will be presented in a summary fashion and the interested reader is encouraged to study the fundamentals of functional anatomy elsewhere.

The constituent parts of the musculoskeletal system are bone, muscle and joints, the latter term denoting all cartilage and fibrous tissue connections between bones. Structural aspects of these constituents will be discussed below, followed by a topographical discussion of different regions, with emphasis on the lower extremity.

For all parts of the human organism, increasing knowledge has led to increasingly complex models of structure and function. The musculoskeletal system is no exception to this rule. This is especially true for the relationship between structure and function, the understanding of which has increased rapidly over the last decades. Three disparate examples of this are:

- the realization that structural elements of joint cartilage (constituent proteins) are important for the amount of cartilage deformation, which in turn influences nutrition
- the understanding of the importance of ligament fibre orientation for liga-

Three-dimensional Analysis of Human Locomotion. Edited by P. Allard, A. Cappozzo, A. Lundberg and C. Vaughan
© 1997 John Wiley & Sons Ltd. ISBN 0 471 96949 4

ment function (ligaments having previously been seen much as straight bundles of parallel fibres)

- the observation that, when analysed with sufficiently accurate methods, the kinematics of different joints will often not conform with the simple patterns previously assumed

This knowledge is continuously being put into clinical practice (in the mentioned examples, knowledge of cartilage structure is used in the tentative development of new treatment strategies for degenerative joint disease, while understanding of the complexity of ligament structure leads to the development of new surgical techniques in, for example, cruciate ligament reconstruction), and movement analysis, by clarifying the basic aspects of individual joint function as well as joint interaction, has provided the major contribution to knowledge underlying the development of joint replacement and modern prosthetic limb replacement. Application of movement analysis information on an individual level, e.g. in planning of surgery in patients with neurological disorders, has proven difficult to implement reliably, probably largely due to a combination of different factors such as lack of method validation, need for large amounts of baseline data for each type of deformity or dysfunction studied and lack of well-defined treatment goals.

There is at present no reason to doubt that the current trend will continue, or that over time the fundamental information needed to use modern techniques such as gait and movement analysis will be gathered. However, as always, this will only happen if the fundamental structural and functional potentials and limitations of the human body are realized.

BONE

STRUCTURE

Bone consists of an inner layer of cancellous bone which is covered by a harder outer layer of cortical bone. The latter shows superior bending strength and is the main supporting structure of diaphyseal bone. The presence of cancellous bone is usually attributed to the better strength–weight relationship of a tube compared to a solid rod.

MECHANICAL PROPERTIES AND FUNCTIONAL ASPECTS

The strength of bone is determined by the relative thicknesses of the cancellous and cortical layers, by the density of each layer and by the orientation of the trabeculae.

The most commonly discussed form of regulation of bone growth is usually referred to as Wolff's law. This law, in simple terms, states that bone will, over time, adapt to load in such a way that its ability to carry the applied load becomes optimal. The background of this adaptation is largely unknown, although current knowledge of cell regulation would suggest a chain combining mechanical and neuroendocrine factors.

MUSCLE

Muscle tissue exists in mammals in three major forms. In this text only skeletal, or striated, muscle will be discussed.

STRUCTURE

Gross Anatomy

To be able to influence movement, a muscle obviously has to span at least one joint. A large number of the muscles involved in locomotion span two (biarticular muscles) or several (multiarticular muscles). In the vicinity of joint complexes such as the wrist–hand and ankle–foot complexes, many muscles span a large number of joints (the maximum numbers being those spanned by the long extensors of the fingers and toes). Considerable effort has gone into studying possible functions for the bi- or multiarticularity of different muscles (Carlsoo and Molbech, 1966; Lieber, 1990; Prilutsky and Zatsiorsky, 1994; Virji-Babul and Cooke, 1995). While the result of isolated contraction of a multiarticular muscle may be unpredictable, it seems obvious that the large number of muscles that have partly overlapping spans represents a probably close to ideal compromise between total number and cross-sectional area of the involved muscles and the amount of control that can be exerted on the other.

The relationship between contraction and joint movement is influenced primarily by:

- the length and cross-sectional area of the muscle
- the distance between muscle insertion and joint axis/joint centre of rotation
- the fibre direction and composition of the muscle

Fibre Composition

Muscle fibres occur in two fundamentally different forms, commonly referred to as slow twitch (red, type I) and fast twitch (white, type II) fibres. These fibre types were originally described in the nineteenth century (Ranvier, 1874), and the understanding of the relative roles of these has shed light on differences both between muscles and between individuals, as well as between species (Barany, 1967; Paukal, 1904). While more recent research has shown the differences between fibre types to be both more diverse and less distinctive than originally assumed, the concept is still helpful in understanding, for example, differences caused by a muscle's phylogenetic adaptation to postural or 'fight or flight' demands (Kelly and Rubinstein, 1994). Within fibres, the contractile actin–myosin elements are responsible for muscle contraction. Regulation is achieved through direct action of efferent nerve fibres in the muscle, and through reflex arcs involving muscle spindles.

JOINTS

GENERAL ASPECTS OF JOINT ANATOMY

Joint Surfaces

The cartilage covering the joint surfaces of synovial joints functions as an impact attenuator, and, together with the synovial fluid, also as a friction attenuator. The main components of the cartilage matrix are collagen and hyaluronic acid. The former is considered to supply the elastic properties of the cartilage, while hyaluronic acid provides viscosity. Cartilage is formed from the layer close to the subchondral bone, and has limited potential for regeneration after damage. Joint cartilage thickness varies between joints but also within the same joint. Thus, in the hip joint, which has comparatively thick cartilage, different parts of the femoral head surface may show cartilage thicknesses from 0.5 to 4 mm (Hodler et al., 1992). Cartilage thickness differences have been postulated to relate to either differences in loading levels or to compensation for surface shape incongruencies (Ateshian, Soslowsky and Mow, 1991; Jonsson et al., 1992; Ljunggren, 1978; Macirowski, Tepic and Mann, 1994). The shape of the joint surfaces, together with ligament tension, is one of the deciding factors for the joint's range of motion.

Intracapsular Ligaments

Examples of intracapsular ligaments are the cruciate ligaments of the knee and the ligament of the femoral head in the hip. These ligaments serve different purposes; the main function of the cruciate ligaments is to maintain sagittal and rotational stability of the knee, whereas the ligament of the femoral head, apart from its stabilizing function, also contains one of the two routes of blood supply to the femoral head.

Capsular Structures

The difference between joint capsule and ligament tissue is often subtle. The joint capsule is generally one of the deciding factors for range of motion, and a fibrotic capsule after trauma or joint disease may severely restrict movement.

Extracapsular Ligaments

Extracapsular ligaments in the strict sense are comparatively few. However, most ligaments that can be dissected away from the joint capsule are commonly seen as extracapsular, making this the largest group of ligaments.

Muscles

The large variation in muscle stiffness is a major element in the differences in range of motion seen in the same joint in different individuals. Examples that are often referred to in the literature are the hamstring and triceps surae muscles. These are also examples of muscles influencing the function of more than one joint. An important consequence of the anatomy of bi- or multiarticular muscles is that they will usually allow more flexion in one joint when the other joint spanned is in extension, and vice versa. In joints with limited structural stability, muscles play an important role in maintaining joint integrity, as is the case, for example, for the supraspinatus and subscapularis muscles in the glenohumeral joint.

JOINT FUNCTION

The basic characteristics of joint function are motion and stability. A trade-off situation is generally considered to exist between these, and this is particularly true for multiaxial joints. In general, joints of the upper extremity are adapted more for mobility than for stability when compared to the corresponding joints of the lower extremity.

JOINT TYPES

Joints are traditionally classified according to mechanical characteristics as hinge, ball and socket (spherical), saddle, plane or condyloid (trochlear) joints. This classification may provide for a general understanding of a joint's properties, but it has to be kept in mind that detailed analysis will usually show the similarities between mechanical design concepts and biological structures to be limited.

'Hinge' Joints

A true hinge joint should only have one possible axis of rotation. There is no joint in the human body that strictly conforms to this description. However, several joints are sufficiently similar to a hinge in function to be usually described as such, notably the elbow and talocalcaneal joints. Even in hinge joints, there may be several possible joint axes, but these are then parallel. This situation occurs when different parts of the convex joint component's articulating surface have different radii.

Spherical ('Ball and Socket') Joints

A spherical joint should have an infinite number of possible axes of rotation, all running through the common centre shared by both the spherical joint surfaces. Again, it should be remembered that joints usually described as ball and socket or spherical are not spherical in a strict geometrical sense. Thus, the diameters of the femoral and humeral heads are different in different directions.

'Saddle' Joints

The classical example of a saddle joint is the first carpometacarpal joint. A saddle joint is characterized by the existence of two natural axes; the combined movement will be similar to that of a universal joint.

Planar Joints

These joints, which should be made up by two opposing flat surfaces with a small amount of possible sliding movement and rotation around an axis perpendicular to the joint surfaces, are not commonly accepted as existing within the locomotor system. As a perfect planar joint can also be described as a spherical joint with infinite radius, it is obvious that even a small difference from a flat surface will tend to put the joint in a different category.

Trochlear and Elliptoid Joints

Trochlear and elliptoid joints share the important property of having two possible fixed joint axes, which are usually more or less perpendicular. From a functional point of view it is important to understand that this means that any joint axis orientation between the two may result from a combined movement and may thus be observed. A pattern of observed joint axes all located in a plane in space thus indicates a trochlear or elliptoid joint. This can be seen in, for example, the knee and the talotibial joint, which are both in this category.

Non-synovial Articulations

Several articulations between parts of the skeletal system that are not synovial joints are important for movement. Perhaps the most obvious example is the articulation between the scapula and the rib cage. Despite the absence of joint surfaces and synovial fluid, this articulation provides a maximum of approximately 90° of movement as well as (together with the sternoclavicular and acromioclavicular joints) stability for the shoulder.

JOINT CENTRES AND JOINT AXES

In mechanical analysis joint centres and joint axes are often studied. Although these concepts may be helpful, the assumption of a simple joint centre pattern or a single axis is often detrimental to the analysis.

Joint Centres

Joint centres were originally described in two-dimensional (2D) analysis of joint motion, and are integral to Reuleux analysis of motion patterns (Sammarco, Burstein and Frankel, 1973). In this context the joint centre largely functions as a 2D counterpart of the joint axis concept, and is subject to the same limitations as other 2D descriptions of three-dimensional (3D) movements. More recently, the term joint centre has been applied to the point of intersection of joint axes in polyaxial joints. In this context, the real-life counterpart of the term is often more consistent, as, in many polyaxial joints (such as spherical, elliptoid or trochlear joints), different axes, if not actually intersecting, will be found to pass within a small volume close to the geometrical centre of the joint. It should still be remembered that knowledge of the joint centre location alone will provide limited information about joint function.

Joint Axes (Figures 1–3)

Joint axes have been postulated for most true articulations. In modelling they are often assumed to be single for any joint that is not spherical. It should be kept in mind that most joint axis information in the literature is derived from work performed on anatomical specimens, and often based on study conditions that may impose certain conditions on the movement studied, thus not allowing for all possible motion types in the joint. In modern kinematical research they are most often calculated as finite or continuous helical axes. When such measurements have been performed under conditions that do not force a certain type of movement to occur, resulting movements have impressed with their diversity rather than their uniformity.

SINGLE JOINTS VERSUS JOINT COMPLEXES

When the function of joint complexes is studied in vivo, examination is often by necessity limited to the segments proximal and distal to the whole complex (such as when ankle/tarsal function is analysed using skin markers on the lower leg and the heel and/or the forefoot). In such cases, the examination rests on the assumption that seeing these joint complexes as a single functional entity will still provide useful information. This assumption may or may not be true in the individual case, depending primarily on the consistency in the behaviour of each individual joint over the studied range of motion. For a joint complex where each individual joint behaves as a single-axis hinge, the assumption will usually not adversely affect the possibility of calculating, for example, moments and forces (provided that soft-tissue factors do not cause unpredictable alterations in joint movement distribution). On the other hand, it has to be kept in mind that, under such circumstances, not even actual determination (rather than a population-based assumption) of the joint complex axis will provide for differences in the relationships between individual joint axes in different individuals.

REGIONAL ASPECTS OF JOINT FUNCTIONAL ANATOMY
(See appendix for terminology.)

Joints of the Upper Extremity and the Trunk

These joints, although, particularly under some circumstances (such as, for the upper extremity, use of walking aids, or, for the spine, a previously fused hip), important for locomotion, will not be covered in this text. The interested reader is encouraged to study their functional anatomy elsewhere. The functional anatomy of, for example, the shoulder joint complex, incorporating the most mobile synovial articulation (glenohumeral) as well as the most mobile non-synovial articulation (thoracoscapular), is quite interesting and well worth considering.

The Pelvis

The pelvic joints are:

1. The sacroiliac joints. These are synovial joints with limited mobility.
2. The pubic symphysis. This is a non-synovial structure which, except in cases of marked connective tissue dysfunction, and in the late stages of pregnancy, where pelvic stability is reduced to facilitate widening of the pelvic ring, has a range of motion close to zero.

The Hip (Figure 1)

The hip joint is anatomically a modified spherical joint. Although the femoral head and the acetabular socket have slightly varying radii in different planes, the main determinants of the pattern of motion are the intra- and extracapsular

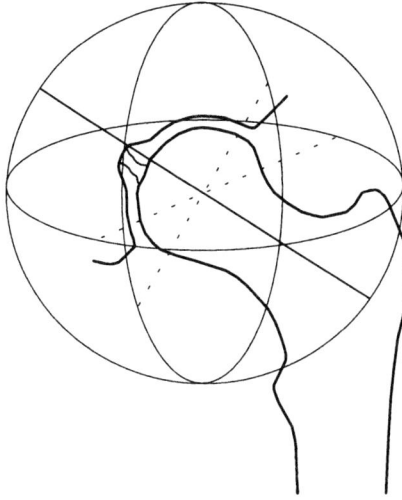

Figure 1. The hip joint as a spherical joint can display rotation about an infinite number of joint axes. The main axis can be defined as an axis close to the direction of the femoral neck; movement about this axis is often preserved even in advanced degeneration

ligaments, and muscles. Of special interest is the function of the main intracapsular ligament (the ligament of the head or lig. capitis femoris). Mobility of the hip is normally restrained by a combination of ligamentous restraints and muscles (such as the short external rotators, the adductors and the iliopsoas muscle). In extreme cases conflict between the anterior or posterior aspect of the proximal femur and the pelvis may restrict movement. Hip movement is extremely difficult to study from the outside, due to a thick layer of soft tissue covering the joint in combination with high soft-tissue mobility over the thigh.

The Knee (Figure 2)

The knee consists of three joints:

1. The tibiofemoral joint again consists of two compartments, lateral and medial. These correspond to the humeroulnar and humeroradial joints in the elbow. The medial compartment is usually slightly larger than the lateral compartment. Each compartment contains a meniscus, the function of which has still not been completely understood (Brown and Shaw, 1984; Bylski-Austrow et al., 1994; Conlan, Garth and Lemons, 1993).
2. The femoropatellar joint is functionally separate from the femorotibial joint (and in many animals this separation is also anatomical). Its main function is to allow the combined distal translation and flexion of the patella, evolved to increase the lever arm of the extensor muscles. The kinematic pattern of the

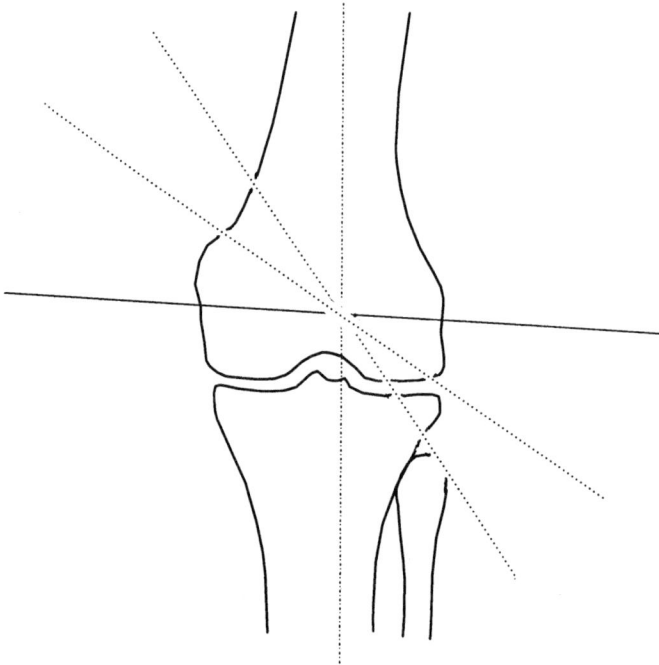

Figure 2. The main axis of the knee joint is the axis of flexion–extension; its direction is a function of the radii of the femoral condyles. In most positions there is an alternate axis which is close to parallel to the long axis of the tibia (dashed line), and combined movements can relate to any axis between these (dotted lines)

    patella is complicated, and crucial in pathological conditions such as patellar instability and post-knee replacement (Brossmann et al., 1993; Goodfellow, Hungerford and Zindel, 1976; Heegaard et al., 1995; Hirokawa, 1993).

3. The proximal tibiofibular joint, which has a limited range of motion and forms part of the knee complex mainly through the fact that the lateral constraints are partly attached to the fibular head (Eichenblat and Nathan, 1983; Ogden, 1974).

The cruciate ligaments are the main intracapsular restraints of the knee joint. The anterior cruciate ligament is wholly intracapsular, while the posterior one is extracapsular near its insertion on the posterior aspect of the tibia. Its proximal portion is intracapsular. Both cruciate ligaments are twisted as well as oblique in all planes, which leads to a pattern of function where different ligament portions are tightened during different parts of the range of motion (Anderson et al., 1992; Arnoczky, 1983; Clark and Sidles, 1990; Hollis et al., 1991; Livesay et al., 1995; Markolf, Wascher and Finerman, 1993; Takai et al., 1993; Woo, Livesay and Engle, 1992; Zavatsky and Connor, 1992). This does not prove the importance of the cruciate ligaments for normal walking, as submaximal activity is often surprisingly unhindered by particularly anterior cruciate ligament injuries.

Extracapsular Structures

The only easily identifiable extracapsular ligament of the knee is the lateral collateral ligament, which runs from the lateral femoral condyle to the head of the fibula. It functions to resist varus stress and as one of several structures resisting anteroposterior stresses of the lateral joint compartment (Ahmed et al., 1992; Fulkerson and Gossling, 1980; Skyhar et al., 1993). The medial collateral ligament is largely integrated with the joint capsule. In general it resists valgus stress but different parts may have somewhat different functions (Ahmed et al., 1992; Arms et al., 1983; Bruns, Volkmer and Luessenhop, 1994).

The popliteus muscle has a fairly small muscle belly and a strong tendon. It runs from the posterior aspect of the proximal tibia to the lateral femoral condyle. A possible function for the stability of the lateral meniscus, and of the knee in general has been postulated; however, this function has not been entirely clarified (Fulkerson and Gossling, 1980; Fuss, 1989; Jones, Keene and Christie, 1995; Skyhar et al., 1993; Tria, Johnson and Zawadsky, 1989; Watanabe et al., 1993).

Kinematics

The knee was generally considered to function as a hinge until the 'screw-home' mechanism was noted in the 1960s (Hallen and Lindahl 1966). This mechanism means that, when the knee is extended, the tibia rotates internally, which leads to a shift in the joint axis inclination. The knee being the most extensively studied and publicized joint over the last 30 years, many opinions on minor or major points in knee kinematics have been published. Most of these relate to states of ligament deficiency.

Commonly Used Simplifications in Modelling

Most modern models take into account the screw-home effect. However, 2D models are still in use. The kinematic pattern of the knee means that these models will tend to be more accurate in the range from 30° flexion upwards, while predictions close to full extension will have to be interpreted with caution.

The Ankle (Figure 3) and Foot

The ankle–foot complex is the joint complex of most importance in locomotion. Study of this region is complicated by several factors:

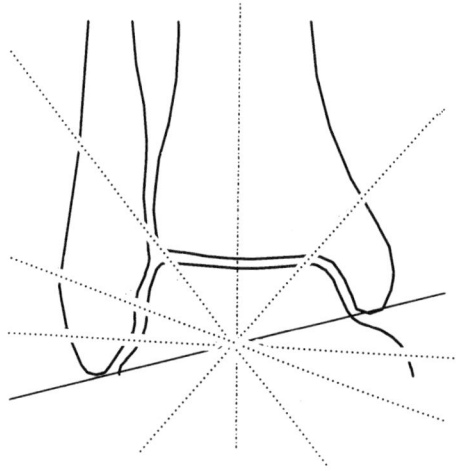

Figure 3. The main axis of the ankle runs close to the tips of the malleoli. There is also rotation about an axis close to parallel to the tibia (dashed line), and combined movements can relate to any axis between these (dotted lines)

1. The conditions under which the foot should be studied often include the use of shoes.
2. The talus has two properties that influence study of its kinematic behaviour:
    (a)   There is no part of the talus that is directly observable from the outside in normal individuals.
    (b)   No muscle or tendon attachments are present on the talus.

A consequence of property (a) is that direct observation of the kinematics of the ankle can only be made with invasive techniques (either in vitro or in vivo), whereas property (b) means that, for example, EMG information cannot be used to make inferences about movements of the talus.

The joints of the ankle–foot complex most often studied are:

1. The talocrural joint. This has traditionally been described as a hinge joint. However, the ankle joint, according to available literature, may vary in kinematic characteristics between individuals, so that some may show a hinge pattern, while others have a kinematic pattern varying between different types and ranges of movement. This is the only possible way in which the large number of publications supporting either a hinge pattern (Sammarco, 1993; Singh et al., 1992), a varying pattern or the possibility of either type of pattern in different individuals (Lundberg et al., 1989) may be interpreted. Several investigations have also shown that the range of rotation around the vertical axis at the ankle may be significant compared to the arc of plantarflexion–

dorsiflexion used in walking (Lundberg et al., 1989; Rasmussen and Tovborg-Jensen, 1982).

2. The talocalcaneal or subtalar joint. This joint is most often described as a hinge (Alexander et al., 1982; Close et al., 1967; Inman, 1976; Isman and Inman, 1969) with an axis running from posterior, lateral, inferior, to anterior, medial, superior. No study indicating major discrepancies from this pattern in normal individuals has been published. However, the hinge axis may not necessarily be constant in position over the range of motion and the differences in axis orientation between individuals may be considerable (Isman and Inman, 1969; Lundberg, 1989). It has also been noted that in certain individuals the inclination of the talocalcaneal axis will vary with load; this, however, does not mean that it changes in orientation in relation to the bone and thus does not contradict the hinge theory.

3. The transverse tarsal joint (or Chopart's joint), made up of the talonavicular and calcaneocuboid joints. Together with the talocalcaneal joint, these joints have been described as a closed kinematic chain (van Langelaan, 1983). Anatomically, the talonavicular joint may be close to spherical in some individuals, and more elliptoid in others. Variations in joint axis orientation occur between different types of movement, but are comparatively small for a spherical joint. It has been postulated that the talonavicular and calcaneocuboid joints should interact in such a way as to lock the midfoot in supination and make it flexible in pronation (Elftman, 1960). It has, in most individuals, a range of motion close to or exceeding that of the tibiotalar joint. The calcaneocuboid joint usually shows a surface shape similar to that of the talonavicular joint, but with the convex surface distally (on the cuboid).

4. The metatarsophalangeal joint range of motion is of importance for locomotion, in that walking demands approximately 60° of dorsiflexion, while running demands approximately 90° (Bojsen-Möller and Lamoreux, 1979).

Extracapsular Restraints

The main ligaments of the ankle joint complex are the following.

Lateral Group.
Anterior fibulotalar ligament This ligament has been described as a rotational constraint in the neutral position (Kjaersgaard-Andersen et al., 1989; McCullough and Burge, 1980; Rasmussen, Jensen and Hedeboe, 1983; Rasmussen and Tovborg-Jensen, 1982) and as a collateral ligament (checking varus forces) in plantarflexion (Inman, 1976). It has been shown to be the ligament most often injured in ankle sprains.

Posterior fibulotalar ligament This ligament has the anatomical characteristics of a rotational constraint (Rasmussen, 1985; Rasmussen, Jensen and Hedeboe, 1983; Stephens and Sammarco, 1992).

Medial Group. This comprises the tibiotalar, tibiocalcanear and tibionavicular portions of the deltoid ligament. Although the functions of different portions of the deltoid ligament have not been fully analysed, the tibiotalar and tibiocal-canear portions seem to function mainly as valgus constraints, while the tibiona-vicular portion may also be part of the arch-maintaining structures (Attarian et al., 1985; Bulucu et al., 1991; Cawley and France, 1991; Colville et al., 1990; Harper, 1987, 1991; Kjaersgaard-Andersen et al., 1989; Mann, 1987; Quiles et al., 1983; Rasmussen, Kromann-Andersen and Boe, 1983; Siegler, Block and Schneck, 1988).

Common Simplifications in Modelling

The ankle complex is probably the region of human anatomy that has been simplified to the largest extent in modelling. A common solution is to depict the joint complex as two oblique hinges, using the data of Isman and Inman (Isman and Inman, 1969) as input data. This to some extent rests on a misinterpretation of that publication, which also gives fairly substantial standard deviations, particularly for the subtalar joint. However, it is also still common to use the concept of a single hinge, even though the risks with calculations following from such a procedure have long been known (Procter and Paul, 1982).

ANATOMICAL ASPECTS ON SURFACE VERSUS BONE-BASED KINEMATICS

In most instances of motion analysis, the aim is to assess the movement of segments of the skeletal system, rather than soft-tissue movement. Therefore, in most kinematic in vivo studies, movements of surface markers are either assumed to directly represent, or used to calculate, using algorithms of varying degrees of sophistication, bone movement. By necessity, the quality of the results of these procedures will vary with local anatomy.

Regional Factors

Hip

The thick soft-tissue layer over the hip and femur makes it difficult to study hip kinematics with external methods. Several ways of overcoming this problem have been suggested, the most promising ones being methods which include weighing of position data from large numbers of markers, taking into account expected patterns of movement. With less sophisticated methods, and in subjects with excessive amounts of soft tissue, it should be remembered that there is no position on the thigh where a skin marker will give data that can be used for analysing any other motion than flexion–extension.

Knee

Movements of the tibia are well reflected by skin movement, particularly of the skin over the anteromedial face of the tibia. As has been stated above, there is considerable skin movement over the distal femur. Anteroposterior skin movement in leg rotation is in the region of 2–3 cm (Karlsson and Lundberg, 1994).

Invasive measurements of femoral movement also encounter difficulties. The large soft-tissue displacements, particularly over the lateral aspect of the distal femur, will tend to limit the measurements utilizing intracortical pins in this region to fairly slow movements within a limited range of motion, while medial pin placement leads to the risk of apprehension of the potential conflict between pin and contralateral knee, which may influence the movement pattern.

Ankle and Foot

The anatomical obstacles to detailed kinematic assessment of the ankle and foot are substantial. The main problems relate to the relatively small segments and intersegmental distances, and to the difficulty of defining the talus from the exterior. There is also large variation in the anatomy of the foot between individuals, which limits the value of universal models of this region. Because of this, in most studies of in vivo function, the foot is seen as connected to the lower leg either through a single hinge joint representing the combined action of all joints between the marker segments, or through two hinge joints (representing the talocrural and talocalcaneal joints).

Individual Factors

In general, soft-tissue movements follow patterns that are predictable. The main problems influencing movement analysis with external markers are variations in the amounts of superficial soft tissues and differences related to the rate (or speed) of movement. All of these factors will influence different individuals differently and lead to problems with most optimization methods suggested for locomotion analysis. A further problem is that, for obvious reasons, the problems will increase with increasing difference from the 'normal' (e.g. movements of neurologically handicapped patients, who have altered soft-tissue action, or patients with joint deformities). These may often be the groups one wishes to study, and it should be understood that, for anatomical reasons, methods validated only on normal subjects or other patient groups may not be applicable.

CONSEQUENCES IN LOCOMOTION ANALYSIS

The functional anatomy of the human musculoskeletal system is, in spite of its fundamental, and largely macroscopic, nature, a rapidly developing field. As in many areas, research findings will sometimes create the impression that uncertainty and confusion is increasing, and that the increased knowledge of, for example, basic joint function will automatically call for increased sophistication in analysis methodology. This is partly true; thus, older methods for 2D movement analysis well served the simple anatomical concepts of their age, while most researchers today will agree that assuming, for example, fixed hinge axes of the joints of the lower extremity will lead to a substantial risk of introducing unacceptable errors. However, the major challenge in locomotion analysis lies in determining the adequate level of accuracy in describing movement, and the optimal methods to provide this information, including optimization of measurement data that only indirectly represent the underlying function of the musculoskeletal system. Once the level of accuracy has been determined, the temptation to overinterpret the data must be resisted. This is particularly important for the type of situation outlined above, where the method used is not validated for the type of individual included in the study. In this case, the functional anatomy of the region has to be considered carefully and objectively to avoid ending up with irrelevant data; in many cases the validation will have to be extended to the group under study.

APPENDIX: TERMINOLOGY

Most of the terms used in this text are uncontroversial. However, as some, particularly those describing movement, are not, it has been considered worthwhile to state how they are used in this text.

Flexion–extension

This is a rotation in the sagittal plane around a transverse axis. It is defined for each joint according to the joint physiology. Thus in hip flexion, the leg is rotated forward, whereas in knee flexion the lower leg is rotated backward. In joint complexes the terms may be contradictory; thus, in the ankle and foot, clarity is usually reached through use of the terms dorsiflexion and plantarflexion. Although these terms are less than consistent, they are difficult to misinterpret, and will be used here.

Abduction–adduction

This is a rotation of the segment distal to a joint outward (abduction) or inward (adduction) around an axis in the sagittal plane which is perpendicular to the long axis of the segment proximal to the joint. This means that as the extremities are mostly vertically oriented in the anatomical position, abduction–adduction mostly occurs around an anteroposterior axis. The exception is the foot distal to the talus where the abduction–adduction axis is vertical.

Internal–external Rotation

These are the rotations in the horizontal plane around the vertical axis. The terms are applied to the movement of the distal segment unless otherwise stated.

Pronation–supination

These terms are specifically applied to the hindfoot and the forearm. Supination means turning up the palm of the hand or the sole of the foot through a clockwise rotation seen from proximally on the right side. In the foot the movement is often not related to the frontal plane but to a plane perpendicular to the axis of the subtalar joint. In this case the terms when applied to the frontal plane are *inversion–eversion*.

The above terms are used to describe physiological movement. Terminology for instability or deformity is often derived from the same basic concepts and may be confusing.

Rotational Instability in the Sagittal Plane

This is denoted hyperflexion–hyperextension. Hyperflexion in the extremities is seldom a problem in locomotion. Hyperextension of the knee may be a significant problem, particularly as it is often combined with neurological conditions that decrease the active stability. The term is also used in the spine.

Rotational Instability in the Frontal Plane

This is usually referred to as valgus (corresponding to hyperabduction) or varus (corresponding to hyperadduction). These terms, when applied to the foot, follow the abduction–adduction convention.

Rotational Instability in the Horizontal Plane

This is referred to as internal or external rotational instability. However, those who describe instabilities are often driven by a desire to sum up not only along or around which axis the increased movement occurs, but also the underlying pathology or a change in position, as well as orientation of the axis of movement, in the name used for the instability. This is the background of such confusing terms as 'anteromedial' or 'posterolateral' rotational instabilities of the knee.

REFERENCES

Ahmed, A. M., Burke, D. L., Duncan, N. A. and Chan, K. H. (1992) Ligament tension pattern in the flexed knee in combined passive anterior translation and axial rotation. *J. Orthop. Res.*, **10**(6), 854–867.

Alexander, R. E., Battye, C. K., Goodwill, C. J. and Walsh, J. B. (1982) The ankle and subtalar joints. *Clin. Rheum. Dis.*, **8**(3), 703–711.

Anderson, A. F., Snyder, R. B., Federspiel, C. F. and Lipscomb, A. B. (1992) Instrumented evaluation of knee laxity: a comparison of five arthrometers. *Am. J. Sports Med.*, **20**(2), 135–140.

Arms, S., Boyle, J., Johnson, R. and Pope, M. (1983) Strain measurement in the medial collateral ligament of the human knee: an autopsy study. *J. Biomech.*, **16**(7), 491–496.

Arnoczky, S. P. (1983) Anatomy of the anterior cruciate ligament. *Clin. Orthop. Related Res.*, **172**, 19–25.

Ateshian, G. A., Soslowsky, L. J. and Mow, V. C. (1991) Quantitation of articular surface topography and cartilage thickness in knee joints using stereophotogrammetry. *J. Biomech.*, **24**(8), 761–776.

Attarian, D. E., McCrackin, H. J., Devito, D. P., McElhaney, J. H. and Garrett, W. E. J. (1985) A biomechanical study of human lateral ankle ligaments and autogenous reconstructive grafts. *Am. J. Sports Med.*, **13**(6), 377–381.

Barany, M. (1967) ATPase activity of myosin correlated with speed of muscle shortening. *J. Gen. Physiol.*, **50**, 197–218.

Bojsen-Möller, F. and Lamoreux, L. (1979) Significance of free dorsiflexion of the toes in walking. *Acta Orthop. Scand.*, **50**, 471–479.

Brossmann, J., Muhle, C., Schroder, C., Melchert, U. H., Bull, C. C., Spielmann, R. P. and Heller, M. (1993) Patellar tracking patterns during active and passive knee extension: evaluation with motion-triggered cine MR imaging. *Radiology*, **187**(1), 205–212.

Brown, T. D. and Shaw, D. T. (1984) In vitro contact stress distribution on the femoral condyles. *J. Orthop. Res.*, **2**(2), 190–199.

Bruns, J., Volkmer, M. and Luessenhop, S. (1994) Pressure distribution in the knee joint. Influence of flexion with and without ligament dissection. *Arch. Orthop. Trauma Surg.*, **113**(4), 204–209.

Bulucu, C., Thomas, K. A., Halvorson, T. L. and Cook, S. D. (1991) Biomechanical evaluation of the anterior drawer test: the contribution of the lateral ankle ligaments. *Foot Ankle*, **11**(6), 389–393.

Bylski-Austrow, D. I., Ciarelli, M. J., Kayner, D. C., Matthews, L. S. and Goldstein, S. A. (1994) Displacements of the menisci under joint load: an in vitro study in human knees. *J. Biomech.*, **27**(4), 421–431.

Carlsoo, S. and Molbech, S. (1966) The functions of certain two-joint muscles in a closed muscular chain. *Acta Morphol. Neerl. Scand.*, **6**(4), 377–386.

Cawley, P. W. and France, E. P. (1991) Biomechanics of the lateral ligaments of the ankle: an evaluation of the effects of axial load and single plane motions on ligament strain patterns. *Foot Ankle*, **12**(2), 92–99.

Clark, J. M. and Sidles, J. A. (1990) The interrelation of fiber bundles in the anterior cruciate ligament. *J. Orthop. Res.*, **8**(2), 180–188.

Close, J. R., Inman, V. T., Poor, P. M. and Todd, F. N. (1967) The function of the subtalar joint. *Clin. Orthop. Relat. Res.*, **50**, 159–179.

Colville, M. R., Marder, R. A., Boyle, J. J. and Zarins, B. (1990) Strain measurement in lateral ankle ligaments. *Am. J. Sports Med.*, **18**(2), 196–200.

Conlan, T., Garth, W. P. J. and Lemons, J. E. (1993) Evaluation of the medial soft-tissue restraints of the extensor mechanism of the knee. *J. Bone Joint Surg. (Am.)* **75**(5), 682–693.

Eichenblat, M. and Nathan, H. (1983) The proximal tibio fibular joint. An anatomical study with clinical and pathological considerations. *Int. Orthop.*, **7**(1), 31–39.

Elftman, H. (1960) The transverse tarsal joint and its control. *Clin. Orthop.*, **16**, 41–45.

Fulkerson, J. P. and Gossling, H. R. (1980) Anatomy of the knee joint lateral retinaculum. *Clin. Orthop. Relat. Res.*, **153**, 183–188.

Fuss, F. K. (1989) An analysis of the popliteus muscle in man, dog, and pig with a reconsideration of the general problems of muscle function. *Anat. Rec.*, **225**(3), 251–256.

Goodfellow, J., Hungerford, D. S. and Zindel, M. (1976) Patello-femoral joint mechanics and pathology. 1. Functional anatomy of the patello-femoral joint. *J. Bone Joint Surg. (Br.)*, **58**(3), 287–290.

Hallen, L. G. and Lindahl, O. (1966) The 'screw-home' movement in the knee-joint. *Acta Orthop. Scand.*, **37**(1), 97–106.

Harper, M. C. (1987) Deltoid ligament: an anatomical evaluation of function. *Foot Ankle*, **8**(1), 19–22.

Harper, M. C. (1991) The lateral ligamentous support of the subtalar joint. *Foot Ankle*, **11**(6), 354–358.

Heegaard, J., Leyvraz, P. F., Curnier, A., Rakotomanana, L. and Huiskes, R. (1995) The biomechanics of the human patella during passive knee flexion. *J. Biomech.*, **28**(11), 1265–1279.

Hirokawa, S. (1993) Biomechanics of the knee joint: a critical review. *Crit. Rev. Biomed. Eng.*, **21**(2), 79–135.

Hodler, J., Trudell, D., Pathria, M. N. and Resnick, D. (1992) Width of the articular cartilage of the hip: quantification by using fat-suppression spin-echo MR imaging in cadavers. *AJR*, **159**(2), 351–355.

Hollis, J. M., Takai, S., Adams, D. J., Horibe, S. and Woo, S. L. (1991) The effects of knee motion and external loading on the length of the anterior cruciate ligament (ACL): a kinematic study. *J. Biomech. Eng.*, **113**(2), 208–214.

Inman, V. T. (1976) *The Joints of the Ankle*. Williams & Wilkins, Baltimore.

Isman, R. E. and Inman, V. T. (1969) Anthropometric studies of the human foot and ankle. *Bull. Pros. Res.*, **10–11**, 97–129.

Jones, C. D., Keene, G. C. and Christie, A. D. (1995) The popliteus as a retractor of the lateral meniscus of the knee. *Arthroscopy*, **11**(3), 270–274.

Jonsson, K., Buckwalter, K., Helvie, M., Niklason, L. and Martel, W. (1992) Precision of hyaline cartilage thickness measurements. *Acta Radio.*, **33**(3), 234–239.

Karlsson, D. and Lundberg, A. (1994) In vivo measurement of body segment motion using external fixator markers. *Biomech. Semin.*, **8**, 179–184.

Kelly, A. M. and Rubinstein, N. A. (1994) The diversity of muscle fiber types and its origin during development. In: *Myology* (ed. A. G. Engel and C. Franzini-Armstrong), McGraw-Hill, New York, pp. 119–133.

Kjaersgaard-Andersen, P., Wethelund, J. O., Helmig, P. and Soballe, K. (1989) Stabilizing effect of the tibiocalcaneal fascicle of the deltoid ligament on hindfoot joint movements: an experimental study. *Foot Ankle*, **10**(1), 30–35.

Lieber, R. L. (1990) Hypothesis: biarticular muscles transfer moments between joints. *Dev. Med. Child Neurol.*, **32**(5), 456–458.

Livesay, G. A., Fujie, H., Kashiwaguchi, S., Morrow, D. A., Fu, F. H. and Woo, S. L. (1995) Determination of the in situ forces and force distribution within the human anterior cruciate ligament. *Ann. Biomed. Eng.*, **23**(4), 467–474.

Ljunggren, A. E. (1978) Articular surfaces of the knee joint. *Acta Morphol. Neerl. Scand.*, **16**(3), 171–197.

Lundberg, A. (1989) Kinematics of the ankle and foot. In vivo roentgen stereophotogrammetry. *Acta Orthop. Scand.* (Suppl.), **233**, 1–24.

Lundberg, A., Svensson, O. K., Nemeth, G. and Selvik, G. (1989) The axis of rotation of the ankle joint. *J. Bone Joint Surg. (Br.)*, **71**(1), 94–99.

Macirowski, T., Tepic, S. and Mann, R. W. (1994) Cartilage stresses in the human hip joint. *J. Biomech. Eng.*, **116**(1), 10–18.

Mann, R. A. (1987) Functional anatomy of the ankle joint ligaments. *Instr. Course Lect.*, **36**, 161–170.

Markolf, K. L., Wascher, D. C. and Finerman, G. A. (1993) Direct in vitro measurement of forces in the cruciate ligaments. Part II: The effect of section of the posterolateral structures. *J. Bone Joint Surg. (Am.)*, **75**(3), 387–394.

McCullough, C. J. and Burge, P. D. (1980) Rotatory stability of the load-bearing ankle. An experimental study. *J. Bone Joint Surg. (Br.)*, **62B**(4), 460–464.

Ogden, J. A. (1974) The anatomy and function of the proximal tibiofibular joint. *Clin. Orthop. Relat. Res.*, **101**(01), 186–191.

Paukal, E. (1904) Die Zuckungsformen von Kaninchenmusckeln verschniedener Farbe und Struktur. *Arch. Anat. Physiol.* 100–120.

Prilutsky, B. I. and Zatsiorsky, V. M. (1994) Tendon action of two-joint muscles: transfer of mechanical energy between joints during jumping, landing, and running. *J. Biomech.*, **27**(1), 25–34.

Procter, P. and Paul, J. P. (1982) Ankle joint biomechanics. *J. Biomech.*, **15**(9), 627–634.

Quiles, M., Requena, F., Gomez, L. and Garcia-Sancho, L. (1983) Functional anatomy of the medial collateral ligament of the ankle joint. *Foot Ankle*, **4**(2), 73–82.

Ranvier, L. (1874) De quelques faits relatifs à l'histologie et la physiologie des muscles striés. *Arch. Physiol. Norm. Pathol.*, **1**, 5–15.

Rasmussen, O. (1985) Stability of the ankle joint. Analysis of the function and traumatology of the ankle ligaments. *Acta Orthop. Scand.* (Suppl.), **211**, 1–75.

Rasmussen, O., Jensen, I. T. and Hedeboe, J. (1983) An analysis of the function of the posterior talofibular ligament. *Int. Orthop.*, **7**(1), 41–48.

Rasmussen, O., Kromann-Andersen, C. and Boe, S. (1983) Deltoid ligament. Functional analysis of the medial collateral ligamentous apparatus of the ankle joint. *Acta Orthop. Scand.*, **54**(1), 36–44.

Rasmussen, O. and Tovborg-Jensen, I. (1982) Mobility of the ankle joint: recording of rotatory movements in the talocrural joint in vitro with and without the lateral collateral ligaments of the ankle. *Acta Orthop. Scand.*, **53**(1), 155–160.

Sammarco, G. J. (1993) Kinematics of the ankle: a hinged axis model. *Foot Ankle*, **14**(2), 113.

Sammarco, G. J., Burstein, A. H. and Frankel, V. H. (1973) Biomechanics of the ankle: a kinematic study. *Orthop. Clin. North Am.*, **4**(1), 75–96.

Siegler, S., Block, J. and Schneck, C. D. (1988) The mechanical characteristics of the collateral ligaments of the human ankle joint. *Foot Ankle*, **8**(5), 234–242.

Singh, A. K., Starkweather, K. D., Hollister, A. M., Jatana, S. and Lupichuk, A. G. (1992) Kinematics of the ankle: a hinge axis model. *Foot Ankle*, **13**(8), 439–446.

Skyhar, M. J., Warren, R. F., Ortiz, G. J., Schwartz, E. and Otis, J. C. (1993) The effects of sectioning of the posterior cruciate ligament and the posterolateral complex on the articular contact pressures within the knee. *J. Bone Joint Surg. (Am.)*, **75**(5), 694–699.

Stephens, M. M. and Sammarco, G. J. (1992) The stabilizing role of the lateral ligament complex around the ankle and subtalar joints. *Foot Ankle*, **13**(3), 130–136.

Takai, S., Woo, S. L., Livesay, G. A., Adams, D. J. and Fu, F. H. (1993) Determination of the in situ loads on the human anterior cruciate ligament. *J. Orthop. Res.*, **11**(5), 686–695.

Tria, A. J. J., Johnson, C. D. and Zawadsky, J. P. (1989) The popliteus tendon. *J. Bone Joint Surg. (Am.)*, **71**(5), 714–716.

van Langelaan, E. J. (1983) A kinematical analysis of the tarsal joints. An X-ray photogrammetric study. *Acta Orthop. Scand.* (Suppl.), **204**, 1–269.

Virji-Babul, N. and Cooke, J. D. (1995) Influence of joint interactional effects on the coordination of planar two-joint arm movements. *Exp. Brain Res.*, **103**(3), 451–459.

Warwick, R. and Williams, P. (1973) *Gray's Anatomy*. W. B. Saunders, Philadelphia.

Watanabe, Y., Moriya, H., Takahashi, K., Yamagata, M., Sonoda, M., Shimada, Y. and Tamaki, T. (1993) Functional anatomy of the posterolateral structures of the knee. *Arthroscopy*, **9**(1), 57–62.

Woo, S. L., Livesay, G. A. and Engle, C. (1992) Biomechanics of the human anterior cruciate ligament. ACL structure and role in knee motion. *Orthop. Rev.*, **21**(7), 835–842.

Zavatsky, A. B. O. and Connor, J. J. (1992) A model of human knee ligaments in the sagittal plane. Part 1: Response to passive flexion. Proceedings of the Institution of Mechanical Engineers, Part H. *J. Eng. Med.*, **206**(3), 125–134.

3

Neuromechanical Concepts for the Assessment of the Control of Human Gait

BRADFORD J. MCFADYEN[1] AND MARC BÉLANGER[2]

[1]Département de Physiothérapie, Faculté de Médecine, Université Laval, Québec, Canada
[2]Département de Kinanthropologie, Université du Québec à Montréal, Montréal, Canada

INTRODUCTION

Locomotion in its most general definition encompasses all movement which displaces the body, whether it be rising from the chair, swimming or even climbing a rope. Walking is the most common form of locomotion and makes up a very large proportion of our normal activities of daily living. Thus, it is understandable that the analysis of locomotion in general and walking in particular has been the subject of numerous review articles, books and book chapters over this century.

The control of walking is necessarily diverse in order to accommodate the varying demands of the environment and changes to locomotor goals. Whether we look at steady-state level walking, gait initiation or the adaptation to uneven environments, a rich system of neuromechanical mechanisms is used for the safe execution of walking. What is seemingly so easy to perform is the result of an intricate coordination of neural substrata simultaneously involved in generating and monitoring locomotion through information processing, pattern generation and reflex control. The subsystem which is the focus of all this attention, the musculotendoskeletal system, is also rich with its own components of multi- and single-function muscles—involving either one or more joints or lines of action— and a network of ligaments and other passive tissues, all usually employed in a

Three-dimensional Analysis of Human Locomotion. Edited by P. Allard, A. Cappozzo, A. Lundberg and C. Vaughan
© 1997 John Wiley & Sons Ltd. ISBN 0 471 96949 4

manner to maintain mechanical integrity within the demanding three-dimensional world (see Chapter 2).

The purpose of this chapter is to provide a theoretical link between the studies of neural control and the biomechanics of human walking in order to introduce some neuromechanical concepts underlying human gait, as well as to raise questions, many of which have already been asked in the literature, but the answers to which are necessary to understand locomotor control in three-dimensional space. It is impossible to include all aspects or references for the neuromechanical control of human walking. This chapter will concentrate on general concepts and considerations for the control of gait by drawing upon examples from two principal categories, anticipatory and reactive control. Chapter 13 of this volume on neural network models of human gait and their comparison to biological neural circuitry should also be consulted. Some brief comments giving consideration to pathological walking will also be offered where appropriate, but readers should refer to Chapter 16 for a more detailed discussion.

MECHANICAL COMPLEXITY OF HUMAN WALKING

The locomotor apparatus detailed in the last chapter provides the flexibility required for general control of movement and equilibrium, but the muscle activity at a joint may not be directly related to the control of its prime movement(s). Activity may arise to control passive contributions from adjacent segments or external forces (such as gravity) and there is little doubt that the central nervous system (CNS) exploits such passive components to coordinate movement along the linked segments (Putnam, 1983). Any assessment of the control of human gait must consider the effects and ultimately the coordination of these passive components in order to fully interpret the active ones.

Because walking has its main effect within the sagittal plane, a number of locomotor strategies are virtually planar in goal (see below). However, over four decades ago, Bresler and Frankel (1950) warned that the mechanical effects of lateral displacement should not be neglected. Interestingly enough, even with the postwar exponential increase in motion analysis technology, and specifically the improvement in technology for three-dimensional analysis of human movement, the concentration on two-dimensional analysis has persisted (present authors included). Such studies involving two-dimensional data are not, of course, necessarily invalid. Winter, Eng and Ishac (1995), for example, have described the need to interpret moments related to body displacement within the overall 'plane of progression' and pointed out that limb rotational orientation (i.e. in the transverse plane) during walking leads to small differences from a given global frame of reference when rotations are less than 8° out of plane. Obviously, each study needs to be scrutinized with respect to the research question asked, as well as the movement and population studied.

As an example, the control of dynamic equilibrium during walking is anything but planar. Even with the small frontal plane displacement of the body centre of mass (less than 4 cm total), MacKinnon and Winter (1993) showed predominant generative power by the hip abductors for lateral control of the head, arm and trunk segments, as well as the swinging limb during normal walking. Foot evertors had comparatively smaller moments, but were no less important. Subjects were found to rely either on foot evertors or hip abductors to compensate for errors in foot placement which threatened to destabilize frontal plane balance. Therefore, although the actual displacements were not large, moments were mechanically significant within the frontal plane. Once the body is in motion, we can easily forget to consider that movements of low velocity and small amplitude can still require significant muscle and neural contributions. Thus, in interpreting muscle function during gait from a neuromechanical point of view, the role in providing medial/lateral support must be considered, perhaps even to the point of partitioning a given muscle's activity into simultaneous functions within multiple planes of movement.

When the nervous system malfunctions, such as in the case of a pathology, the three-dimensional effects can be greatly pronounced and, even though the plane of progression may be maintained, there often exists increased variability about other movement axes. Increased movement in the frontal and transverse planes may be introduced to offset some threat to upright walking such as loss of equilibrium or unwanted foot contact arising from such problems as hyperactive reflexes (e.g. spasticity), structural changes (e.g. amputation) or decreased strength (e.g. hypotrophy). Of course, changes or compensations in particular joints or limbs must be assessed with respect to the whole linked segment chain. Contractures of the hip extensors, for example, could cause decreased step lengths that may be more responsible for forefoot or toe walking than any problems at the ankle. When this is added to the specific dysfunctions that may occur due to trauma, disease or deformities, the speed of gait can also be significantly affected. As the speed of walking decreases, there exists a greater amplitude of movement within the frontal plane (Thorstensson et al., 1984). This is obviously a factor for all forms of slower walking, including that of normal subjects of different ages.

Finally, the mechanical subsystem which is the object of the CNS's control is, fortunately, ripe with many redundant degrees of freedom. However, depending on the research question at hand, a reductionist assumption that muscles perform a two-dimensional, unifunctional role may misrepresent the underlying neuro-mechanical goal of a given activity. Therefore, researchers need to juggle the great amount of information within the physical limitations of the recording systems and the time constraints imposed for reduction and analysis of the ever-increasing data sets.

NEURAL COMPLEXITY OF HUMAN WALKING

Some of the first references dealing with the neural control of locomotion date back to the end of the nineteenth century (Freusberg, 1874; Sherrington, 1899). While the main aim of much of the research has been to understand human locomotion, animal models have mostly been used to determine the neural substrata involved. Based mainly on these animal models, three control mechanisms of locomotion have been postulated (see Figure 1): central pattern generator

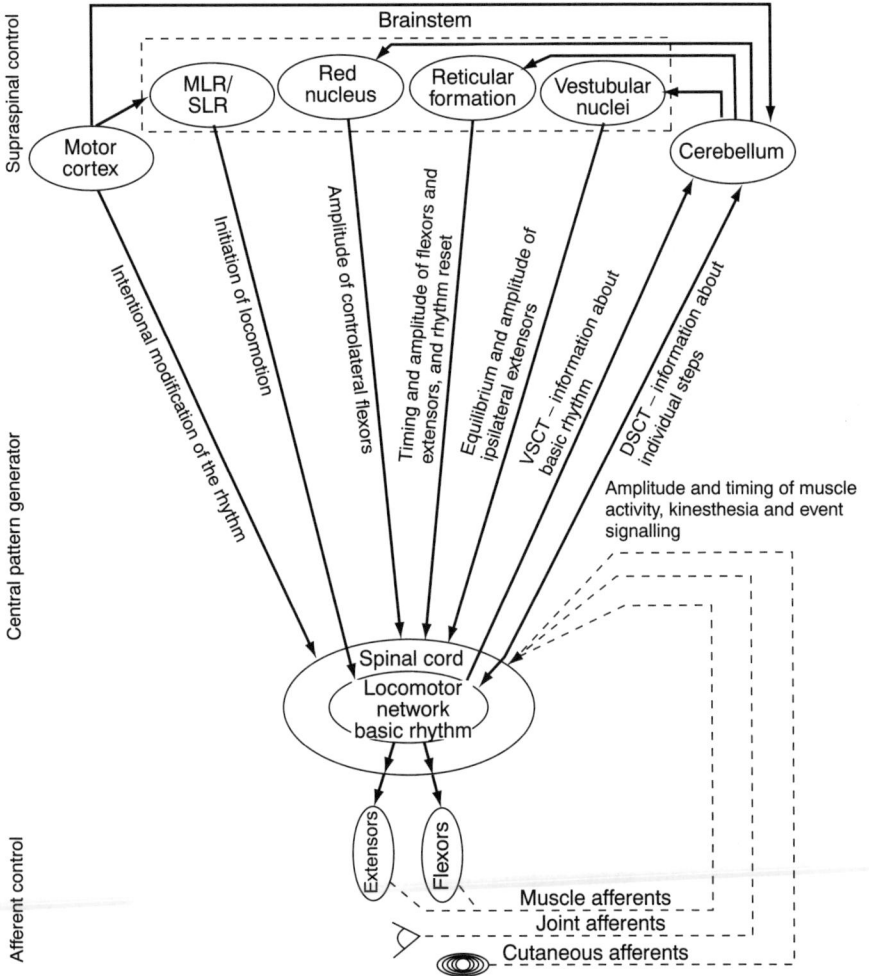

Figure 1. Schematic of the three levels of control of locomotion (afferent, central pattern generator and supraspinal) based on animal models, along with various related neural centres and their basic functions and interactions

(CPG); afferent or peripheral controls; and descending or supraspinal controls. Three different approaches have been used to investigate these control mechanisms: direct stimulation of a structure; recording the activity of a structure during locomotion; and removal of a structure. While the last two approaches have seldom been used in human locomotor studies, the first approach has been employed extensively, particularly in the last 15–20 years (see pp. 59–61). In order to be able to interpret the biomechanical aspects of locomotor tasks (i.e. output variables), one also needs a good understanding of the neural control mechanisms (i.e. input variables) shown in Figure 1. A brief introduction is offered below, but the reader is encouraged to seek out further work.

GENERAL CONCEPTS OF CENTRAL PATTERN GENERATION

As early as 1911, Brown (Brown, 1911) has shown that in a decerebrate spinalized cat with ligations of the dorsal roots, rhythmic locomotor activities in the hindlimbs could still be observed. With a paralysed spinal cat preparation, Grillner and Zangger (1974) clearly demonstrated that the isolated spinal cord was capable of generating rhythmic activity with 'locomotor-like' properties in the hindlimb flexor and extensor nerves (this is often referred to as 'fictive locomotion'). Other work (e.g. Grillner, 1976, 1982; Shik and Orlovsky, 1976; Delcomyn, 1980; Selverston, 1980) has demonstrated that a vertebrate's spinal cord has a network of neurons capable of generating rhythmic behaviours, recognized collectively and functionally as a CPG. Since the exact nature of the CPG is not known, numerous models of varying complexity have been proposed to explain the production of timing cues for the basic rhythmic activity of muscles during locomotion (see also Chapter 13). A simple but interesting model has been proposed by Grillner (1981) involving interconnected subunits (unit burst generators), one for each joint capable of producing the rhythmic patterns. Timing of activity for the different muscle groups, including bifunctional muscles, is the result of having interconnections between the generators such that it is possible to obtain either forward or backward locomotion by switching the connections between units from excitatory to inhibitory (Grillner, 1981).

In contrast to the work noted above, the direct evidence of a CPG in primates, including humans, is much less clear. The evidence of CPGs in humans is mainly derived anecdotally from spinal lesions which occurred during the World War I (Holmes, 1915) and World War II (Kuhn, 1950). Several patients from these studies were reported to be capable of generating alternating lower limb movements which resembled walking despite having lesions of the spinal cord that were made complete by the presence of shrapnel or grenade fragments. More recently, Calancie et al. (1994) provided EMG and kinematic records of what they believe were locomotor rhythms produced by a CPG in a patient with spinal cord injuries. On the other hand, the authors could not fully discount the possibility that parts of the generated movements and activity patterns were due to the effects of some of the descending systems. Other evidence of CPGs in humans comes from work in infants which demonstrates some innate capabilities of the spinal cord to generate locomotor movements (Forssberg, 1985) before the descending systems have matured and reached their targets in the spinal cord. This issue of insufficient maturation has, however, been questioned (Thelen, 1992). Thus, although the presence of a rhythmic spinal generator in humans can be postulated from a phylogenetic point of view, direct evidence is still missing.

AFFERENT CONTROL

It was initially believed that afferent inputs were responsible for producing basic locomotor rhythm as dictated by Sherrington's 'chain reflex hypothesis' (Sherrington, 1910). This idea persisted even with Brown's demonstration of a basic locomotor rhythm in the absence of peripheral feedback, as discussed above (Brown, 1911). Afferent control is now commonly understood to represent feedback from the periphery (e.g. the limbs) in order to control locomotion.

Based on the animal models, afferent inputs have been shown to have several important functions (see Rossignol, Lund and Drew, 1988). For example, recordings of afferent discharge demonstrate that signals from cutaneous (Loeb, 1981), muscle (Prochazka, Westerman and Ziccone, 1976) and joint (Loeb, Bak and Duysens, 1977) receptors are involved in kinaesthesia or for signalling particular events such as foot contact or the transition from one gait phase to another (Grillner and Rossignol, 1978). Deafferentiation experiments have shown the importance of afferent input in regulating muscle force, the amplitude of the muscular activity, and the extent of limb excursions (e.g. Sprong, 1929; Wetzel et al., 1976; Grillner and Zangger, 1984). Moreover, the removal of the afferent inputs also produces significant effects on the timing of muscular activity (Grillner and Zangger, 1975) and the interlimb coordination (Shik, Orlovsky and Severin, 1966).

Cutaneous afferent stimulation experiments in animals (Forssberg, Grillner and Rossignol, 1975) and in humans (Bélanger and Patla, 1984) have shown that the reflex responses are modulated according to the time of occurrence of the perturbations and the task being performed. Modulations have also been shown for muscular responses in humans (Capaday and Stein, 1986). These modulations represent gains of the spinal circuitry due to incoming peripheral information that are phase and task dependent. One caveat of these experiments is that the responses occurred at the muscular level, with little or no effects on the kinematics of the movements being performed. However, this does not mean that this lack of local movement will have no effect on the system as a whole. For example, a perturbation which occurs as a result of uneven terrain may trigger muscular responses which allow for the stabilization of the ankle and thus prevention of unwanted movements in all three planes.

Thus, although afferent inputs may not be necessary for generating the basic locomotor pattern, they are important for the coordination within and between limbs (including muscle coordination and joint coupling) and the regulation of the force and amplitude of movements. In brief, afferent information is important for both the maintenance of walking and its adaptation to changes in environmental conditions (e.g. changing floor grades) and to unexpected perturbations.

DESCENDING CONTROL

Figure 1 also provides details of certain levels of the supraspinal control. Data obtained from animal models, particularly in the cat, suggest that the roles of the descending system are to initiate locomotion, to make postural and gait adjustments (i.e. anticipatory and directional) and to mould the rudimentary patterns into movements which are fine-tuned to the locomotor task being performed. Stimulation and isolation experiments have identified two brain regions capable of initiating the locomotor functions: the mesencephalic locomotor region (MLR) (Shik, Severin and Orlovsky, 1966) and the subthalamic locomotor region (SLR) (Hinsey, Ranson and McNattin, 1930). Once activated, these regions send their messages to the spinal locomotor networks via noradrenergic descending systems (Jankowska et al., 1967) and the propriospinal polysynaptic pathways (Shik, 1983). The motor cortex appears to be unnecessary for the initiation of locomotion and also for some goal-directed behaviours such as feeding (Bjursten, Norrsell and Norrsell, 1976). In contrast, recent recordings from motor cortical cells during obstructed locomotion in the cat suggest that the motor cortex is involved in the modification of limb trajectory (Drew, 1993). The limb modifications in these cat experiments resemble those seen during obstructed walking in humans (McFadyen, Lavoie and Drew, 1995).

Other supraspinal structures which influence the locomotor activities include the red nucleus, the reticular formation, the vestibular system and the cerebellum. The first three structures, along with the motor cortex, can act directly on the spinal motoneurons or via interneuron circuitry to influence the locomotor rhythm (Orlovsky, 1972a). The rubrospinal neurons have been suggested to modulate the amplitude of the flexor muscle activity in the contralateral limb swing phase (Orlovsky, 1972b). The reticulospinal systems have been reported to regulate both the timing (Drew, Dubuc and Rossignol, 1986) and the amplitude of the ongoing flexor and extensor muscle activity (Orlovsky, 1972a). More importantly, Drew and Rossignol (1984) have shown that the locomotor rhythm can be reset by stimulation of the medullary reticular formation.

With lesions and stimulation of the lateral vestibular (Deiter's) nucleus, Orlovsky (1972c) has shown that the vestibulospinal system can increase the extensor activity on the ipsilateral side. Of course, the vestibulospinal system also provides equilibrium in the frontal plane, as evidenced by chronic spinal cat preparations where these cats have a tendency to lose their balance after several steps during treadmill locomotion (e.g. Bélanger et al., 1996). In addition to receiving vestibular inputs, the cerebellum receives information about the individual step cycles via the spinocerebellar tracts (Grillner, 1981). The dorsal spinocerebellar tract (DSCT) transmits exteroceptive and proprioceptive inputs, while information about the spinal interneuronal circuitry (e.g. CPG) is relayed via the ventral spinocerebellar tract (VSCT). The cerebellum has no direct connection to the spinal locomotor networks and must act through the other descending systems (see above) to fine-tune and coordinate the locomotor movements (Orlovsky and Shik, 1976).

Although the descending systems play an important role in fine-tuning the locomotor movements, the animal literature suggests that they are not essential for the generation of the basic coordinated walking movements, since such movements have been observed after complete transection of the spinal cord, as described earlier.

One of the shortcomings of the studies on the neural control of locomotion is that it has been examined most often in two dimensions. Some postural (e.g. MacPherson, 1988) and arm-reaching movement studies (e.g. Georgopoulos, Schwartz and Kettner, 1982) are among the few which have attempted to explain how such movements are controlled in three-dimensional space. MacPherson (1988) has shown that the cat's hindlimb muscles are activated maximally in a preferred direction for static control of equilibrium. Similarly, cortical recordings during arm movements in the monkey have shown that cells can discharge to a preferred direction in three-dimensional space (Georgopoulos, Schwartz and Kettner, 1982). If we are to understand locomotion fully, then neuroscientists, along with biomechanists, will have to study the underlying mechanisms which control such movements in three dimensions.

NEUROMECHANICAL STRATEGIES FOR THE CONTROL OF WALKING

The study of human walking is, for the most part, still predominated by protocols involving steady-state behaviour. Because the environment and our locomotor goals are rarely constant, we are required to continuously adjust and adapt. In order to maintain safe and effective (i.e. meeting the movement goal) gait, such changes must be, whenever possible, controlled in a proactive way, anticipating the needs and mutual effects of both the displacement and equilibrium goals for the specific environment involved. However, not all environmental factors are foreseeable, and nor is locomotor control always exact. Thus, reactive control, by decreasing the effects of a given perturbation and continuing the locomotor goal if possible, is equally as important to effective walking. This section will briefly introduce some fundamental aspects related to both anticipatory and reactive locomotor control, but is in no way meant to be an exhaustive discussion.

ANTICIPATORY ADJUSTMENTS OF MOTOR PATTERNS

Body displacement requires anticipatory locomotor adjustments (ALAs), which were defined by McFadyen, Magnan and Boucher (1993) as 'the change in the present locomotor dynamics associated with foreseeable environmental changes or redefined movement goals. ALAs are changes within the patterns focal to displacing the body' (pp. 260–261). In contrast, balance requires anticipatory postural adjustments (APAs), which, as pointed out by Belen'kii, Gurfinkel and Pal'tsev (1967), are involved when 'the anticipatory activation of the muscles as preparation for voluntary movement is connected with the need to maintain balance' (p. 161). Massion (1992) suggested two different ways in which focal and postural control may be organized together. A hierarchical mode was dominated by the movement to be executed, while a parallel mode involved both movement and postural control levels receiving information from higher centres and then being executed in parallel with one another.

Figure 2 is inspired by the parallel configuration presented by Massion (1992).

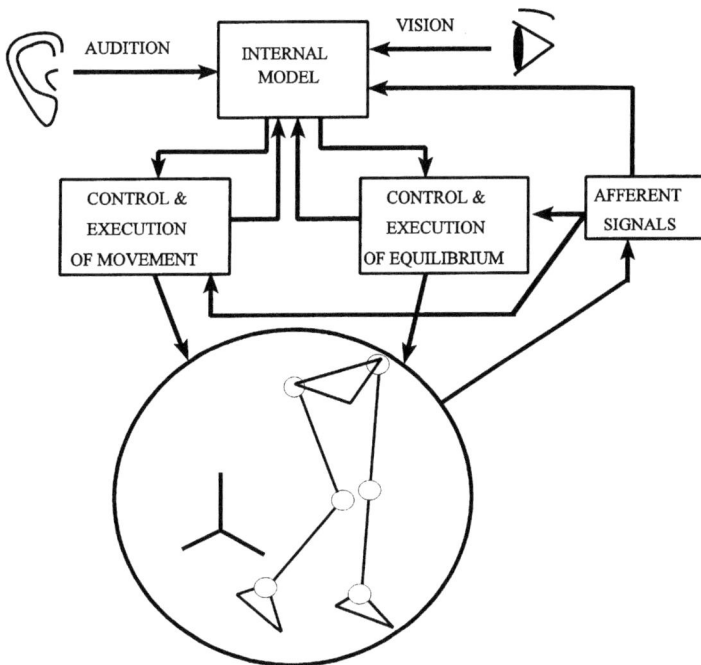

Figure 2. Simplified schematic of the control and execution of lower limb movement and equilibrium presented in a parallel configuration with coordination from a higher-level internal model. Exteroceptive information is represented by visual and auditive inputs, and proprioceptive information by afferent inputs. The three bars beside the lower limb cartoon represent a general three-dimensional reference system

An internal model has been included at the higher level to provide the assimilation of exteroceptive information, such as from vision and audition, with knowledge of the postural and focal control signals and of the present mechanical state of the body. With this information, predictions about the changes required to meet or maintain movement goals can be made and used to modify descending control.

The concept of the internal model provides an effective metaphor for discussing the information processing involved in motor planning and adaptation. It has been presented in simulation studies in general (Rosen, 1985) as well as with respect to human movement control of the upper (e.g. Jordan, 1990) and lower (e.g. McFadyen, Winter and Allard, 1994) limbs. Although it is not possible at this point to locate the exact neural centres (including the ones indicated in Figure 1) making up the internal model for locomotor control, such a model must contain some kind of knowledge base of the three-dimensional mechanical characteristics of both the body and the environment. Only in this way can the CNS effectively generate anticipatory adjustments for both adapting the patterns focal to displacing the body (ALAs), and assuring the necessary equilibrium before, or in conjunction with, movement execution (APAs). Of course, the two modes of control (ALA or APA) are not mechanically exclusive and each goal may even be superimposed on the muscular activity of natural movements (see pp. 50–51).

For the remainder of this section we will concentrate on two examples of ALAs. One is obstructed walking, which is environmentally related, while the other, initiation of walking, is related to a redefined movement goal.

It has been recently shown (McFadyen and Winter, 1991; McFadyen, Magnan and Boucher, 1993) that there exists a robust reorganization of lower limb patterns by the CNS in order to avoid obstacles of various relative heights and proximity to the leg. In particular, there are increases in both the hip and knee joints' flexion angles caused by the introduction of a knee flexor generation power at the point of toe-off. At the same time, the normal hip pull-off energy decreases in magnitude with the introduction of the obstacle.

Such reorganized patterns in the lower limb dynamics of ongoing gait exemplify the ability of the CNS to exploit the multiarticular nature of the lower limb and maintain continual locomotion despite the environmental constraints. Such patterns show a predominance of a neuromechanical strategy directed within the plane of progression. However, although the control by the unilateral limb is probably the most important direct control for immediate obstacle clearance in young healthy adults, the control of the pelvis and the contralateral limb is also important for added hip hiking and equilibrium during the destabilization of obstacle clearance and may play an increasingly important role, depending on the age of the person (Patla, Prentice and Gobbi, 1995).

With respect to changing a movement goal, the initiation of walking has been argued to involve predominantly sagittal plane control at the ankle (Herman et al., 1973) for forward progression of the body. However, there is an important weight-transfer process where ipsilateral abductors pull the body weight over to the swing leg initially and then unload to complement the increased contralateral abductor activity to move the weight in the opposite direction during the first step (e.g. Jian et al., 1993).

In the case of the initiation of walking in hemiparetic subjects, St-Vincent (1994) has shown that frontal plane control is increasingly important. Hemiparetic subjects showed a larger medial–lateral amplitude of the total centre of pressure trajectory coupled with a pulling by the stance leg (seen as a medially applied force), which implies hip adductor activity. This pulling force was absent in the age- and speed-matched control group (although the control subjects did display a decrease of the applied lateral ground force). In this way the hemiparetic subjects also preserved balance while restraining the total body centre of mass forward progression during its transfer over the stance side. These findings for hemiparetic subjects have been repeated by others (Malouin et al., 1995).

Thus, the CNS provides proactive control for the maintenance and adaptation of human locomotion. Such anticipatory intervention is essential, but not sufficient, in a world fraught with many unknowns. The ability to update the internal model, react to the unexpected and correct for system errors is also necessary.

REACTIVE CONTROL OF WALKING

Until quite recently, reflexes, which represent automated responses that serve to protect us, were thought to be constant irrespective of the conditions, whether static (e.g. lying supine, sitting and standing) or dynamic (e.g. walking, running and cycling). Influential work in the cat by Forssberg, Grillner and Rossignol (1975) demonstrated that cutaneous reflexes, evoked either by a mechanical or a weak electrical stimulation of the skin of the paw, were modulated and dependent on the phase of the cycle of locomotion in which the stimulation was applied. Typically, a stimulus applied during the swing phase resulted in the excitation of the ipsilateral flexors and contralateral extensors. The same stimulus applied during the stance phase had no effect on the ipsilateral flexors but produced a large response in the ipsilateral extensors and a small burst of activity in the contralateral flexors.

Phase-dependent responses to cutaneous stimulation during human locomotion were first demonstrated in studies by Bélanger and Patla (1984) and Crenna and Frigo (1984). Duysens et al. (1990) reported that a complete reflex reversal as described in the animal studies was only observed when the stimulation was of low intensity. These authors not only observed a reversal in the muscular activity of flexor and extensor muscles but also demonstrated a movement reversal at the ankle joint (i.e. from flexor to extensor responses). Thus, in addition to showing that the cutaneous reflex is modulated according to the phase of the step cycle, these studies also indicated that stimulation intensity is important for reflex reversal.

Phase-dependent reflexes are not restricted to inputs arising from cutaneous sources but could also be observed in the reflexes from muscular origins. Capaday and Stein (1986) demonstrated that, while keeping a relatively constant stimulation intensity as assessed by the amplitude of the motor (M) response, the soleus H-reflex amplitude was strongly modulated during the step cycle. The H-reflex amplitude increased nearly in parallel with the motor activity of the muscle during the stance phase, while it was quite low during the swing phase. It was demonstrated that this modulation also occurred in other types of movements such as running, walking on a narrow beam and cycling (Capaday and Stein, 1987; Llewellyn, Yang and Prochazka, 1990; Brooke and McIlroy, 1990). Interestingly, Brooke, McIlroy and Collins (1992) found that it was not necessary to have active movements in order to see phase dependency in the soleus H reflex. Phase modulation was observed by simply having the subjects' lower limb fixed at different positions and during passive cycling moments (McIlroy, Collins and Brooke, 1992).

Cutaneous and muscular reflexes are dependent not only on the phase of the movement cycle (i.e. phase dependent) but also on the particular task (i.e. task dependent). The first study to show task dependency in humans was that of Lisin, Frankstein and Rechtmann (1973), who reported a weakening of the flexor reflex during the walking cycle compared to the static condition. More recently, Duysens et al. (1993) showed the opposite effect when comparing standing and running. This type of response is more consistent with that reported in animals (Rossignol and Drew, 1986). The reflex amplitude (reflex gain) of the muscular response was also shown to be altered in different tasks despite similar muscle contraction levels (Capaday and Stein, 1986). These authors have shown a significant drop in soleus reflex gain when going from standing to walking and a further decrease when the gait progressed from walking to running (Capaday and Stein, 1987), and this depression in the H-reflex amplitude was independent of the speed of locomotion (Edamura, Yang and Stein, 1991). Brooke and his colleagues also demonstrated a relationship between the speed of the passive cycling movements and the reflex gain reduction (e.g. McIlroy, Collins and Brooke, 1992).

In daily real-life situations, the CNS is bombarded by inputs arising from different sources, including skin, joint and muscular afferents. Several studies have started to examine the effects of conditioning stimuli on the muscular responses during cyclical movements (e.g. Fung and Barbeau, 1994). These authors reported that the soleus H reflex was most inhibited at foot contact and midswing of the locomotor step cycle and that the inhibition was greater during walking compared to static conditions which mimicked the dynamic postures.

The corrective responses recorded during stumbling studies have been pro-posed to involve both spinal and supraspinal reflexes (Berger, Dietz and

Quintern, 1984). Moreover, it was proposed that such responses may involve preprogrammed mechanisms which generate ipsilateral and contralateral adjustments. Hence, the responses can be considered as muscular locomotor synergies similar in concept to those obtained following standing platform perturbations (Horak and Nashner, 1981).

From this short summary of reflexes during locomotion, one can see the complexity of the responses which are influenced by the phase of the cycle, the task and the stimulation parameters. These complex responses are also paramount for continual and adaptable locomotion.

CONCLUSIONS AND FUTURE CONSIDERATIONS

Locomotor control requires a coordination of passive and active forces across the musculotendoskeletal linkages. Such forces must maintain the integrity of the intended displacement within the plane of progression (or planes in the case of directional changes) coupled with equilibrium within all directions. We have only briefly begun to uncover the supraspinal, peripheral and pattern generation networks associated with such control, as well as the mechanical characteristics underlying the multiple degrees of freedom system of connected segments. It has been suggested here that neuromechanical strategies for such control can be generally divided into anticipatory or reactive strategies. We are still a long way from fully understanding all that underlies the control of human gait, and can empathize with Hemami and Stokes (1983) when they referred to the study of locomotor control as a 'challenging, difficult, multifaceted, and multidisciplinary' problem.

As the future unfolds, we will be increasingly forced to acknowledge that gait—all movement for that matter—is manifested from an intricate blending of neural, psychological and mechanical substrata within the human body that is oblivious to these scientific disciplines. The study of the control of gait will require multidisciplinary teams with top-down approaches, each with a list of what they feel are the pressing issues requiring study and clarification. Norman (1986) has already made an attempt to list the substantive issues facing biomechanics. We would like to close with our own list for understanding the control of locomotion (recognizing both its biases and tendency to borrow from others). We have classified this list into two general categories of needs: conceptual and technical.

CONCEPTUAL NEEDS

- Discover the neural source of basic locomotor patterns (both the super-imposed progression and equilibrium goals) and their adjustments in humans (i.e. elucidate on the levels of control and their interactions as indicated in

Figures 1 and 2).

- Understand the mechanisms of dysfunction of such patterns.
- Detail and understand the multifunctionality of muscle, including:
  the three-dimensional lines of action of muscle at a joint
  the role of multijoint muscles
  functional compartmentalization of muscle
- Blend research strategies within and across disciplines, including:
  human versus animal works
  top-down versus bottom-up versus reductionist approaches

TECHNICAL NEEDS

- Seek more direct and easier measures of internal forces and moments of force.
- Develop faster and less labour-intensive algorithms for the collection and reduction of large data sets.
- Standardize data reporting and collection for easier exchange, including:
  common coordinate system (mathematical or anatomical) or agreed-upon differences
  common terminology
- Decrease problems of skin movement during motion recording.

ACKNOWLEDGEMENTS

The authors wish to acknowledge the financial support of the Natural Sciences and Engineering Research Council of Canada and the assistance of Mme Carole Roy.

REFERENCES

Bélanger, M. and Patla, A. E. (1984) Corrective responses to perturbation applied during walking in humans. *Neurosci. Lett.*, **49**, 291–295

Bélanger, M., Drew, T., Provencher, J. and Rossignol, S. (1996) A comparison of treadmill locomotion in adult cats and after spinal transection. *J. Neurophysiol.*, **76**, 471–491.

Belen'kii, V. Ye., Gurfinkel, V. S. and Pal'tsev, Ye. I. (1967) Elements of control of voluntary movements. *Biofizika*, **12**, 154–161.

Berger, W., Dietz, V. and Quintern, J. (1984) Corrective reactions to stumbling in man: neuronal co-ordination of bilateral leg muscle activity during gait. *J. Physiol. (Lond.)*, **357**, 109–125.

Bjursten, L. M., Norrsell, K. and Norrsell, U. (1976) Behavioural repertory of cats without cerebral cortex from infancy. *Exp. Brain Res.*, **25**, 115–130.

Bresler, B. and Frankel, J. (1950) The forces and moments in the leg during level walking. *Trans ASME*, **72**, 27–36.

Brooke, J. D. and McIlroy, W. E. (1990) Movement modulation of a short latency reflex linking the lower leg and the knee extensor muscles in humans. *Electroencephalogr. Clin. Neurophysiol.*, **75**, 64–74.

Brooke, J. D., McIlroy, W. E. and Collins, D. F. (1992) Movement features and H-reflex modulation. I. Pedalling versus matched controls. *Brain Res.*, **582**, 78–84.

Brown, T. G. (1911) The intrinsic factors in the act of progression in the mammal. *Proc. R. Soc. Lond. (Biol.)*, **84**, 308–319.

Calancie, B., Needham-Shropshire, B., Jacobs, P., Willer, K., Zych, G. and Green B. A. (1994) Involuntary stepping after chronic spinal cord injury. Evidence for a central rhythm generator for locomotion in man. *Brain*, **117**, 1143–1159.

Capaday, C. and Stein, R. B. (1986) Amplitude modulation of the soleus H-reflex in the human during walking and standing. *J. Neurosci.*, **6**, 1308–1313.

Capaday, C. and Stein, R. B. (1987) Difference in the amplitude of the human soleus H reflex during walking and running. *J. Physiol. (Lond.)*, **392**, 513–522.

Crenna, P. and Frigo, C. (1984) Evidence of phase-dependent nociceptive reflexes during locomotion in man. *Exp. Neurol.*, **85**, 336–345.

Delcomyn, F. (1980) Neural basis of rhythmic behaviour in animals. *Science*, **210**, 492–498.

Drew, T. (1993) Motor cortical activity during voluntary gait modifications in the cat. I. Cells related to the forelimbs. *J. Neurophysiol.*, **70**, 179–199.

Drew, T., Dubuc, R. and Rossignol, S. (1986) Discharge patterns of reticulospinal and other reticular neurons in chronic, unrestrained cats walking on a treadmill. *J. Neurophysiol.*, **55**, 375–401.

Drew, T. and Rossignol, S. (1984) Phase-dependent responses evoked in limb muscles by stimulation of medullary reticular formation during locomotion in thalamic cats. *J. Neurophysiol.*, **52**, 653–675.

Duysens, J., Trippel, M., Horstmann, G. A. and Dietz, V. (1990) Gating and reversal of reflexes in ankle muscles during human walking. *Exp. Brain Res.*, **82**, 351–358.

Duysens, J., Tax, A. A. M., Trippel, M. and Dietz, V. (1993) Increased amplitude of cutaneous reflexes during human running as compared to standing. *Brain Res.*, **613**, 230–238.

Edamura, M., Yang, J. F. and Stein, R. B. (1991) Factors that determine the magnitude and time course of human H-reflexes in locomotion. *J. Neurosci.*, **11**, 420–427.

Forssberg, H. (1985) Ontogeny of human locomotor control: I. Infant stepping, supported locomotion and transition to independent locomotion. *Exp. Brain Res.*, **57**, 480–493.

Forssberg, H., Grillner, S. and Rossignol, S. (1975) Phase dependent reflex reversal during walking in chronic spinal cats. *Brain Res.*, **85**, 103–107.

Freusberg, A. (1874) Reflexbewegungen beim Hunde. *Pflugers Arch. Ges. Physiol.*, **9**, 358–391.

Fung, J. and Barbeau, H. (1994) The effects of conditioning cutaneomuscular stimulation on the soleus H-reflex in normal and spastic paretic subjects during walking and standing. *Exp. Brain Res.*, **72**, 2090–2104.

Georgopoulos, A. P., Schwartz, A. B. and Kettner, R. E. (1982) Neuronal population coding of movement direction. *Science*, **233**, 1416–1419.

Grillner, S. (1976) Some aspects of the descending control of the spinal circuits generating locomotor movements. In: *Neural Control of Locomotion* (ed. R. M. Herman, S. Grillner, P. S. G. Stein and D. G. Stuart), Plenum Press, New York, pp. 351–375.

Grillner, S. (1981) Control of locomotion in bipeds, tetrapods and fish: In: *Handbook of Physiology*, section 1: *The Nervous System*, Vol. II, Part 2 (ed. J. M. Brookhart and V. B. Mountcastle) American Physiological Society, Bethesda, pp. 1179–1236.

Grillner, S. and Rossignol, S. (1978) On the initiation of the swing phase of locomotion in chronic spinal cats. *Brain Res.*, **146**, 269–277.

Grillner, S. and Zangger, P. (1974) Locomotor movements generated by the deafferented spinal cord. *Acta Physiol. Scand.*, **91**, 38–39A.

Grillner, S. and Zangger, P. (1975) How detailed is the central pattern generation for locomotion? *Brain Res.*, **88**, 367–371.

Grillner, S. and Zangger, P. (1984) The effect of dorsal root transection on the efferent motor pattern in the cat's hindlimb during locomotion. *Acta Physiol. Scand.*, **120**, 393–405.

Hemami, H. and Stokes, B. (1983) A quantitative discussion of mechanisms of feedback and feedforward in the control of locomotion. *IEEE Trans Biomed. Eng.*, 681–688.

Herman, R., Cook, T., Cozzens, B. and Freedman, W. (1973) Control of postural reactions in man: the initiation of gait. In: *Control of Posture and Locomotion* (ed. R. B. Stein, K. G. Pearson, R. S. Smith and J. B. Redford), Plenum Press, New York, pp. 363–388.

Hinsey, J. C., Ranson, S. W. and McNattin, R. F. (1930) The role of the hypothalamus and mesencephalon in locomotion. *Arch. Neurol. Psychiatry*, **23**, 1–43.

Holmes. G. (1915) Spinal injuries of warfare. II. The clinical symptoms of gunshot injuries of the spine. *B. Med. J.*, **2**, 815–821.

Horak, F. B. and Nashner, L. M. (1981) Central programming for postural movements: adaptation to altered support-surface configurations. *J. Neurophysiol.*, **55**, 1369–1381.

Jankowska, E., Jukes, M., Lund, S. and Lundberg, A. (1967) The effect of DOPA on the spinal cord. 6. Half-centre organization of interneurones transmitting effects from flexor reflex afferents. *Acta Physiol. Scand.*, **70**, 389–402.

Jian, Y., Winter, D. A., Ishac, M. G. and Gilchrist, L. (1993) Trajectory of the body COG and COP during initiation and termination of gait. *Gait Posture*, **1**, 9–22.

Jordan, M. (1990) Motor learning and the degrees of freedom problem. In: *Attention and Performance XIII* (ed. M. Jeannerod), Erlbaum, Hillsdale, pp. 796–836.

Kuhn, R. A. (1950) Functional capacity of the isolated human spinal cord. *Brain*, **73**, 1–51.

Lisin, V. V., Frankstein, S. I. and Rechtmann, M. B. (1973) The influence of locomotion on flexor reflex of the hindlimb in cat and man. *Exp. Neurol.*, **38**, 180–183.

Llewellyn, M., Yang, J. F. and Prochazka, A. (1990) Human H-reflexes are smaller in difficult beam walking than in normal treadmill walking. *Exp. Brain Res.*, **83**, 22–28.

Loeb, G. E. (1981) Somatosensory unit input to the spinal cord during normal working. *Can. J. Physiol. Pharmacol.*, **59**, 627–635.

Loeb, G. E., Bak, M. J. and Duysens, J. (1977) Long-term unit recording from somatosensory neurons in the spinal ganglia of the freely walking cat. *Science*, **197**, 1192–1194.

MacKinnon, C. D. and Winter, D. A. (1993) Control of whole body balance in the frontal plane during human walking. *J. Biomech.*, **26**, 633–644.

MacPherson, J. M. (1988) Strategies that simplify the control of quadrupedal stance. II. Electromyographic activity. *J. Neurophysiol.*, **60**, 218–231.

Malouin, F., Richards, C., Dumas, F., Bonneau, F., Comeau, F. and Lavoie, S. (1995) Effects of paresis on gait initiation after stroke. *Proc. Soc. Neurosci.*, **21**, 2083.

Massion, J. (1992) Movement, posture and equilibrium: interaction and coordination. *Prog. Neurobiol.*, **38**, 35–86.

McFadyen, B. J., Lavoie, S. and Drew, T. (1995) Unilateral hindlimb anticipatory locomotor adjustments for near and far obstacles in the cat. *Proc. Soc. Neurosci.*, **21**, 419.

McFayden, B. J., Magnan, G. A. and Boucher, J. P. (1993) The anticipatory locomotor adjustments for avoiding visible fixed obstacles of varying proximity. *Hum. Move. Sci.*, **12**, 259–272.

McFadyen, B. J. and Winter, D. A. (1991) Anticipatory locomotor adjustments during obstructed human walking. *Neurosci. Res. Commun.*, **9**, 37–44.

McFadyen, B. J., Winter, D. A. and Allard, F. (1994) A computer model of the control of unilateral anticipatory locomotor adjustments for obstructed gait. *Biol. Cybern.*, **72**, 151–160.

McIlroy, W. E., Collins, D. F. and Brooke, J. D. (1992) Movement features and H-reflex modulation. II. Passive rotation, movement velocity and single leg movement. *Brain Res.*, **582**, 85–93.

Norman, R. W. (1986) A barrier to understanding human motion mechanisms: accurate estimation of individual muscle and ligament force-time histories. In: *Proceedings of the Symposium on Future Directions in Exercise and Sport Research* (ed. J. Skinner, C. Corbin, D. Landers, P. Martin and C. Wells), Human Kinetics Publishers, Champaign, pp. 1–21.

Orlovsky, G. N. (1972a) The effect of different descending systems on flexor and extensor activity during locomotion. *Brain Res.*, **40**, 359–371.

Orlovsky, G. N. (1972b) Activity of rubrospinal neurons during locomotion. *Brain Res.*, **46**, 99–112.

Orlovsky, G. N. (1972c) Activity of vestibulospinal neurons during locomotion. *Brain Res.*, **46**, 85–98.

Orlovsky, G. N. and Shik, M. L. (1976) Control of locomotion: a neurophysiological analysis of the cat locomotor system. In: *International Review of Physiology Neurophysiology II*, Vol. 10 (ed. R. Porter), University Park Press, Baltimore, pp. 281–317.

Patla, A. E., Prentice, S. D. and Gobbi, T. (1995) Visual control of obstacle avoidance during locomotion: strategies in young children, young and older adults. In: *Changes in Sensory-Motor Behaviour in Aging* (ed. A.-M. Ferrandez and N. Teasdale), Elsevier Science BV, Amsterdam, pp. 257–277.

Prochazka, A., Westerman, R. A. and Ziccone, S. P. (1976) Discharges of single hindlimb afferents in the freely moving cat. *J. Neurophysiol.*, **39**, 1090–1104.

Putnam, C. A. (1983) Interaction between segments during a kicking motion. In: *Biomechanics VIII-B* (ed. H. Matsui and K. Kobayashi), Human Kinetics Publishers, Champaign, pp. 688–694.

Rossen, R. (1985) *Anticipatory Systems: Philosophical, Mathematical and Methodological Foundations*, Pergamon Press, Oxford.

Rossignol, S. and Drew, T. (1986) Phasic modulation of reflexes during rhythmic activity. In: *Neurobiology of Vertebrate Locomotion* (ed. S. Grillner, P. S. G. Stein, D. G. Stuart and H. Forssberg), MacMillan, London, pp. 517–534.

Rossignol, S., Lund, J. P. and Drew, T. (1988) The role of sensory inputs in regulating patterns of rhythmic movements in higher vertebrates. A comparison between locomotion, respiration and mastication. In: *Neural Control of Rhythmic Movements in Vertebrates* (ed. A. Cohen, S. Rossignol and S. Grillner), Wiley and Sons, New York, pp. 201–283.

Selverston, A. F. (1980) Are central pattern generators understandable? *Behav. Brain Sci.*, **3**, 535–571.

Sherrington, C. S. (1899) On the spinal animal. *Med. Chir. Trans.*, **82**, 449–486.

Sherrington, C. S. (1910) Flexion-reflex of the limb, crossed extension-reflex, and reflex stepping and standing. *J. Physiol. (Lond.)*, **40**, 28–121.

Shik, M. L. (1983) Action of the brainstem locomotor region on spinal stepping generators via propriospinal pathways. In: *Spinal Cord Reconstruction* (ed. C. C. Kao, R. P. Bunge and P. J. Reier), Raven Press, New York, pp. 421–434.

Shik, M. L. and Orlovsky, G. N. (1976) Neurophysiology of locomotor automatism. *Physiol. Rev.*, **56**, 465–500.

Shik, M. L., Orlovsky, G. N. and Severin, F. V. (1966) Organization of locomotor synergism. *Biophysics*, **11**, 1011–1019.

Shik, M. L., Severin, F. V. and Orlovsky, G. N. (1966) Control of walking and running by means of electrical stimulation of the mid-brain. *Biophysics*, **11**, 756–765.

Sprong, W. L. (1929) A study of reflexes in the deafferented leg of the cat and their relation to tonus. *Bull. Johns Hopkins Hosp.*, **45**, 371–395.

St-Vincent, G. (1994) The initiation of hemiplegic gait. MSc Thesis, Université du Québec à Montréal, Montréal, Québec.

Thelen, E. (1992) Development of locomotion from a dynamic systems approach. In: *Movement Disorders in Children: Proceedings of the International Sven Jerring Symposium* (ed. H. Forssberg and H. Hirschfeld), Karger, Basel, pp. 169–173.

Thorstensson, A., Nilsson, J., Carlson, H. and Zomlefer, M. R. (1984) Trunk movements in human locomotion. *Acta Physiol. Scand.*, **121**, 9–22.

Wetzel, M. C., Atwater, A. E., Wait, J. V. and Stuart, D. G. (1976) Kinematics of locomotion by cats with a single hindlimb deafferented. *J. Neurophysiol.* **39**, 667–678.

Winter, D. A., Eng, J. J. and Ishac, M. G. (1995) A review of kinetic parameters in human walking. In: *Gait Analysis: Theory and Application*, (ed. R. L. Craik and C. A. Oatis), Mosby, St Louis, pp. 252–270.

4

The Gait Laboratory

RACHID AISSAOUI[1,2], PAUL ALLARD[1,2], RÉGIS LACHANCE[1,2]
AND MORRIS DUHAIME[1,3,4]

[1]Laboratoire d'Étude du Mouvement, Research Center, Sainte-Justine Hospital,
Montreal, Canada
[2]Department of Physical Education, University of Montreal, Montreal, PQ, Canada
[3]Orthopedic Surgery, Shriner's Hospital, Montreal, PQ, Canada
[4]McGill University, Montreal, PQ, Canada

At one time or another, those involved in three-dimensional analysis of human locomotion have faced the challenge of setting up a gait laboratory. Some, perhaps more fortunate, begin their work in an existing laboratory. Nonetheless, we eventually have to modify the existing facilities to meet our growing needs. These can take the form of larger facilities, upgrading of equipment or completely changing a system, buying new equipment to complement that in place, installing new furniture, etc. This chapter deals first with the planning of a gait laboratory and should be useful to those planning a new gait laboratory and to the many who will eventually be called to substantially modify their installations.

The basic requirements for starting or running a gait laboratory can be summarized in terms of facilities, equipment and personnel. These aspects, detailed in Table 1, are all equally important. The physical aspects are presented first, since once the space is allotted for a gait laboratory it may decrease over the years but seldom expands. However, technological advances lead to the upgrading of equipment and often to the replacement of other equipment which is either obsolete or discontinued. A few years ago, high-speed film cameras were widely used to document gait patterns. Now, this technology is being rapidly replaced by video camera systems. Though electromyography is often used in the quantification of able-bodied locomotion and in the evaluation of pathological gait, the instrumentation and software related to muscle assessment will not be reviewed here. Similarly, the expertise required in fundamental

Three-dimensional Analysis of Human Locomotion. Edited by P. Allard, A. Cappozzo, A. Lundberg and C. Vaughan
© 1997 John Wiley & Sons Ltd. ISBN 0 471 96949 4

Table 1. Basic requirements for setting up or operating a gait laboratory

Facilities		Equipment		
Physical aspects	Utilities	Hardware		Software
		Acquisition	Peripherals	
Space Layout Access	Electrical Communication Protection Air conditioning	Computer Data capture	Storage Printers	Operating systems and platforms Data capture Data processing Software tools Communication

(locomotion studies), applied (orthotics and prosthetics) or clinical (pathological gait assessment) research will not be addressed.

FACILITIES

The facilities consist of the physical aspects of the room where the laboratory will be located and its utilities. For example, the lack or inadequacies of certain facilities in a larger room may be too expensive to remedy. A room can be large enough but the installation of the air-conditioning system can be prohibitive, whereas expanding a small room by moving a wall and installing a new telephone line could be less expensive. The total renovation cost must be considered when seeking an adequate room size to carry out the gait evaluations.

PHYSICAL ASPECTS

Some of the important physical aspects to consider in setting up a gait laboratory or when expanding an existing facility are related to the working space, the laboratory layout and accessibility to the patients as well as to other services.

Space

Unfortunately, gait laboratories have traditionally had a reputation for taking up a lot of space compared to other more traditional laboratories. Indeed, in some universities, professors and students do part of their experiments in confined areas and even in corridors! Though they can do their research without too much difficulty, gait evaluations require space for the subject to walk in and to maintain a reasonable camera-to-subject distance. This leaves plenty of unoccupied and valuable space when the laboratory is not in full use, making it sometimes difficult to justify all this space.

Walkway

Some of the physical considerations are detailed in Table 2. The size of the room is most important. You will need about a 10 m long by 2 m (approx 33 ft by 6 ft 6 metres) wide unobstructed area where the walking trials will be performed. There should be a distance of 3 m at the beginning and the end of the walkway, allowing the subject to accelerate and decelerate (Jian et al., 1993), though constant walking speed can be reached within 1 s (Nakurama, Yamamoto and Tsuji, 1994). The cameras are usually located at a distance of about 4 m from the middle portion of the walkway. Thus for a unilateral 3D gait analysis, the minimal room size should be about 5 m by 9 m (16 ft by 30 ft). For simultaneous bilateral gait analyses, the minimal unobstructed dimensions would be about 8 m by 9 m (26 ft by 30 ft).

These values correspond to the floor area of the Laboratoire d'Étude du Mouvement at Sainte-Justine Hospital where the eight video cameras are set against two opposite walls to cover a common field of view of 3 m long by 1.5 m wide and 2 m high. In addition to natural-speed walking, the subjects are able to perform slow jogging of 2.1 m/s. Those wishing to study different running conditions other than on a treadmill will need a walkway of length at least 15–18 m. In some motion laboratories, doors can be opened to make use of hallways as a run-up distance. Others more fortunate have a direct access to a running track. Naturally, a 15–20 m stopping distance may also be required!

Floors

Force plate installation can also be a problem. For these devices to be flush with the walkway, force plates have to be embedded into the floor. If the laboratory is

Table 2. List of facilities for setting up or operating a gait laboratory

Physical aspects	Utilities
Space	Electrical
Walkway	Power requirements
Floors	Power surcharge
Ceiling	Lighting
Columns and pillars	
Windows	
Layout	Communication systems
Work space	Telephone
Storage	Ethernet
Access	Protection
Main entrance	Fire protection
Clinic	Alarm system
Washroom	
	Air conditioning

not located in the basement of a building, drilling the floor may weaken the concrete slab. It may then be necessary to build an elevated floor around the force plates.

Floor tiles supported at each corner by a vertical post come in different materials, colour and finishes. Those having a high reflective index, 0.8 and above, can become a source of camera glare. Aluminium tiles are more expensive than those made of wood and are not normally necessary. The gait laboratory equipment is not exceedingly heavy, and walking and running are not too damaging to a laminated wooden surface. Since ground reaction forces are usually measured from uncovered force plates, the surrounding floor should not be carpeted, to ensure no abrupt changes in ground surfaces.

Usually, the tiles are insulated to reduce noise when walking over them. An elevated floor made of plywood is much less expensive but is not recommended because it can give out a hollow sound as subjects walk over it. Regardless of the floor structure, it must be rigid and vibration-free, which is best for the cameras as well as the subjects. Nearby heavy machinery may transmit vibrations through the floor to computers, causing the chips to vibrate out of their sockets, the boards to dislodge out of their slots, and the bearings to move out of their races.

Ceiling

If you decide on an elevated floor, make sure you have enough ceiling height. A 3-m (approx 10 ft) ceiling in an open floor makes the room appear flat and cavernous. A 3.30-m (approx 11 ft) ceiling gives you more headroom and does not make the room too gloomy. More ceiling height is needed if you intend to hang the cameras from the ceiling, or wish to study stair ascent or throwing activities. If electromyographic data are also collected, then a rail should be fixed to the ceiling so that the cables can slide on teflon rollers and unfold as the subject walks, assuming that a telemetry system is not used.

Columns and Pillars

Sometimes, one or two columns may occlude a large area. Since these columns support the upper floors or may contain air ducts for heating and ventilation, they cannot be removed or displaced. If you are fortunate, they may not disturb the cameras' field of view. If not, you may need additional cameras to adequately cover the test volume, reorganize the camera placement accordingly or orientate the walkway along the diagonal of the room.

Windows

Incoming light from a window is another disturbance. Daylight or ambient light sources from an adjacent corridor are additional sources of infrared light, and the visible light may create glare in the video cameras. Black venetian blinds can be pulled down to block the undesirable light during data acquisition, while maintaining some privacy for the subjects. The blinds can be raised at other times.

Layout

Work space and storage needs should be considered in terms of growing needs and efficient use of the laboratory area. Graduate students using the facilities may wish to have a temporary or permanent work space. Additionally, under-graduate students may have access to the laboratory for demonstrations or during a work term. During those peak periods, it may be difficult to satisfy everyone's needs, but good planning combined with a good layout can minimize part of the space-requirement problem.

Work Space

Once the size of the room has been determined, the floor area has to be divided into different areas for different functions. The walkway and the cameras' field of view take most of the floor area of the gait laboratory. If the gait evaluations are performed throughout the day and every working day, the offices, the additional computers and their peripherals as well as the storage space must be located outside the laboratory. One or two workstations can be located within the gait laboratory so that routine tasks such as data verification can be performed. If the professionals processing and analysing the data are those who take care of the evaluation, their work area can be located within the laboratory. The ergonomic aspects of office working posture (Corlett, Wilson and Manenica, 1986) and workstation design (Kvalseth, 1983) should be carefully considered, especially when planning a new office layout.

With the minimal room size of 4.75 m by 9 m for unilateral gait analyses or 8 m by 9 m for bilateral gait analyses, the workstations can be spread along the wall, leaving the room centre unobstructed for the gait assessments. The data capture system and its associated computers and peripherals can occupy one of the corners.

There must also be a place where the subject can have some privacy for changing. Curtains can be drawn to close off a section of the laboratory for this purpose. The same area can also be used for applying surface markers and taking the anthropometric and transformation measurements (external markers to joint centres). Curtains can divide the room into different sections when the laboratory is used for assessments other than gait, such as spinal mobility evaluations. This leaves part of the laboratory free for those processing data and filing reports. Offices for the other professionals are better located outside the gait laboratory.

If the laboratory is used for undergraduate teaching (and depending on the number of students), the teaching space should be separated from the data capture systems and processing units. Teaching, research and clinical activities must be tightly scheduled to avoid conflicting situations.

Storage

A storage area is required for miscellaneous small instruments, such as calipers, as well as basic electrical (voltmeter, soldering iron, etc.) and mechanical tools (screwdrivers, clamps, etc.). Precision instruments requiring careful storage and handling should be easily visible and accessible to eliminate the risk of damage. Expensive video components should be put in a separate locker. You cannot usually store anything under an elevated floor because of fire regulations, since the ceiling's sprinkler system is inadequate to stop a fire underneath the floor.

For different camera set-ups such as for gait analyses and spinal evaluations, the video cameras and cables must be stored in an orderly manner. Cables can be easily twisted and break at the connector end. Adequate storage is required for the data capture material. Sometimes, the cameras are left on their stands until the next data collection. If this is done, then the cameras should be close to the wall, well away from any passageway and mobile wall dividers which can accidentally fall on the cameras.

Access

Finally, the location within the building is also important. Ideally, it should be easily accessible from the building entrance, close to the clinic or to the rehabilitation services and not far from the washroom. It is already difficult to find an adequate room for the gait laboratory, and so these considerations are often neglected.

UTILITIES

The principal utilities you need in a gait laboratory are electricity, communication services, basic protection systems and air conditioning. Twenty or even 15 years ago, the electrical power requirements were enormous. Four or five 1000-W and even 1500-W photographic flood-lamps would be used to balance the lighting conditions prior to high-speed film acquisition. A uniform illumination of about 500–1000 lux (50–100 foot-candles) was recommended to film the subject. The high-speed cameras and the computers at that time also consumed significant electricity. During those days, our laboratory had as much power as a small house with its 200 A and 16 different electrical circuits. Each of the 1500-W flood-lamps was connected to its own electrical circuit. Today, energy-efficient systems reduce the energy demands significantly.

Electrical Needs

The main electrical requirements are usually those related to the electrical power consumption, the prevention of power surcharge, which can damage expensive equipment, and the appropriate lighting conditions for those using a visual display terminal to collect and process data.

Power Requirements

The power outlets should be numerous and located around the room. Ceiling-mounted receptacles with hanging or retractable cords reduce the number of cables on the floor. Today, video cameras using ambient lighting conditions and fitted with ring lights to increase marker detection reduce significantly the power requirements compared to 1500-W flood-lamps. Our laboratory is now equipped with an eight-camera system, two force plates, nine SUN Sparc workstations and personal computers (PCs) and other peripherals. The total amperage has dropped by half to 100 A, and we only use 86 A spread over seven electrical circuits. The SUN workstations and the two video processors consume about half our power needs.

Protection against Power Surcharge

Often in a hospital, the electrical power can be inadvertently cut during maintenance or renovations. Most computer systems, three-dimensional (3D) kinematic video-based systems and power bars have a built-in fuse which will protect the equipment in case of a power surcharge. Computers sensitive to high-voltage peaks and 3D kinematic systems which are connected to a built-in power bar are in our opinion vulnerable and insufficiently protected against power surcharges.

To protect your valuable equipment, we recommend the use of a voltage regulator. These devices have two functions. They will prevent voltage peaks or surcharges which can damage your equipment, and allow you 5–15 min to save your files after the power has been cut off. The length of time depends on the type of battery in the voltage regulator. They are not expensive compared to the investment in a 3D system and computers. If your motion analysis system and computers are connected by a network, then modern voltage regulators can shut down all the workstations automatically. During electrical storms, machines should also be switched off until the storm passes. Abrupt power cycling of computer hardware is hard on the machine, and while it may not break down immediately, it can cause damage in the long run.

Lighting

Screen glare from fluorescent luminaires is one of the major problems. Various reflective grids are used to diffuse the light and reduce the glare on the screen or in the video cameras. Ambient lighting is also decreased, since the light loses its luminance as it bounces off several reflective surfaces of the grid. Task lighting at the operator's workstation may be needed to compensate for a lower ambient illuminance. Indeed, when grids were installed in our laboratory to replace the translucent plastic cover, five or six new luminaires were added to increase the ambient lighting.

Care must be taken to avoid a gloomy or cavernous effect, which can occur in a low-ceilinged room when the floor is raised. The overhead lighting can be on different circuits. In this way, part of the laboratory can be dimmed if necessary to reduce glare and reflection during data acquisition, while the other part can be normally lit for data processing. The ambient lighting level in the laboratory should be about 150 lux, while for visual tasks of high contrast, such as reading and writing, it should be between 200 and 500 lux. For more information on this topic, the reader is referred to Grandjean and Vigliani (1982) and Kaufman and Christensen (1987).

Communication Systems

Data collection and processing is usually a two-step procedure. Separating the data capture system from the area processing units enables an efficient use of the laboratory space. When purchasing data capture systems, allow part of the budget for additional workstations. Data can then be transferred by network communication.

A computer network is a collection of computers connected together by both hardware and software. Ethernet, a high-speed local area networking technology, was developed in 1973. An Ethernet installation has a tap, which makes the electrical connection to the inner and outer conductors of the network coaxial cable. Attached to the tap is the transceiver which relays the signal on the network cable of the Ethernet interface board on to the host computer via the drop cable. The actual Ethernet communication rate is 10 megabyte/s.

Serial communication is the process of converting an 8-bit byte to a stream of eight single data bits travelling consecutively through a single wire. More often, a modem receives the serial data and converts them into a form that can be transmitted on a telephone line; however, the speed of transmission is limited to a baud rate of 19 200 (bits per second). For higher-speed flow, the optical cable replaces the traditional modem with special numerical lines.

Protection

In some hospitals, there is an emergency code in case of emergency situations such as fire, theft, flood, suspicious persons or patient behaviour. At worst, the telephone number of house security should appear somewhere on the telephone so that appropriate measures can be taken immediately.

Fire, theft and emergency situations must be seriously considered. We have already mentioned that equipment, tools and other material cannot be stored underneath the walkway because of fire regulations in some states or provinces. Cables running under an elevated floor must also be protected and enclosed in a metallic conduit. Power bars may also be considered as a fire hazard in some states or provinces and consequently cannot be used. If the laboratory has ventilation ducts, it may be necessary to have a fire detection device within the ducts. In case of a fire, shutting off the ventilation would immediately prevent further damage. One or two argon fire extinguishers must be within easy reach. Argon, being an inert gas, will replace the oxygen and stop the fire, and will not damage the electronic equipment, as would carbon dioxide extinguishers.

Before installing formal protection devices, such as security locks, alarm and fire detection systems, consult the person responsible for house security. That person will not only give you valuable information but will also inform you as to what can be done. For example, if you control access to the laboratory by replacing the standard lock with a high-security lock requiring special keys, housekeeping will be unable to come during their working hours in the evening. Then, you have to clean the laboratory or make alternative arrangements with housekeeping so that someone is present in the laboratory to unlock the door for the cleaners.

At Sainte-Justine Hospital, we have two locks. A coded key is given to the personnel who usually open and close the laboratory during the daytime. An extra key is kept by house security, which has the names of those authorized to use the laboratory facilities at any time. The second lock uses a regular key. This lock is used when one leaves the laboratory unattended for a short period of time.

A coded key lock can be supplemented with movement detectors (not for gait analysis). Again, the chief of security will want to be informed, especially if the detector is hooked up to a 120-dB horn! Our system is connected to a private security company which immediately reports any incidents to both the chief of security at the hospital and the person in charge of the laboratory. In this way, house security can check out the laboratory and additional procedures instituted by the director of the gait laboratory, if necessary.

Usually, hospitals and universities have an insurance policy which covers them against theft and damage. According to the contract, the amount deductible can vary widely. You should consult the chief of security of your institution to find out what is covered, how much is deductible and who assumes any of the expenses resulting from theft or loss.

Air Conditioning

Another important consideration in the gait laboratory is the heating and ventilation system. Though much can be said about these systems, more often than not, less can be done about them! The ventilation system should be connected to the fire detection system. In case of fire, the ventilation in your area should cease immediately to reduce the influx of fresh air and stop the propagation of toxic fumes to other areas of the building.

It is difficult to control cold drafts from the air-conditioning system in a large, open space. We use baffles to block or redirect the air coming out from the ventilation ducts. Furthermore, the thermostat is located in the laboratory, where we can adjust the temperature to our needs.

In Canada and the northern USA, air-conditioning systems may be difficult to adjust correctly during the change of seasons. Fresh air coming into the laboratory may be warm one day and cold the next. Once the weather is stable, the ambient temperature in the laboratory becomes comfortable.

The ideal operating room temperature for computer equipment is between 18 °C and 20 °C (Nemeth, Snyder and Seebass, 1989). This temperature ranges below the ideal office temperature, which is between 20 °C and 24 °C (Grandjean, 1983). An ambient room temperature of 27 °C corresponds to about 49 °C inside the computer (Nemeth, Snyder and Seebass, 1989).

Electronic components are very sensitive to static electricity and can easily be damaged by it. The ideal relative humidity for most computer hardware is in the range 40–60%. If the humidity is too low, static electricity becomes a problem. It can come from carpeted floors, which generate more static electricity than hard-surface floors. If the humidity is too high, condensation can form on the boards, which causes oxidation and subsequent short-circuits.

In summary, the physical aspects of the laboratory must be well thought out at the beginning so that you have sufficient space to carry out your evaluations in a comfortable environment.

EQUIPMENT AND ACCESSORIES

Most turnkey gait analysis systems are built around three elements: the data capture system which collects analog signals from video cameras, force plates, etc. and stores them on the computer's hard disk; the accompanying software, which is used to calibrate the cameras and transforms video data into 3D coordinates; and peripherals, such as printers, so that the results can be presented in a hard-copy form. The basic configuration is relatively adequate for a single user as a start-up system. As more people use the system and more data are being processed, the needs become diversified. Some of these are listed in Table 3.

Table 3. List of equipment for setting up or operating a gait laboratory

Hardware		Software
Acquisition	Peripherals	
Computers	Data storage	Operating systems and
Data capture systems	Hard disks	platforms
Kinematics systems	Optical disks	MS-DOS
Video	Magnetic tapes	UNIX
Goniometers	Printers	Data capture software
Accelerometers		Data-processing software
Kinetic systems		3D reconstruction
Force plate		Software tools
Pressure sensors		Programming
Force transducer		Signal processing
A/D converters		Statistics
		Word processing
		Graphics
		Drawing
		Communication
		Internet
		World Wide Web
		Biomch-1

DATA CAPTURE SYSTEMS AND PERIPHERALS

A gait evaluation can take many forms and can include a combination of kinematic, kinetic and electromyographic assessments. Depending on your objectives, you may prefer one type of assessment over another and build your gait laboratories around it.

Kinematic Data Capture Systems

Most of the 3D kinematic systems come with two or four video cameras. Due to the colinearity and coplanarity conditions, we are able to perform a 3D reconstruction of an object point (marker) with a minimum of two cameras (Allard, Blanchi and Aissaoui, 1994). However, the estimation of the marker's coordinates can be slightly inaccurate because of measurement errors, resulting in a phantom point. Using three or more video cameras to track the same marker or marker set can reduce the difference in the position of the phantom point and the marker. For unilateral gait analysis, a minimum of three video cameras is needed, and preferably four, since not all markers are seen by all the cameras.

A basic understanding of a few photogrammetric parameters is useful before building a calibration object which will cover the useful volume for gait analysis. The field of view width W and height H corresponding to the portion of the field which can be digitized by the video processor can be determined by:

$$W = 0.006\,827(U - f)/f$$
$$H = 0.006\,408(U - f)/f$$

where f stands for the focal length (mm) of the lens, and U for the object distance (the distance from the focal point of the lens to the subject, in mm). For example, if the object distance is set at 4 m and an 8-mm lens is selected, then the field width and field height are 3.4 m and 3.2 m respectively (Motion Analysis Corporation, 1989).

Goniometers and accelerometers can be added to complement the video assessment. Ladin (1994) has described some of these devices in terms of their principle of operation and outlined their advantages and limitations.

Kinetic Measurement Systems

Force platforms should be solidly fixed on their base to avoid vibrations which occur during loading. This can be a concrete floor or a steel base onto which the force platform is bolted. It should be positioned in the middle of the cameras' field of view, midway along the walkway. The second force platform should be similarly secured. However, if the distance between the force platforms has to be adjustable to the subject's step length (children and adults) or to the type of gait trials (slow and fast walking, running, etc.), then the second platform obviously has to be moveable.

Our second force platform is fastened to a sliding steel bed which can be turned from aft to fore by rotating a universal screw. Once in position, the bed is bolted into place. Ground reaction forces from the moveable force plate are noisier than those from the fixed one. This noise has a high-frequency content associated with the vibrations, but a low-pass digital filter can eliminate most of that noise.

Analog-to-digital (A/D) Converters

The number of additional devices that can be synchronized with the video system depends largely on the number of channels available in the analog-to-digital (A/D) converter. For example, a 16-channel A/D card enables the data collection of one AMTI force plate and up to 10 EMG signals. If a second force plate is used, the number of available channels is four, but a 32-channel A/D converter allows for more possibilities.

Peripherals

Data storage devices are probably the most useful peripherals. Depending on the type of 3D analysis system, a single four-camera video data acquisition and corresponding force plate data represent about 1 Mbyte of information per gait trial. If several acquisitions are performed, the hard disk of the computer will soon become saturated with information. This is why we recommend the purchase of a 2–5-Gbyte hard disk when acquiring a 3D kinematic system. To make additional space on the hard disk, the data will have to be processed, keeping only the reconstructed coordinates of the markers. At one time or another, and sooner than one imagines, the hard disk will be full and some data will have to be stored elsewhere.

Data storage can be done on different media: nine-track tape, 1/4-inch cartridge tape, 8- or 4-mm video tape, floppy disk and rewritable optical disk. These devices can be divided into two categories. The first three are of the sequential access type, whereas the remaining two are of the random access type.

A gait evaluation consisting of several gait trials and including video data from eight cameras and information from two force plates may use 32–49 Mb of memory. Neither the standard floppy disk (1.44 Mb) nor the new floppy disks (20 Mb) have sufficient memory capacity to store multiple gait evaluations. Although the optical disk system can store 600 Mb, it suffers from its unreliable technology and lack of robustness. The most powerful medium to our knowledge is the Exabyte tape, which is an 8-mm video tape. Exabyte systems are relatively fast and allow a complete dump which can be performed without operator intervention because of their large capacity (5 Gb). The drive itself can handle data at up to 15 Mb/min, but dumping by Ethernet reduces the overall transfer rate to about 2 Mb/min. The transfer of the data from a single subject can take 1.5 s. After decompression, the time required to retrieve a single file is about 30 s.

Data storage devices can also be used for periodic backups. The question remains: why are backups essential? No matter how the system's hardware is maintained and how efficient the users are, every system manager will eventually be confronted with lost files. Lost files range from a single file inadvertently deleted to an entire file system destroyed by hardware or power failure. For more information about backup and restore procedures, the reader is referred to Nemeth, Snyder and Seebass (1989) and Frisch (1991).

SOFTWARE

The basic software provided with a 3D kinematic system takes care of the data capture, which usually includes some form of tracking, digitization and labelling. Additionally, there are procedures for calibration and 3D reconstruction. These software packages are system dependent. When purchasing a 3D system, both the accuracy and resolution of the equipment and the modelling techniques used in the software must be considered and compared.

According to the application, you may want to use or develop your own software for further data processing. Today, most commercial software packages allow you to process the data as you wish and return to the principal program. For example, you can apply your own filtering technique to the reconstructed coordinates and still be able to continue your analysis with the commercial software. The flexibility of access to and exit from commercial software and the ease of operation are some aspects to consider when purchasing data-processing software. There are a number of commercial software packages available for force and pressure measurements, EMG signal processing, muscle modelling and computer animation.

Operating Systems and Platforms

The operating system is software that is a native part of each computer; it manages the hardware and software of the machine. The most common one is MS-DOS, which is dedicated to PCs. It is a single-user operating system. In the world of workstations, the equivalent one and the most common is probably UNIX, which is a multi-user and multi-task operating system. UNIX is available in different versions, namely SunOS, Solaris, Irix, etc., where the primary difference is the graphical user interface.

Environment platforms such a Windows, Sunview, Motif and Open Windows are user-friendly operating systems which allow complex operations to be performed by clicking the mouse buttons. Because a motion analysis system always comprises both hardware and software, the native or host software is usually defined by the manufacturer. Make sure that the software developed by the manufacturer, which makes use of the environment platform, is robust. If not, you may have some difficulties during data collection or in processing video and force plate data. Error messages may be related to poor management by the user but may also arise from poorly defined instructions by the manufacturers. When selecting a system, purposely make mistakes to determine how the software will respond and see if the system crashes.

Data Capture Software

The software that allows you to collect data from a single or multiple video camera system is closely related to the hardware. With the software, the video images are digitized. Then, the significant information is extracted and stored in real time. The latter step corresponds to a contour extraction by the operation of segmentation. These two operations are performed within the internal sampling duration (i.e. $1/50$ s, $1/60$ s or less). The segmented images are used in the calibration of the cameras and in the tracking of surface markers (Greaves, 1994).

Data-processing Software

Photogrammetry generally deals with the mathematical representation of the geometrical relations between objects in space, based on their projected images. This is also known as the calibration problem (Yuang, 1989). The estimation of the internal (focal point, principal point, etc.) and external (position and orientation) camera parameters is followed by the 3D reconstruction of surface markers (Allard, Blanchi and Aissaoui, 1994). The data capture and data-processing software modules are organized around black box executable files and are usually supplied by the manufacturer. One must not base the choice of the motion analysis system on the user-friendly software only but also on the quality of the hardware and the performance of the software in providing accurate data.

Software Tools

Many manufacturers provide basic to advanced data analysis software as complementary tools to their video-based systems. Nonetheless, the user may need to develop specific software applications. These include filtering methods to reduce noise in the data to inverse dynamic analyses. The Matlab programming language is a powerful and flexible tool for software development. Other application software packages include those for statistical analyses (Statistica, Sigma stat), word processing (Microsoft Word, WordPerfect) and graphical representation (Sigmaplot) and animation (Musculographics, Softimage).

Communication Software

Many confuse the terms Ethernet, TCP/IP (Transmission Control Protocol/Internet Protocol) and Internet. Ethernet is the hardware system previously discussed that enables communication between different computers, while TCP/IP is the software that manages this communication. Internet is the standard protocol that the software uses to communicate and is the world's largest network. It grew from fewer than 6000 computers at the end of 1986 to more than 600 000 computers five years later (Hunt, 1992).

The common thread that ties the enormous Internet group together is the TCP/IP network software. TCP/IP refers to an entire set of data communication protocols known as Transmission Control Protocol/Internet Protocol. The popularity of the TCP/IP protocol on the Internet grew rapidly because of its several important features:

- Open protocol standards, freely available and independently developed from specific computer hardware or operating systems.
- Independence from specific physical network hardware. This allows TCP/IP to integrate many different kinds of networks.
- A common addressing scheme allowing any TCP/IP device to uniquely address any other device in the entire network, even if the network is as large as the world-wide Internet.
- Standardized high-level protocols for consistent, widely available user services.

The World Wide Web (WWW) is one of the fastest growing parts of the Internet, with 6 million to 8 million users. For example, you can visit the International Society of Biomechanics WWW at http://www.kin.ucalgary.ca/isb. Gopher, for example, is a document retrieval program which provides a menu-driven interface to aid the user in his or her research, while MOSAIC and NetScape are graphical interfaces which allow the user to navigate within Internet. One must not forget the biomechanics list server called Biomch-l, developed by Herman J. Woltring in 1988. Though it is limited to subscribers, it is an active forum on biomechanics for open discussions and information dissemination. To date, there are more than 2585 subscribers in 49 countries (van den Bogert, 1996). You can join by sending an E-mail message to LISTSERV@nic.surfnet.nl with the following message: SUBSCRIBE BIOMCH-L forename, name (affiliation).

LABORATORIES OF THE FUTURE

In the last 10–15 years, gait laboratories have developed into complex systems enabling simultaneous 3D bilateral human locomotion studies. Most of the video-based camera systems have many hundreds of feet of cable, which too often lie on the floor or are fastened to the ceiling during data collection. These cables are connected to a PC system equipped with a large-screen monitor and force platforms having their own amplifiers and connecting cables. The standard procedures, from camera calibration to the calculation of the marker 3D coordinates, are still time-consuming and involve too much manual input. A complete 3D inverse dynamic analysis also involves too many manual operations. Though this is a marked improvement from the days of high-speed cinematography and manual digitization, locomotion studies are undermined by tedious tasks such as opening and closing numerous files and windows, and clicking the appropriate instruction. These tasks are usually performed by highly skilled professionals.

These drawbacks, namely the bulkiness of the data collection systems and the numerous manual tasks, should be eliminated or at least greatly reduced in the next few years. In fact, the industry is already moving in this direction, due in part to the demands from the animation industry.

One vision of the gait laboratory of the future could include the following aspects. To film the subject, tiny cameras fitted with distortion-free wide-angle lenses would transmit the data by radio waves, thus eliminating all cables. Portable PC computers would reduce the bulkiness and the weight. Data cards would need to be more compact. The calibration process could be simplified by using a single marker (Sabel, 1994) rather than a calibration frame which needs to be carefully measured. The accuracy of the data capture systems can be improved at different levels: instrumentation, software for instrument correction plus faster and simpler 3D reconstruction techniques.

To substantially reduce some of the tiresome tasks, quasi-real-time 3D coordinates calculation is highly desirable. This implies: (1) high-accuracy data collection system; (2) automatic marker recognition and labelling; (3) noise-free video files (no file editing); (4) fast 3D coordinate calculations and representations; and (5) some form of data validation to avoid problems resulting from marker occultation and lost markers, etc.

Good laboratory practice (GLP) must be addressed. Equipment care and proper procedures must be used to ensure quality data acquisition. In many laboratories, the equipment is handled by numerous persons who have their own interest in the data being collected. They can be professionals (engineers, therapists, doctors, research assistants, prosthetists, etc.), undergraduate and graduate students, professors, etc. Most laboratories have one or more in-house protocols which should be in written form. Some manufacturers of video-based kinematic systems also provide useful instructions on equipment setting-up, calibration and marker placement. They are sometimes so good that they are used as textbooks. This information could be adapted and included in a GLP manual. Other items such as maintenance, instrument breakdowns, safety procedures, lists of suppliers, etc. would complete the GLP manual. The GLP manual allows everyone to follow approved procedures and serves as a useful document for new staff or students.

Adequate space and facilities, good equipment and a GLP manual should provide and ensure numerous quality data for human movement analysis.

ACKNOWLEDGEMENTS

Though the outline of this book chapter was written in October 1994, the authors are grateful to those who have contributed to the ongoing discussions on Biomch-1 with respect to the following topics: setting up a gait laboratory, gait equipment and discussions on GLP. We also wish to extend our gratitude to Mr Jean É. Acé for his constructive remarks.

REFERENCES

Allard, P., Blanchi, J.-P. and Aissaoui, R. (1994) Bases of three-dimensional reconstruction. In: *Three-dimensional Analysis of Human Movement* (ed. P. Allard, I. A. F. Stokes and J.-P. Blanchi), Human Kinetics, Champaign, IL, pp. 19–40.

Corlett, N., Wilson, J. and Manenica, I. (1986) *The Ergonomics of Working Posture.* Taylor and Francis, London.

Frisch, A. (1991) *Essential System Administration.* O'Reilly and Associates, Inc., Sebastopol, CA.

Grandjean, É. (1983) *Précis d'Ergonomie.* Les Éditions d'Organisation, Paris.

Grandjean, É. and Vigliani, E. V. (1982) *Ergonomic Aspects of Visual Display Terminals.* Taylor and Francis, London.

Greaves, J. O. B. (1994) Instrumentation in video-based three-dimensional systems. In: *Three-dimensional Analysis of Human Movement* (ed. P. Allard, I. A. F. Stokes and J.-P. Blanchi), Human Kinetics, Champaign, IL, pp. 41–56.

Hunt, G. (1992) *TCP/IP Network Administration.* O'Reilly and Associates, Inc., Sebastopol, CA.

Jian, Y., Winter, D. A., Ishac, M. G. and Gilchrist, L. (1993) Trajectory of the body COG and COP during initiation and termination of gait. *Gait Posture*, **1**, 9–22.

Kaufman, J. E. and Christensen, J. F. (1987) *IES Lighting Handbook: Application Volume.* Illuminating Engineering Society of North America, New York.

Kvalseth, T. O. (1983) *Ergonomics of Workstation Design.* Butterworths, London.

Ladin, Z. (1994) Three-dimensional instrumentation. In: *Three-dimensional Analysis of Human Movement* (ed. P. Allard, I. A F. Stokes and J.-P. Blanchi), Human Kinetics, Champaign, IL, pp. 3–18.

Motion Analysis Corporation (1989) *User's Guide Manual.* Motion Analysis Corporation, Santa Rosa, CA.

Nakurama, T., Yamamoto, T. and Tsuji, H. (1994) Distance and velocity meter for human motion analysis. *Med. Biol. Eng. Comput.*, **32**, 115–118.

Nemeth, E., Snyder, G. and Seebass, S. (1989) *UNIX System Administration Handbook.* Prentice Hall Software Series, New Jersey.

Sabel, J. C. (1994) Camera calibration with single marker. In: *Proceedings of the Third International Symposium on Three-dimensional Analysis of Human Movement,* Stockholm, pp. 7–9.

van den Bogert, T. (1996) *International Society of Biomechanics Newsletter,* Issue Number 61, February/March, 4–5.

Yuang, J. S. C. (1989) A general photogrammetric method for determining object position and orientation. *IEEE Trans. Robotics Automation*, **5**, 129–142.

5

Real-time Motion Capture Systems

HANS FURNÉE

Delft University of Technology, Motion Study Laboratory, GA Delft, The Netherlands

INTRODUCTION AND EARLY HISTORY

PHOTOGRAPHIC AND CINEFILM RECORDING OF MOTION EVENTS

Direct visual observation was, until the mid-nineteenth century, the sole tool used to provide the database for scientific description of, and enquiry into, the nature of animal and human movement. Supplemented by palpation or dissection, it was only detailed visual observation of the human and animal body in motion that was available to a long line of scientists from Aristotle, through da Vinci, Harvey, Borelli and Boerhaave. These earliest gait analysts nonetheless compiled meticulous, often comparative, descriptions of muscle action, and performed geometric and kinematic analyses on locomotion, jumping and other motion patterns.

The German brothers, physiologist and physicist Wilhelm and Eduard Weber (1836), were perhaps the first to attempt some quantitative measurement of locomotion. They used clocks, measuring tapes and optical surveying instruments, and evolved a theory in which, during the swing phase of gait, the leg was treated as a pure pendulum. As opposed to such non-contacting methods, mechanical contraptions like the myograph or kymograph, originally developed by Helmholz, transmitted the length change of contracting isolated muscle to a pointer. This inscribed the time pattern on a rotating drum, blackened with a thin layer of soot: this was the primordial movement recorder, and operated in real time. Expanding the degrees of freedom by attaching more writing pointers to

Three-dimensional Analysis of Human Locomotion. Edited by P. Allard, A. Cappozzo, A. Lundberg and C. Vaughan
© 1997 John Wiley & Sons Ltd. ISBN 0 471 96949 4

the drum recorder, the French engineer–physiologist Etienne-Jules Marey in 1868 recorded the wing beat of a dove flying in a carousel with a pneumatic tube linkage to the apparatus.

It was not until the advent of photography that significant advances were made in the quantification of locomotion and, in due course, many other aspects of the human and animal motor act. The methods developed were basically non-contacting and capable of providing lasting records for diligent scrutiny, archiving and subsequent manual processing. But evidently, if only because of the time lapse of chemical processing, the photographic methods are not in the real-time category, as those crude mechanical devices were.

It was the British photographer Eadward Muggeridge (later Muybridge) who, in Sacramento California, USA, demonstrated the ability of photography to arrest motion by a sequence of pictures. With an array of 12 cameras, with trip-wired shutters to secure properly ordered delay, he, in 1872, demonstrated the unsupported transit phase of the horse in trot: a wager had been laid on the number of hooves in contact with the turf. Muybridge later incorporated electro-magnetic shutters and sequential triggering of up to 24 cameras by a rotary commutator. His heritage of serial photography consists of some 100 000 plates of animal and human motion recorded at Palo Alto California and, from 1884, at the University of Pennsylvania.

By 1888, Dercum had already presented displacement–time curves for normal and pathological gait which were derived from Muybridge's recordings.

Some improvements to Muybridge's multi-camera single-exposure system were due to the contributions of Marey, mentioned previously, such as his version of the photographic gun (Marey, 1885). Here, once it is triggered, a chamber loaded with 12 pieces of photographic emulsion rotates at high speed behind a barrel-like lens holder. The chamber came to intermediate full stops for exposure. This gave serial exposures within 1 s, with aperture times of $1/720$ s. Using three such cameras, Marey obtained three perpendicular projections of 10 phases of a seagull in free flight.

Marey (1885) was aware of stroboscopy, as already used by Savart in the optical subsampling of periodic phenomena by exploiting the eye's image retention. Marey extended the method to photography with what he called photochronography, where he incorporated a rotary disc shutter with one or more slits inside the single-plate camera. In discussing stroboscopy, Marey was well aware of the possibilities offered by periodic illumination by what was then the technology of spark discharges.

With his shuttered photochronography instrument he recorded human movement with subjects dressed in white contrasted against dark backgrounds, thus obtaining a time decomposition of, for example, pole-vaulting as in his 1894 book (Marey, 1894). To obtain pictures less crowded with unmanageable detail, Marey went on to dress his subjects in black, with limbs and head brightly marked by stripes and buttons shining in the sun. Thus his subject was reduced to a more abstract representation of complex progressive motion. This is how photochronography produced the first so-called stick diagrams as reported in Marey's 1885 and 1894 books.

Observing that the pictures, taken at successive time instants, tended to crowd on top of each other if the subject did not move enough, Marey installed a moving mirror inside the camera to spread the images out across the sensitive surface. He then thought of moving the plate itself. In his chronophotographic box, presented in 1888 to the Paris Academy of Sciences, Marey indeed had a strip of photographic paper transported at 20 exposures/s behind the lens and rotary shutter mechanism. Again, Marey was aware of the necessity of the film being at rest during each of the successive exposures, and he described the transport mechanism employed.

In his 1894 book, Marey mentions a sensitive celluloid film, employed at a 50 exposures/s rate with 1/4000 s aperture times: the film camera was thus born. It was originally a scientific research instrument, developed and reported years earlier than W. K. L. Dickson's kinetoscope at the Edison factory in the USA. The Dickson camera in turn pre-dated by three years the much more publicized cinematograph of the French Lumière brothers—the birth of cinema as an entertainment and educational medium.

COMPUTER PROCESSING OF PHOTOGRAPHIC MOTION DATA

Within a decade of photographic motion capture, the German investigators Braune and Fischer (1895) were among the first to perform exhaustive manual calculations on human gait, including 3D reconstruction from central projections utilizing four cameras. These monumental studies even included double differentiation and the consideration of forces acting on the estimated centres of gravity of the relevant body segments.

These milestone efforts were followed by half a century of analytical processing of motion data on cinefilm or multiple-exposure strobed photography, before the advent of the digital computer.

The first computer generation was best utilized by Paul (1967); the University of Strathclyde (Glasgow) Biomechanics Laboratory had a manually operated Spectro XY film analyser available for numerical coordinate extraction from cinefilm, obtained with two orthogonal Bolex cameras running synchronously at 50 frames/s. Two operators were occupied for 9 h transcribing 60–70 film frames into tabular numerical form, representing about one stride in stereo view. Another five person-hours were spent punching paper-tape for computer input.

Developments in the automation of this tedious and error-prone post-processing of film frames, such as the flying-spot scanner initiated by Ledley (1965), can safely be left aside in this chapter on real-time motion capture. Indeed, cinefilm recording has been replaced by either the medium of taped video or by the technology of on-line, sensor/computer-based motion capture devices. The exceptions are found in very high-speed applications, which arise more in certain industry segments than in biomechanics.

In biomechanics, it may be safely stated that, currently, cinefilm is used in only a few places for purposes other than archiving and visual inspection.

ON-LINE MOTION CAPTURE SYSTEMS

The video-to-digital coordinate converter (Furnée, 1967) was the first documented image-sensor-to-computer interface to constitute an on-line motion capture system that encoded the coordinates of multiple markers in real time. The original system was designed for the analysis of human arm movement, as a research tool in a project aimed at the control of externally powered prostheses at the Delft University of Technology. Basically a low-cost hardware preprocessor, it was hooked up to an IBM-1130 by an in-house input channel. The input sensor was a commercial 50-Hz closed-circuit television camera.

The coordinate converter was and still is based on the principle that the television scan inherently constitutes a coordinate grid which can be recovered by counting the line number for the vertical (Y) coordinate, and on each line, the sweep time increments for the horizontal (X) coordinate. With later charge-coupled device (CCD) cameras replacing the image tube, the X coordinate is just the column (or pixel) number.

The coordinate conversion system is essentially only two counters, one for Y, and one faster counter for X; the coordinate conversion of imaged markers is no more than the instant read-out of these counters for each above-threshold video value. The counter read-outs are input in real time to the computer: the (X, Y) coordinates of all bright dots (pixels). Early prototypes already had a marker contour detector for suppressing computer input of points interior to the contour, and which also, for data reduction, could select upper and lower points on the contour.

The 1967 prototype used 256 non-interlaced television lines, and eight-bit counters for a 256 $\times$ 256 coordinate grid. Camera synchronization was, in the next prototypes, derived from the same two counters. Markers changed from active pinlights to passive reflective paper disks, and follow-up software included elementary tracking for the purpose of marker identification. A rotary shutter was introduced, and in the course of development was replaced by stroboscopic scene illumination. The first documented locomotion analysis by real-time acquisition of digital coordinates was reported Furnée et al. (1974), and this included multichannel EMG recorded in synchronous analog-to-digital (A/D) conversion.

The Delft prototypes as published and demonstrated led to interest at and cooperation with other institutes, as witnessed by Paul, Jarrett and Andrews, (1974); at the University of Strathclyde (Glasgow) Biomechanics Laboratory, a similar system was developed. This was later expanded to include multiple cameras for three-dimensional (3D) work and the system was interfaced to the

DEC-PDP computer family. This, with credit due to Dr J. R. W. Morris of Oxford Medical Systems for recognizing the market potential, was the origin of VICON (1980) as the first commercial on-line television/computer-based motion analysis system.

Several similar academic systems (Cheng, Koozekanani and Fatehi, 1975; Lappalainen and Tervonen, 1975) had no follow-up other than a few published applications. A one-time product at Newington Children's Hospital (Connecticut, USA), developed by United Technologies and reported by Taylor et al. (1982), has made its mark both in the literature and in practice. The VICON system adopted AMASS software (by Adtech, Maryland, USA) for marker midpoint estimation and has subsequently become a major source of customer-oriented analytical software. From the mid-1980s major additional players have been the ELITE system marketed by BTS (Milan, Italy) and the ExpertVision system by Motion Analysis Inc. (Santa Rosa, California, USA). Each of these contributed some distinctive features, as reported in Ferrigno and Pedotti (1985) and Greaves (1986), while essentially belonging to the same optoelectronic sensing family.

No review of early optoelectronic motion capture systems would be complete without reference to the radically different Selspot system (Gothenburg, Sweden). This, again, had an academic origin (Lindholm and Oeberg, 1974) but, in contrast to the above systems, made a rapid entry into the commercial market in 1975.

The sensor is a position-sensitive device (PSD), a planar, large-area silicon device which uses the lateral photoeffect. The two-dimensional (2D) application is based on the duo-lateral PSD, where, with two pairs of lateral electrodes along the square's sides, and concomitant analog electronics, the (X, Y) coordinates are extracted as the (intensity-weighted) centroid or median of the impinging light. Because of this light-averaging effect across the sensor surface, in practical applications the image must correspond to a single focused lightspot, as projected from a single marker.

In multimarker applications, the standard requirement in motion analysis, Selspot uses time multiplex switching of individual markers. Consequently, markers are in the form of tiny light-emitting diodes (LEDs). An important feature is that, being active in a multiplex mode, markers are self-identified.

A contribution to PSD design for linearizing the (X, Y) response was made by Woltring (1975).

Developments in 3D calibration by Woltring (1981) were also applied in early television-based motion analysis such as the University of Strathclyde system used at the Dundee Limb Fitting Centre in Scotland.

COMPUTER PROCESSING OF VIDEOTAPED MOTION DATA

A different and essentially off-line family of motion data acquisition systems, though optoelectronic and using the same sensor (a television camera), is represented by the video tape and framegrabber concept.

As a camera-to-computer input device, its publicly documented origin was again as a research tool in medicine, for the automated analysis of dynamic images in morphology or physiology, as recorded by video X-ray or fluoroscopy (Winter, Malcolm and Trenholm, 1968).

This early system, with a 20 × 20 sample window of limited spatial resolution (every fifth television line) and a four-bit grey-scale resolution, was succeeded by the Winnipeg group's more flexible and higher-resolution computer interface for television (CINTEL), as reported by Dinn, Winter and Trenholm (1970).

A first CINTEL application in the analysis of videotaped human gait was fully reported in Winter and Reimer (1972). Large reflective markers were used, and their midpoint was calculated by unweighted averaging, as in this application the grey-scale resolution was set to one bit. This accommodated to CINTEL's then maximum spatial resolution of 96 × 96 points within a 192 × 192 window.

If a major instrument and computer company had recognized the potential of CINTEL (D. Winter, personal communication), commercial framegrabbers would have reached the market a decade before they did.

In fact, from the mid-1980s framegrabber PC boards have been used for the off-line input and analysis of pre-recorded video as a medium for the capture of human motion (this section ends with a mention of on-line framegrabbing of real-time camera video).

Much of the development effort in automated analysis is based on marker acquisition and tracking, using identical principles to those used in real-time sensor-based motion capture.

Evidently the video tape approach with its full view of subjects' movement is more supportive for manual operator intervention in resolving possible ambiguities in marker association. Also, video recording is a convenient medium in outdoors capture sites, such as in sports.

One other distinction of on-line CCD camera-based motion capture systems is that a majority of these have the feature of pixel-synchronous capture and marker coordinate extraction, for the utmost in sub-pixel definition and stability. This intrinsic quality of pixel clock availability is absent in standard video tape. Since this affects the overall performance, the extent of clock jitter should be considered when contemplating the purchase or use of a video tape-based motion analysis system. One of the solutions is using fiducial marks for position reference within each video frame. Digitizing these additional points adds to error. Another solution is in prospect in the form of clocked, digital video formats in this application area. In any case, in a 3D system, all cameras should have a rigorous common frame synchronization.

The video tape/framegrabber approach has been further developed by consultancy and manufacturing companies; two pioneers are Ariel Life Systems, originally of La Jolla, California, USA, and Peak Performance Technologies Inc. of Englewood, Colorado, USA.

A specific niche is occupied by high-speed video recording (presently on CD-ROM) and analytical post-processing, exemplified by the product range of Kodak Ektapro, which originates from the 2000 frames/s Spin Physics imager and multitrack video recorder of the mid-1980s.

But technology as evident in current camcorders may not necessarily stop at the tape barrier: with the advent of on-line, moving-image framegrabbers, we may expect new, affordable systems to emerge with multiple cameras, and real-time framegrabbers in a power host platform. This should support a sustained throughput, and the in-line processing of high-resolution, high-frame-rate data: a convergent approach in camera-based real-time motion capture.

PRINCIPLES OF ON-LINE MOTION CAPTURE

TELEVISION COMPUTER SYSTEMS

On-line television computer systems were introduced in the historical review earlier in this chapter. The basic principles of operation have remained much the same, in so far as the real-time video-to-digital coordinate conversion is essentially a counting operation.

The video signal, when crossing a preset threshold level, triggers the instant read-out of the counter contents. The counters are one for Y, which increments at horizontal synchronization and resets at vertical synchronization, and one for X, which increments at pixel clock (or at a multiple of pixel clock, in an oversampling mode advocated by some manufacturers) and resets at horizontal synchronization.

The importance of rigorous synchronization between camera(s) and converter system, down to the pixel clock level, is evident, whether this is by direct synchronization drive or phase-locking of local synchronization. Phase jitter is an error source which translates into coordinate noise by shifting the video threshold crossing with respect to the defining pixel boundaries.

Most current manufacturers adhere to non-interlaced video, and in this case the active line count is half the count of the sensor's unblanked lines, while image rate is the field rate which is twice the standard full frame rate. Including the blanked or non-integrating lines, standard (non-HDTV) line count is 625 at a 25-Hz frame rate in the PAL format, and 520 at 30 Hz in NTSC.

Where interlaced video is used, the alternate video fields are shifted by half a line distance. In this mode (Macleod, Morris and Lyster, 1990), an average value may be produced of the vertical Y read-outs which then implies the full sensor resolution but at the expense of a two-sample time averaging. Users should be aware of this limited-bandwidth mode, which may be justified in certain applications and can also be of advantage in the static capture phase of 3D system calibration.

Once the real-time coordinates of the imaged marker contours are obtained, it is further processing that yields the estimated midpoints of the markers at sub-pixel resolution, and with minimal error, assuming that the markers as imaged do not depart significantly from a circular shape. Non-circular images can result from partly obscured markers, or from using non-spherical markers like planar disks if oblique to their line of sight.

Some manufacturers remain dedicated to marker processing on the host computer platform. In some instances, powerful algorithms, like those based on AMASS, are provided which also cater for the imperfectly imaged markers as above, if necessary augmented by operator intervention. Originally developed as non-real-time, post-processing algorithms, these and similar procedures are speeding up on the present powerful and affordable platforms, so that we may see a convergence to the concept of real-time marker centroid processing.

Real-time marker centroid processing was introduced by Furnée (1984) in a system where sub-pixel resolution as in Taylor et al. (1982) was obtained in real time, with a microprocessor embedded in the coordinate converter hardware. At the same time, this extreme of on-line marker data reduction unburdened both the transfer link to the host PC and the storage requirements, which at the time were critical design items.

The MacReflex system introduced in 1990 by Qualisys A/B (Gothenburg, Sweden) was the first to adopt the same approach, as expanded by Furnée (1986), of real-time, microprocessor-based, marker midpoint extraction. The MacReflex product is also marketed by Adaptive Optics Associates, a United Technologies subsidiary.

Now, considering the CCD sensor and passive markers in an otherwise minimally restricted environment, such as within a well-illuminated or outdoor observation space, a major problem is the contrast of the (retro-)reflective markers in ambient light or against background light clutter. An original solution to the contrast problem was presented by Ferrigno and Pedotti (1985) with the ELITE system by BTS (Milano, Italy). The assumption is that, ideally, the circular marker images are in shape and size range distinct from clutter high-lights. Dedicated A/D conversion and digital signal processing (DSP) hardware is used in cross-correlating the video signal with a reference template which is defined on a small grid of, say, 6×6 pixels. Only a matching video sequence is accepted as a marker for further processing. The possibility of enhanced, sub-pixel resolution by using the correlates in computing the weighted marker centroids is mentioned.

Another approach to increasing marker contrast was the electronically shut-tered CCD camera presented by Furnée (1986) as part of the PRIMAS system. Here, the short integration or aperture time, a fraction of the frame period, is synchronous with the (infrared) scene illumination pulse. The twin stroboscopy concept reduces the apparent intensity of ambient light and background clutter versus the full intensity of the retro-reflective marker. The same PRIMAS system introduced a 100 Hz frame rate for cameras and real-time marker processing. Similar camera speeds and/or a short integration mode are currently found with a majority of motion capture systems.

To return to trends in marker processing, some manufacturers are introducing total PC-based solutions with proprietary video/coordinate boards on the expansion bus (ISA, and moving to PCI, to name the Intel bus standard architectures). This requires only a lookalike box for camera power and controls, in case even these are not moved to the PC enclosure. This approach was introduced in new hardware brands like Kinemetrix from Orthodata (Lüdenscheid, Germany).

The middle position is represented by the marker acquisition/coordinate converter system, which serves multiple cameras at one end and has a one-channel interface to the host platform at the other end. The division of labour, such as marker centroid computing, between host and preprocessor has, in summary, several solutions. These correspond to higher and lower data rates, respectively, and manufacturer-specific solutions for data transfer and buffering into the host. These system-specific differences merit user awareness only if they are meaningful for improving the capabilities, performance and cost-effectiveness.

It is estimated that more television computer-based units are installed worldwide, and by more companies, including some Japanese manufacturers serving their domestic market, than any other optoelectronic systems.

OTHER OPTOELECTRONIC SENSOR SYSTEMS: PSD, LINEAR ARRAYS

The Selspot system, based on one or more 2D PSD sensors, was mentioned in the historical review earlier in this chapter. As detailed there, coordinate detection allows only one marker to be imaged at a time, so the markers are active, wired and are switched in time-division multiplex. The *XY* output per marker is essentially in real time, and the individual markers are inherently self-identified by their order in the switching sequence.

The switching period of 0.1 ms between successive markers, or the 10 kHz sampling rate per marker, implies an ensemble sampling rate of 10 kHz divided by the number of markers. For any reasonable number of markers, this results in a comparatively high (ensemble) sampling rate. Nevertheless, it must be kept in mind that even though individual markers are sampled equidistantly in time, the markers are not sampled simultaneously, as is the case with strobed-mode CCD motion capture systems.

For proper work where, for example, angles are implied by three markers, it is the job of the analytical software to de-skew the marker coordinates as sampled. Consequently, this requires some form of interpolation to a common time instant within each ensemble sample.

The ideal, error-free interpolation can only be approximated in practical applications, with additional difficulty in real time. Moreover, any approximate interpolation implies a low-pass filter operation. A statement on this error source, and the method and extent of compensation, should form part of the system specification.

These considerations apply to all time-multiplex systems, including the twin linear-array systems such as CoSTEL (LOG.In, Rome) and OPTOTRAK (Northern Electric, Waterloo, Canada), i.e. to all systems using optoelectronic sensors other than sequential-address area scanners like the CCD. The CCD area scanners must, moreover, be operated in a strobed or electronically shuttered mode for the concept of simultaneous marker sampling to apply.

The advantage of PSD as a non-addressable total area integrator is that as an *XY* sensor it exhibits continuity rather than discretization of coordinates. Depending on design details, and marker light intensities, the absence of quantization noise may, however, be offset by other sources of electronic noise. Total noise performance is summarized in the specified precision.

The last and by no means least significant source of possible error with PSD-based systems is the fact that the *XY* output corresponds to the intensity-weighted centroid of impinging light. Two possible sources of spurious light are ambient or background light, and the projection via spurious pathways of reflections from the active high-intensity marker itself. The errors are manifest in a shift of the intensity-weighted centroid, a bias offsetting the apparent marker coordinates. The first error category may be minimized by applying high-intensity markers, by optical filtering, and by subtracting sensor output measured in-between active-marker pulses. The latter approach was used in the WATSMART system (Northern Electric, Waterloo, Canada), a Selspot lookalike introduced in 1986 and superseded in 1988 by the OPTOTRAK system to be reviewed below.

The second error source, of often intractable spurious reflections, can to some extent be combated by elaborate *ad hoc* measures such as special lining and interceptive baffles in observation space. Precautions like these can be complemented by a judicious positioning of the PSD cameras to obviate the most damaging reflective pathways. On the premise that reflections from the moving subject can be excluded, a rigorous calibration of floor/walls and/or ceiling reflections using a single marker at known positions is recommended practice.

Following a different concept, the sequential-access sensor system which uses at least two uniaxial or linear CCD arrays (Leo and Macellari, 1979) is as immune to spurious marker reflections as are the area-scan CCD systems of the preceding section. This was further developed in 1983 as the CoSTEL system and marketed by LOGIn (Rome, Italy). Up to six uniaxial cameras are supported. Each linear CCD array has a 2048 pixel count and, by necessity, the single active marker is projected by a toroidal or anamorphic lens as a line of light perpendicular to the linear array. Where the light line intersects the array, at, ideally, one, or, in practice, a small number of pixels, this yields the read-out of one coordinate value along the array. Again, with CoSTEL and similar systems, the markers are active, wired and switched in time-multiplex. If more than a single marker were active, each compound of two uniaxial cameras would be defeated in ambiguities of multiple X and Y values. One or more calibrated, and preferably orthogonal, pairs of these uniaxial cameras constitute a 3D sensor system.

Evidently, the anamorphic lens has a low efficiency with regard to the fraction of marker light projected on the line array, and this necessitates high active marker intensities and excellent low-noise properties of sensor and associated electronics.

This is emphasized by the OPTOTRAK efforts, with correspondingly high performance figures. Distinct features of early OPTOTRAK systems (Krist et al., 1990) are the attention to detail, such as the size and isotropy of LED markers, and the rigid encasing of two uniaxial arrays with their fixed and precalibrated anamorphic lenses. This implies that each unit is a 'metric' 2D camera. This facilitates 3D work with two or more of these cameras, and a real-time 3D option is provided. In another version, a fixed-baseline oblong unit with three uniaxial arrays performs as a highly accurate 3D camera, within fixed observation space dimensions.

MAGNETIC SENSOR SYSTEMS

Magnetic sensing systems were pioneered by B. Polhemus, who in 1970 founded his namesake company (Colchester, Vermont, USA). Early applications were in the military sector, such as tracking of air pilots' head motions. Applications have diversified to fields such as computer-aided design, computer graphics and computer-generated video animation, biomedicine and biomechanics, including gait analysis and rehabilitation, and remote robotic control in industry.

In many respects, the system concept is the inverse of optometric systems. With magnetic systems, the reference metric is actively imposed upon the observation space in the form of a magnetic field distribution which is emitted by three orthogonal base coils.

Sensors are called targets, as they are not at the emitting but at the receiving end. Targets contain a triaxial assembly of miniature pickup coils, which are wired to the system box. The wire hookup tends to be bulky, but inevitably the low-frequency signals require high transmission quality, such as freedom from mechanically induced noise (microphony) while moving about.

Specific to these 3D magnetic systems is that any one target produces six degree-of-freedom (6-DOF) values, representing both the target position and attitude. The equivalent in optoelectronic (and electroacoustic) systems is a cluster of three markers.

Magnetic sensing systems are essentially real time, and sample rates are achieved in the order of 100 Hz. Increasing the number of targets may increase time latency in some basic system solutions; other options include parallel processing to maintain low response times.

Originally, the magnetic fields are generated as sinusoidal time functions. It is straightforward to neutralize the static earth magnetic field.

A derivative system by Ascension (Burlington, Vermont, USA), established in the early 1980s, relies on pulsed DC magnetic fields, and subtracts the earth magnetic field between pulses.

One disadvantage is the influence on the magnetic fields of iron appliances or construction parts within or near the observation space. Careful calibration across the whole of observation space is mandatory to compensate for local warping, and repeatability is only ensured if no iron substances are moved around.

An important difference from optometrics is that magnetic sensing does not suffer from 'shadowing', in that there is no need for a direct 'line of sight' between the emitting and the sensing coils. Cost is a last category which separates magnetic systems from the more expensive optometrics.

ULTRASOUND LOCALIZATION SYSTEMS

Sonic systems, characterized by having ultrasound emitters on the subject and microphone receivers fixed in observation space, were reported by Andrews and Youm (1979). Measurement volume was restricted to wrist and ankle analysis. For larger volumes, the subject is presented in Steffny and Schumpe (1990) and early commercial products are exemplified by Zebris (Munich, Germany) and Ultrametrix by Orthodata (Lüdenscheid, Germany).

The landmarks are active ultrasound transmitters, miniaturized and necessarily wide-beam for allowing pickup while attached to the body in motion. As the sensor, a fixed set of microphones is used for spatial triangulation, which is essentially based on the measurement of pulse propagation delay.

Again, this concept requires the markers to be pulsed individually for the triangulation algorithm. As the system is based on pulse propagation, the chosen solution is time-division multiplex switching. This, again, implies that the markers are inherently self-identified. For synchronization of the markers, an additional infrared link is one option for replacing the tethered wiring to the subject. The system is equipped with a dedicated PC-board, and provides 3D coordinates in real time.

The Ultrametrix system originally specified up to 16 markers, ensemble sample rate equals 100 Hz divided by marker count, and measuring errors are given as between 1 and 10 mm, depending on position in a $5 \times 4 \times 3$ m observation space.

No more than a modest penetration has yet been achieved in biomechanics or other applications.

PROPERTIES OF ON-LINE MOTION CAPTURE SYSTEMS

From the preceding sections, we can summarize some characteristics of the various capture system families.

ACTIVE AND PASSIVE MARKERS

Passive, unwired markers are found with CCD-based systems. Common shapes are disks of (semi)spheres. These markers are made from or covered by retro-reflective adhesive sheet.

Active, wired markers are found with most other systems, such as those based on the PSD, uniaxial arrays and the magnetic and electroacoustic sensing systems. In PSD and uniaxial optical systems, the markers are LEDs or infrared-emitting diodes.

In both optical marker types, reflection or emission should best approximate the ideal of isotropy, for the measured coordinates to be independent of the line-of-sight angle. Small size and low inertia are preferred. Spurious movement such as skin slip must be minimized. In a majority of 3D biomechanics applications, involving rigid-body segment approximations, the use of marker clusters is advocated. This chapter does not deal with skin slip as an error source or with proposed compensation methodologies.

MARKER IMAGING AND EXTRACTION: CONTRAST ENHANCE-MENT, RECOGNITION TECHNIQUES

Problems of reflection by spurious pathways are specific to PSD-based active-marker systems.

CCD-based systems commonly use electronically shuttered, short-aperture-time cameras for the enhancement of marker contrast under synchronous strobed illumination.

In view of the time-multiplexed, active markers the shuttering approach is not supported in uniaxial camera systems.

In the ELITE CCD camera system, selective marker extraction from background is further enhanced by a cross-correlation template-matching method.

TIME SAMPLING: SIMULTANEOUS, TIME-OFFSET OR SKEW, SAMPLE RATES

CCD-based systems, if operated in the shuttered mode or at least in stroboscopic illumination, have simultaneous sampling of all markers as imaged.

Sample rates, equal to camera field or frame rate as given by the manufacturer, do not depend on the number of markers.

In capture cases for a multitude of markers, such as for a calibration acquisition, the sample rate may be set at reduced values by skipping frames. Other applications for subsampling are slow-motion phenomena.

PSD-based systems or uniaxial cameras are characterized by time-multiplexed switching, so that the position samples of individual markers are, by definition, time-offset.

Even at marker speeds well below those dictated by sampling theory, the individual time offset is an error source if not accounted for in the kinematic fitting model, or if data are not de-skewed before further analysis. Correction algorithms, other than the impractical ideal interpolation, are preferably checked for bandwidth-limiting consequences.

With PSD or uniaxial cameras, the per marker switching period is generally fixed, so that the maximum sample rate for the ensemble of all markers depends on the number of markers. Nevertheless, the ensemble sample rate is generally higher than frame rates of common CCDs.

SPATIAL SAMPLING: DISCRETE OR CONTINUOUS

CCD and uniaxial array systems are in the category of discrete spatial sampling. Discrete pixel positions are implied, even though the per pixel light intensity is represented by a continuous-time and bandwidth-limited video signal.

The geometric accuracy and stability of the pixel grid or array is at the basis of overall accuracy and precision, given the further condition of appropriate signal processing.

The array-based systems inherently exhibit quantization noise, even in the presence of sub-pixel-definition algorithms, as these are intended only to minimize the quantization step.

The PSD-based system and also, as non-projective sensing or triangulation systems, the magnetic and electroacoustic systems are representative of position sampling on a spatial continuum. The, not always relevant, difference is that any additive quantization noise is introduced by the signal processing only.

INTENSITY SAMPLING: THRESHOLD CROSSING OR GREY SCALE

The uniaxial concept, as used in CoSTEL and OPTOTRAK, applies A/D digitized processing of the pixel array video signal for the highest degree of sub-pixel resolution.

Most other CCD-based systems extract the coordinates of marker edges on the pixel grid by threshold-level crossings. These involve no A/D conversion but a voltage comparator, equivalent to a single-bit discrimination.

With PSD systems, the coordinate extraction is essentially the digitized amplitude processing of the four electrode signals.

COORDINATE GENERATION: MARKER OUTLINE, MIDPOINTS, WEIGHTED MIDPOINTS

With PSD systems, the coordinate generation is equivalent to obtaining the difference signal between two opposing electrodes, divided by their sum signal for normalization or independence of marker image intensity.

As reviewed, PSD coordinates represent the weighted average (centroid) of the total projected light. As such, coordinate values, which should only represent the focused marker image, are susceptible to error caused both by the surface integration of ambient and background light and by marker projections along spurious reflective pathways.

With area or line-scan cameras the markers are individually imaged on the pixel grid or array. With area sensors using the threshold-crossing concept, the coordinates of the marker outline are retrieved as line number Y and horizontal clock count X, respectively.

The horizontal clock, generally the pixel clock, may in other cases be a multiple of the pixel clock. As advocated by Morris (1990), the VICON system features this form of video oversampling. In any case, the horizontal clock must be rigidly synchronized to the camera pixel clock, as any time jitter of clock or sampling aperture is a possible source of coordinate error.

All area-scanning systems as reviewed here include data reduction, from the marker contour coordinates to a single coordinate pair per marker. These are the marker midpoints, as estimated by algorithms, whether off-line or in real time, and it is these algorithms which achieve sub-pixel definition. Geometric centroid and circle-fitting methods are among those implemented in current systems.

Marker size, as imaged on the pixel grid, is a determining factor for attainable resolution and accuracy, as first pointed out by Winter and Reimer (1972). A review and a proposed ring-fitting method, with attention to distorted marker images, is presented in Jobbágy and Furnée (1994).

Instead of threshold crossings, digitized grey-scale values of the video signal are used with some image-based systems (Ferrigno and Pedotti, 1985).

Grey-scale weighted midpoint estimation in the method of choice with the line-scan camera systems of CoSTEL and OPTOTRAK.

CALIBRATION AND PERFORMANCE

SENSOR CALIBRATION

All image-based 3D systems are based on multiple sensors to determine the XYZ coordinates of any individual marker attached to the relevant landmark on the subject. Line-scan sensors yield one coordinate, while area scanners and PSD devices yield two coordinates each. For the XYZ reconstruction from multiple 1D or 2D sensor data, the individual sensors must be calibrated as to their disposition, roughly speaking their position and attitude, in 3D observation space. If the sensors are moved, the external parameters need recalibration.

The sonic transducer triangulation systems are included in this same category, in that the sensing microphones must be calibrated for position if not assembled on a rigid carrier, and for such issues as individual divergencies in propagation delays. After calibration, environmental factors such as air humidity must remain stable, to maintain the propagation time constants.

Magnetic 3D sensing systems require calibration with regard to the spatial distribution of the magnetic fields as produced by the driving master coil set. With AC magnetic fields, no modelling is available for the usually unavoidable permeability disturbances, such as the presence of iron construction parts in the proximity of observation space. This requires a full calibration to be completed over observation space. An appropriately dense correction matrix is established by positioning a known distribution of sensors serving as probes at a succession of reference points.

In image-based systems, the sensors (cameras) must also be calibrated individually for their internal parameters of the optical system as referred to the sensor imaging surface. Among these are the focal length and the position of the principal point, the intersection of sensor and the optical axis. Scale factors can account for the conversion from metric to pixel-based coordinate values; for generality these may differ in X and Y.

Additionally, lens distortion must be incorporated in the calibration procedure. Assuming appropriate electronics in the sensor-based coordinate conversion, lens distortion, if not linearized or otherwise unaccounted for, is the major source of systematic error.

An unwieldy, 3D calibration rig was required in the classic calibration procedures, based on the direct linear transform (DLT) for solving the colinearity equations (Abdel-Aziz and Karara, 1971; Marzan and Karara, 1975; Hatze, 1988). Woltring (1978, 1980) introduced the biomechanics community to a photogrammetric calibration technique that allows the use of a planar calibration object. Equipped with a dense grid of control points, the SMAC (simultaneous multiframe analytical calibration) plane has to be viewed in a number of attitudes and/or positions: one full view per camera, not necessarily perpendicular to the optical axis, and a limited number of views common to two or more cameras.

For lens correction, Morris (1990) also advocates the use of a very dense planar calibration object, in the form of a stable film with control points held perpendicular to the optical axis. Despite the theoretical and practical advantages of SMAC, its application is limited to a camera configuration where the calibration plane can be viewed by all cameras. In a many-camera, say bilateral, set-up, SMAC can be applied to each of the camera pairs or triplets.

The external parameters of opposing camera sets may then be linked by an unrestricted view of a centrally located marker cluster; this could also define the laboratory coordinate system. As an alternative by Sabel (1994), full calibration can also be performed with a single large marker that is moved around so as to cover a large part of observation space. The internal camera parameters are assumed to be known from the previous unilateral SMAC calibrations.

As starting values, the algorithm requires a rough estimate of the external parameters of two of the cameras to produce the final estimates of all external parameters. If internal camera parameters are unchanged, the single-ball method provides an easy approach to recalibration.

No reference other than in the trade literature was found to an extension of the method utilizing a single rod equipped with two or more markers. Moving the rod around the observation space should add scaling in a straightforward way and could also be relevant to lens distortion correction.

Lens correction, as applied by many authors, is done using a polynomial equation where a radial symmetric and an asymmetric part are characterized by three and two parameters, respectively. Usually, without adverse effects, fewer than these five terms are entered in the linearization.

MARKER IDENTIFICATION

The subject of marker identification is only relevant to image-based, area-scanning systems like those depending on CCD cameras. These are the only systems to use unwired, passive markers, but this implies that no distinguishing features exist between markers.

There will certainly be a future for coloured markers, in systems where the expense of colour cameras and the drawbacks of using wide-spectrum, visible illumination are justified.

As noted earlier, all line-scan camera and PSD-based systems use time-multiplex-switched, active markers which are inherently self-identifying.

Marker identification with CCD camera systems is relevant in two senses. The first is matching, or the assessment of which of the $(X, Y)_{mi}$ coordinate pairs from i cameras to associate with one $(X, Y, Z)_m$ triplet as representing one—as yet unidentified—marker m. The second is the identification or association of any specific marker m or triplet $(X, Y, Z)_m$ to the corresponding landmark L.

If landmarks are numbered or named, this implies that the identified markers are assigned the same number or name, and that marker number or name is maintained along the trajectory. This task, supported by tracking and prediction, is currently by preference carried out in three dimensions because this is less sensitive to ambiguities caused by the apparent merging of marker tracks in single-camera 2D projection. Tracking of individually assigned markers can also be used in support of 3D marker matching.

Matching in the above terminology can be achieved by using per-camera 2D coordinates to produce the inverse ray, which is the line of sight from the marker image point through the lens' projection centre back into observation space. Points of (near) intersection of inverse rays from two or more cameras are likely to correspond to an actual 3D marker position. As pointed out by Macleod, Morris and Lyster (1990), such a geometric self-identification method depends on highly dependable 2D data to avoid ambiguities in the ray-tracing process. At the same time, the ray or forward intersection method is part of the 3D reconstruction algorithm used in providing the estimates for the marker 3D coordinates.

Another approach to multiple 2D coordinate matching is by stereometric image matching (Baltsavias and Stallmann, 1991; Maas, 1991), where the inverse ray from a specific $(X, Y)_{m1,i}$ point on one camera is projected as the epipolar line onto the other camera images. There, any $(X, Y)_{m,j}$ closest to the epipolar line is likely to correspond to $(X, Y)_{m1}$ and thus to marker $m1$.

A computationally fast alternative, applied in real-time image matching by Sabel, van Veenendaal and Furnée (1993), relies on the advance calculation of auxiliary vectors, derived from the calibration parameters. This allows the real-time matching problem to be reduced to a classification of values resulting from a few simple equations, which relate to the slope in a predefined bundle of epipolar lines.

With all methods reviewed, however, ambiguities may occur if two markers are coplanar with the two cameras, or more precisely with their lens projection centres. This results in two additional ghost intersections with the inverse ray-matching method, and to multiple points on the epipolar line.

Marker tracking, or the use of a third camera, can help to resolve the ambiguity. In more remote cases, the three-camera method can still fail if one of the markers, coplanar with two cameras, gets obscured and cannot be seen by the third.

Evidently, marker occlusion will also present a problem for marker association in the second sense. Ideally, on the marker re-emerging into sight, the tracking software should automatically reassign the previous landmark number. It is more likely, in all but the more elaborate, model-based software, that a new track may be started which must be manually corrected. This poses a challenge to the software developer for real-time optometric applications.

THREE-DIMENSIONAL RECONSTRUCTION

In summary of the above, once the multiple-camera 2D coordinates are matched, marker reconstruction in three dimensions is either an application of the DLT, or an extension of the forward intersection method. This is computationally more efficient than the DLT.

Marker (X, Y, Z) is at an estimated point on the common perpendicular to the two inverse rays, previously determined as having a match because this perpendicular is the shortest of all.

For solid bodies, an extension from three dimensions to 6-DOF estimates can be given by applying a rigid cluster of more than two point markers.

This is, then, equivalent to the inherent 6-DOF coordinates offered by magnetic sensing systems with the standard triple-coil targets.

PERFORMANCE (PRECISION, RESOLUTION, ACCURACY);
STANDARDS FOR ASSESSMENT

Some selected reviews on performance parameters are Walton (1986), Morris (1990), Furnée (1990) and Furnée et al. (1992). To a large extent based on Morris (1990), the CAMARC Consortium under the Advanced informatics in Medicine program of the European Union (CAMARC, 1994) has agreed a performance test protocol for 3D kinematic measurement systems.

In the preamble this protocol distinguishes first the product test, to be performed at the manufacturer's premises or elsewhere, where there is access to sophisticated facilities. These should have specifications surpassing the generally high accuracies of the systems under test.

The specimen test, subject of the protocol, involves less complex test devices and is designed so that it can be performed for acceptance testing and for validation at the user's premises. The simple spot check is intended for a regular validation of the user's system performance. As such, the spot test reveals the preservation of length in the relevant part of the observation volume, essentially as a measure of the maintenance of calibration quality.

Quantities involved in the performance specifications and in the performance tests are the usual statistical figures of merit, expressing accuracy and noise.

Effects of finite resolution are reflected in the quantization noise contributing to noise. The resolution in this sense denotes the just-discernible difference—the just-discernible marker displacement in this position-sensing equipment. Resolution is emphatically not defined by the fraction of measurement range known as the least significant bit (LSB), unless the LSB is really significant, i.e. standard deviation below one LSB. For reasons reviewed above, absolute positional accuracy cannot be assessed without very specialized equipment, which is the reason why the tests refer to measurement of distance, except for static noise (precision), which refers to the standard deviation of coordinate values.

If Value$_i$ stands for a measured value, and RMS for root mean square, then as a shorthand reminder the statistical performance figures are:

Average	$= \mu =$ Mean (Value$_i$)
Bias	$= B =$ Average $-$ True value
Standard deviation	$= \sigma =$ RMS (Value$_i$ $-$ Average)
RMS error	$= A =$ RMS (Value$_i$ $-$ True value)

such that the RMS error

$$A = (\sigma^2 + B^2)^{1/2}$$

and A encompasses both the measure of accuracy, expressed as bias, and the extent of noise error, expressed as standard deviation.

Precision expresses noise as standard deviation, relative or normalized by division by the measurement range. Measurement range is commonly taken as the field of view (FOV) diagonal (2D or 3D). Similarly, the other average, bias and error figures are often specified relative to FOV. It is emphasized that precision is a figure of merit for repeatability (noise), not for accuracy.

OTHER PERFORMANCE ISSUES

The sample rates with various systems have been reviewed in the preceding sections. The sample rate should be checked for possible violation of the Whittaker–Shannon sampling theorem, with respect to the highest non-zero frequency components in the assumedly band-limited movement phenomenon. The theory is textbook material in the area of signal processing. Violation of the sampling theorem is at the expense of aliasing or folding of any non-zero frequency components from above half the sample rate. These then appear as indistinguishable spurious additives in the frequency band below half the sample rate.

For any particular (motion capture) system, the sample rate f_s and the noise performance as expressed in the precision p can be combined into a figure of merit Q (Furnée, 1990). The quality factor Q indicates the spatio-temporal resolution, with $Q = \sqrt{(f_s/p)}$, and should be maximized, as shown below.

With the above expression for Q, the equation

$$p_k \geqslant p_{k,\min} = \sigma_2 (2/\omega_s) \omega_0^{2k+1}/(2k+1) \quad \text{(Gustafsson and Lamshammar, 1977)}$$

can be expressed as

$$p_k \geqslant p_{k,\min} = (1/Q^2) \omega_0^{2k+1}/\pi(2k+1) \quad \text{(Furnée, 1990)}$$

Here, $f_s(\omega_s)$ is the sample (radian) frequency, p_k is the precision of the undistorted linear estimate of the kth-order derivative, and the input signal is assumed to be strictly band-limited, with ω_0 the cut-off radian frequency, and obeying the Whittaker–Shannon theorem with $\omega_0 \leqslant \omega_s/2$. The equality, where p_k equals its minimum attainable value $p_{k,\min}$, holds only for the case when the kth-order differentiator has a transfer function of zero for $\omega \leqslant \omega < \omega_s/2$.

The equations also hold for $k = 0$, the signal estimate or zero-order derivative, and they show that sampled-data systems, as in motion capture, must be designed to maximize spatio-temporal Q. It follows from its definition above that, for maximizing Q, the abatement of system noise is quadratically more effective than raising the sample rate.

The effects of marker size, relative to the FOV dimensions, were mentioned above in connection with their relevance to attainable accuracy, resolution and precision.

Resolution in the sense of minimally needed separation between the markers, or their outlines, as imaged in projection, is an issue to be considered especially where significant limb rotations are foreseen.

Similarly, the capability to cope with marker images connected or merging, again only observed in projection, is an issue in evaluating marker midpoint algorithms. The capability to cope with partly occluded markers was mentioned in preceding sections.

Trade-offs are shifting with available computing technology, such that sophisticated algorithms like ring-fitting and grey-scale processing are moving from off-line into real-time systems.

REAL-TIME MOTION CAPTURE

The motion capture systems reviewed can be summarized with respect to their real-time capabilities in the several dimensions.

Ultrasound localization systems have not found wide application. They use active, wired transducers. By virtue of the multi-microphone triangulation concept, these systems can be classified as real-time 3D systems.

Among optoelectronic systems, the active, wired marker systems like those based on position-sensitive devices and twin uniaxial photo-arrays are inherently real-time 2D systems. As the time-multiplexed markers are self-identified, real-time 3D is achieved with little additional computational effort.

Corrections for the inherent time-skew between the individual marker samples account for some additional processing burden, both in this segment and with ultrasound sensing.

Magnetic sensing systems use wired targets with triple sensing coils, and as such are real-time 3D systems. Inherently, more than that, the magnetic targets are essentially 6-DOF sensors. This means that on any rigid-body segment each of the, admittedly more bulky, targets can replace a marker cluster, to provide attitude additional to the position coordinates. With parallel processing, marker coordinates can be generated as simultaneous samples. Magnetic systems are the only ones where there is no shadowing, in that is there is no line-of-sight requirement between moving targets and the fixed-field driver subsystem. This attractive feature is conditional on the intervening body or substance not altering the magnetic fields.

Simultaneous marker sampling is also inherent with the area-scan, CCD image-based systems. These are the only systems to use passive, unwired markers, generally in the form of retro-reflective disks or (semi)spheres. Several systems provide straight, real-time 2D marker coordinates. Other manufacturers have optional real-time 2D. Algorithmic trade-offs, such as to cope with imperfect markers, were reviewed above. With increasing processing power, the trend is to real-time 2D.

The main concern in this class of systems is marker identification. Once the 2D coordinate pairs from multiple cameras are assigned, associated or matched as representing one and the same marker, 3D reconstruction is no less straightforward than with the other systems. When a marker becomes temporarily occluded, the particular problem exists of reassigning this marker to the original track or corresponding landmark. Only in part can multiple cameras contribute to maintaining real-time marker track continuity. While identification software gets more robust, user demands increase in new imaginative applications.

With the PRIMAS system (Delft, The Netherlands), an early real-time CCD system for full-body 3D was exhibited at the Paris, 1993, Congress of the International Society of Biomechanics. Real-time 3D potential with non-wired optometric systems is offered in systems now marketed by several manufacturers. For the more demanding situations, 'near real-time' options are offered.

Real-time applications are the subject of several of the following chapters. Real-time visualization is bound to make its mark on such areas as patient exercise, biofeedback and medical staff training. Real-time motion capture in the control loop is extending research in functional electrical stimulation. Real-time motion capture can control industrial processes. Animation and virtual reality driven by real-time motion capture is pervading the entertainment industry.

REFERENCES

Abdel-Aziz, Y. L. and Karara, H. M. (1971) Direct linear transformation from comparator coordinates into object-space coordinates. In: *Proceedings of the ASP/UI Symposium on Close-range Photogrammetry*. American Society of Photogrammetry, Falls Church, Urbana, Illinois, pp. 1–18.

Andrews, J. G. and Youm, Y. (1979) A biomechanical investigation of wrist kinematics. *J. Biomech.*, **12**, 83–89.

Baltsavias, E. P. and Stallmann, D. (1991) Triangular vision for automatic and robust three-dimensional determination of the trajectories of moving objects. *Photogramm. Eng. Remote Sensing*, **57**(8), 1079–1086.

Braune, C. W. and Fischer, O. (1895) Der Gang des Menschen I. In: *Abhandlunge Math.-Physische Classe der Kön. Sachs. Ges. für Wissenschaft*, **21**, 153–322.

CAMARC (1994) Performance test protocol for 3-D kinematic measurement systems. In: *Standards for Instrumentation and Specifications* (ed. J. P. Paul), Deliverable no. 24 to the Commission of the European Communities, programme AIM, project A-2002, 30 August, pp. 65–76.

Cheng, I. S., Koozekanani, S. H. and Fatehi, M. (1975) Simple computer–television interface system for gait analysis. *IEEE-BME*, May, 259–260.

Dercum, F. X. The walk in health and disease. *Trans. Col. Physicians*, **10**, 308–338.

Dinn, D. F., Winter, D. A. and Trenholm, B. G. (1970) CINTEL: computer interface for television. *IEEE Trans. Comp.*, November, 1091–1095.

Ferrigno, G. and Pedotti, A. (1985) A digital dedicated hardware system for movement analysis via real-time TV signal processing. *IEEE Trans. BME*, **32**(11), 943–950.

Furnée, E. H. (1967) Hybrid instrumentation in prosthetics research. In: *Proceedings of the 7th International Conference on Medical and Biological Engineering*, Stockholm, p. 446.

Furnée, E. H. (1984) Intelligent movement measurement devices I: video converter with real-time marker processor. In: *Proceedings of the 8th International Symposium on External Control of Human Extremities.* ETAN Belgrade, supp., pp. 39–41.

Furnée, E. H. (1986) High-resolution real-time movement analysis at 100 Hz. In: *Proceedings of the North American Conference of Biomechanics*, Montreal, pp. 273–274.

Furnée, E. H. (1990) Opto-electronic movement measurement systems: aspects of data acquisition, signal processing and performance. In: *Proceedings of the International Symposium on Gait Analysis* (ed. U. Boenick and M. Näder), Berlin, pp. 112–129.

Furnée, E. H., Halbertsma, J. M., Klunder, G., Miller, S., Nieukerke, K. J., van der Burg, J. and van der Meché, F. G. A. (1974) Automatic analysis of stepping movements in cats by means of a television system and digital computer. *J. Physiol.*, **240**, 3–4.

Furnée, E. H., Sabel, J. C., van Veenendaal, H. L. J. and Jobbágy, A. (1992) 3-D Performance characterization in motion analysis systems: an example. In: *Proceedings of the European Symposium on Clinical Gait Analysis* (ed. E. Stüssi), Zürich, pp. 280–283.

Greaves, J. O. B. (1986) State of the art in automated motion tracking and analysis systems. *Proc. SPIE* 693.

Gustafsson, L. and Lamshammar, H. (1977) ENOCH—An integrated system for measurement and analysis of human gait. PhD thesis, Uppsala University, Sweden. UPTEC 7723R.

Hatze, H. (1988) High precision three-dimensional photogrammetric calibration and object space reconstruction using a modified DLT approach. *J. Biomech.*, **21**(7), 533–538.

Jobbágy, A. and Furnée, E. H. (1994) Marker centre estimation algorithm in CCD camera-based motion analysis. *J. Med. Biol. Eng. Comput.*, **32**, 85–91.

Krist, J., Melluish, M., Kehl, L. and Crouch, D. (1990) Technical description of the OPTOTRAK 3D motion measurement system. In: *Proceedings of the International Symposium on Gait Analysis* (ed. U. Boenick and M. Näder), Berlin, pp. 23–39.

Lappalainen, P. and Tervonen, M. (1975) Instrumentation of movement analysis by raster-scanned image source. *IEEE Trans. IM*, **24**(3), 217–221.

Ledley, R. S. (1965) High-speed automatic analysis of biomedical pictures. *Science*, **146**, 216–223.

Leo, T. and Macellari, V. (1979) On-line microcomputer system for gait analysis data acquisition based on commercially available optoelectronic devices. In: *Biomechanics VV-B*, PWN-Polish Scientific Publishers, Warsaw, pp. 163–169.

Lindholm, L. -E. and Oeberg, K. E. T. (1974) An optoelectronic instrument for remote on-line movement monitoring. In: *Proceedings of the 2nd International Symposium on Biotelemetry* (ed. P. A. Neukomm), Basel.

Maas, H. G. (1991) Automated photogrammetric surface reconstruction with structured light. In: *International Conference on Industrial Vision Metrology*, SPIE Proceedings Series, Vol. 1526.

Macleod, A., Morris, J. R. W. and Lyster, M. (1990) Highly accurate video coordinate generation for automatic 3D trajectory generation. In: *Proceedings on Image-Based Motion Measurement* (ed. J. S. Walton), SPIE Proceedings Series, Vol. 1356, pp. 12–18.

Marey, E. -J. (1885) *La Méthode Graphique dans Les Sciences Expérimentales et Particulièrement en Physiologie et en Médecine*, 2nd edn, with supplement: *Le Développement de la Méthode Graphique par l'emploi de la Photographie*. Masson, Paris.

Marey, E. -J. (1894) *Le Mouvement*. Masson, Paris.

Marzan, G. T. and Karara, H. M. (1975) A computer program for direct linear transformation solution of the collinearity condition, and some applications of it. In: *Proceedings* of the *Symposium on Close-Range Photogrammetric Systems*, American Society of Photogrammetry, Falls Church, pp. 420–426.

Morris, J. R. W. (1990) A standard test protocol for assessment of 3D kinematic system accuracy exemplified on its application to the VICON system. In: *Proceedings of the International Symposium on Gait Analysis* (ed. U. Boenick and M. Näder), Berlin, pp. 40–50.

Paul, J. P. (1967) Forces transmitted by joints in the human body. *Proc. Inst. Mech. Eng.* **181**(3J), 8–15.

Paul, J. P., Jarrett, M. O. and Andrews, B. J. (1974) Quantitative analysis of locomotion using television. In: *Abstracts of the World Congress of the International Society for Prosthetics and Orthotics*, Montreux, p. 43.

Sabel, J. C. (1994) Camera calibration with a single marker. In: *Proceedings of the 3rd Conference on 3D Analysis of Human Movement* (ed. A. Lundberg), Stockholm, pp. 7–9.

Sabel, J. C., van Veenendaal, H. L. J. and Furnée, E. H. (1993) PRIMAS, a real-time 3D motion analysis system. In: *Proceedings of the 2nd Conference on Optical 3D Measurement Techniques* (ed. A. Grün and H. Kahmen), Zürich, pp. 7–9.

Steffny, G. and Schumpe, G. (1990) Ultraschalltopometrisches System der Universität Bonn. In: *Proceedings of the International Symposium on Gait Analysis*, (ed. U. Boenick and M. Näder), Berlin, pp. 102–106.

Taylor, K. D., Mottier, F. M., Simmons, D. W., Cohen, W., Pavlak Jr, R., Cornell, D. P. and Hankins, B. (1982) An automated motion measurement system for clinical gait analysis. *J. Biomech.*, **15**(7), 505–516.

Walton, J. S. (1986) The accuracy and precision of a video-based motion analysis system. *Proc. SPIE*, **693**.

Weber, W. and Weber, E. (1836) *Mechanik der Menschlichen Gehwerkzeuge*. Dieterich, Göttingen.

Winter, D. A., Malcolm, S. A. and Trenholm, B. G. (1968) Real-time conversion of video fluoroscopic images. In: *Digest of the 3rd Canadian Medical and Biological Engineering Conference*, Toronto.

Winter, D. A. and Reimer, G. D. (1972) Quantization errors in calculation of volumes, areas and coordinates in medical images. In: *Proceedings of the 16th SPIE Conference*, San Francisco.

Woltring, H. J. (1975) Single- and dual axis lateral photodetectors of rectangular shape. *IEEE Trans. ED*, **22**, 581–590.

Woltring, H. J. (1978) Simultaneous multi-frame analytical calibration (S.M.A.C.) by recourse to oblique observations of planar control distributions. In: *Nato Conference on Applications of Human Biostereometrics*, SPIE Proceedings Series, Vol. 166, pp. 124–135.

Woltring, H. J. (1980) Planar control in multi-camera calibration for 3-D gait studies. *J. Biomech.*, **13**, 39–48.

Woltring, H. J. (1981) 3-D Utility software for S.M.A.C. at the Dundee Limb Fitting Centre. Internal report on behalf of Dundee Limb Fitting Centre/Scottish Home & Health Dept., 28 October 1981.

6

Technology and Application of Force, Acceleration and Pressure Distribution Measurements in Biomechanics

EWALD M. HENNIG[1] AND MARIO A. LAFORTUNE[2]

[1]Biomechanik Laboratoire, Universität Essen, Essen, Germany
[2]Nike Sport Research Laboratory, Beaverton, Oregon, USA

INTRODUCTION

Movements originate from forces acting on the human body. For events of short duration, cinematographic techniques are normally not sufficient to estimate the forces and accelerations experienced by the body's centre of mass (CoM) or any one of its parts. Therefore, mechanical sensors are necessary to register forces, accelerations and pressure distributions that occur during human locomotion. Although transducers can be used to quantify the loads within the human body, for ethical reasons external forces are recorded and internal forces are computed in most locomotor applications.

Force transducers rely on the registration of the strain induced in the sensing element by the force to be measured. Accelerometers are measuring devices which are based on the determination of the forces produced by a known mass m. Pressures are calculated from recorded forces across a known area A. Because all three measurements are based on the registration of the same phenomena—strain—similar technical characteristics are possessed by the three types of transducers. Desirable transducer characteristics for biomechanical applications may differ from the characteristics which are advantageous for engineering

Three-dimensional Analysis of Human Locomotion. Edited by P. Allard, A. Cappozzo, A. Lundberg and C. Vaughan
© 1997 John Wiley & Sons Ltd. ISBN 0 471 96949 4

usage. Measurement of pressure during sitting or lying on a bed requires a soft and pliable transducer mat that will adapt to the shape of the human body. However, such a transducer will not incorporate good technical specifications.

In this chapter, we will focus on the aspects of measuring techniques that are most pertinent to biomechanical applications. After an initial description of general sensor characteristics, a discussion of commonly used transducer technologies will follow. Subsequently, we will examine the prime requirements for the measurement of force, acceleration and pressure during locomotion. Finally, running data are presented to illustrate the type of information that can be collected with force platform, accelerometer and pressure distribution sensors.

GENERAL TRANSDUCER CHARACTERISTICS

There are several characteristics to consider when buying a transducer for biomechanical research. It should be suitable for a wide range of applications, from low forces during standing and walking to high loads during running and jumping. A high-resolution transducer allows the detection of small fluctuations, even when a large signal is present. Static as well as high-dynamic-frequency responses are desirable to allow the recording of slowly fluctuating loads as well as short impacts. Linearity and hysteresis of a transducer are primarily dependent upon the deformation and the elastic properties of the sensing element. Yet, they can also be influenced by the electronic processing. Furthermore, transducer characteristics may also vary with environmental conditions (temperature and humidity). Therefore, technical data sheets should be interpreted with care, as they most often refer to transducer response obtained under standard laboratory conditions that seldom replicate the utilization environment. The specifications of a 40 cm by 60 cm glass-top piezoelectric 'Kistler' Type 9285 force plate (Kistler Instrumente AG, Winterthur, Switzerland) will be presented for illustrative purposes.

SENSITIVITY AND RANGE

The sensitivity of a force transducer refers to its change in output with an increment of applied force. For a piezoelectric force plate (KIAG 9285) measuring vertical forces, an electrostatic charge of approximately 3.8 picocoulomb (pC) is generated for a force of 1 newton (N). The sensitivity may vary across the surface of a transducer. Variation of sensitivity is given as a percentage of the mean value (e.g. $\leq \pm 2\%$ for KIAG 9285). The output signal of a transducer after electronic processing exhibits a certain level of noise. This noise can be either low- or high-frequency fluctuations of the output signal without corresponding changes in the mechanical input. These fluctuations relate to the stability of the transducer. The maximum deviations from the mean signal, divided by the transducer sensitivity, represent the error band. This is commonly expressed in mechanical units. The range of a sensor refers to the minimal and maximal load at which the transducer operates according to the data sheet specifications ($F_z = 0-5$ kN, F_x, $F_y = -2.5$ to $+2.5$ kN for KIAG 9285). The overload values refer to the loads that the transducer can sustain without being damaged ($F_z = 7.5$ kN, F_x, $F_y = 3.75$ kN for KIAG 9285).

To obtain sufficiently high output signals for low accelerations, accelerometers often have a limited range of 0 to 5, 10 or $25\,g$ with over-range values of 25, 50 and $125\,g$ ($g = 9.81$ m/s^2; acceleration due to gravity). These over-range values can easily be exceeded by inadvertently dropping the unit onto a rigid surface. Thus, to shield accelerometers from damage, some manufacturers offer over-load-protected units. These units are highly recommended for biomechanical applications that typically involve frequent mountings to the body. Over-range stops are mechanical limits that prevent damaging displacements of the cantilever beam of strain gauge sensors. Transducers with overload protection can typically withstand peak accelerations of more than $20\,000\,g$.

FREQUENCY RESPONSE

Ideally, a transducer should respond to both slow- and fast-changing mechanical input without phase distortion. The frequency response of a transducer depends upon the type of sensing element, its geometric dimensions and the electronic characteristics of the processing unit. The natural frequency of a sensor refers to the free oscillation of a fully assembled transducer system when exposed to a short mechanical pulse (> 300 Hz for KIAG 9285). To elicit free oscillation, the pulse duration must be considerably shorter than the period of the natural frequency. The frequency at which a forced sinusoidal oscillation causes the highest amplitude of the vibrating system is called the resonant frequency. Most important to the measuring properties of a transducer is its frequency response. This provides information about changes in transducer output, gain or attenuation across a frequency band. The graphical representation of the frequency response is typically shown on percentile or logarithmic (dB) scales.

LINEARITY

Ideally, a transducer should provide constant increases in its output signal with equal increments in force application. However, depending on the quality of the transducer, greater or lesser deviations from the ideal straight-line relationship between mechanical input and transducer output will be present. Different methods exist to assess the linearity of transducer output. The most commonly used method to quantify linearity deviation depends upon the maximum deviation of the output signal from the best straight line. The best straight line is drawn so that the transducer response curve shows equal maximum deviations at zero, one-half of full-scale and full-scale output. The largest positive and negative percentile deviation from that best straight line is expressed relative to the full-scale output (FSO) of the transducer. A typical linearity value for the vertical transducer output of a high-precision force platform (KIAG 9285) is $< \pm0.3\%$ for a calibration range of 0–2500 N. In the absence of hysteresis, transducer output deviations from linearity do not diminish measurement accuracy. Modern computer data-processing techniques allow numerical corrections of nonlinear transducer behaviour.

HYSTERESIS

The hysteresis of a transducer describes its response as a function of previous loading history. For a force transducer, it is defined as the largest difference in output between two identical forces during loading followed by unloading. The hysteresis is typically expressed as the percentage of this largest difference towards full-scale output. Hysteresis values for rigid and elastic transducers are low ($\leq 1\%$ for KIAG 9285). However, transducers exhibiting large and visco-elastic deformations (e.g. capacitive sensors) may show substantially higher hysteresis values. In addition, hysteresis is dependent on the rate of loading and it normally increases with higher-frequency mechanical inputs. A simple model of a force plate, including a spring and a dash-pot, is shown in Figure 1.

The oscillating mass on top of the force plate exerts identical forces on both a soft and a rigid force transducer. The response of the rigid force platform follows the mechanical input in a linear fashion, whereas the soft transducer exhibits a different behaviour during loading as compared to unloading. Because loading history and load amplitude influence the shape of the curve, measurement accuracy is substantially reduced in the presence of large hysteresis. A numerical correction of transducer hysteresis is only possible for periodic sinusoidal events. In locomotor activities, non-sinusoidal impacts occur. Therefore, numerical corrections of hysteresis effects are, for most biomechanical applications, impossible. The hysteresis value of a transducer is a key variable when judging its quality as a measuring tool.

CROSSTALK

Apart from the general transducer characteristics, there is a specific property that applies to transducer assemblies. A purely normal force F_z on top of a measuring force platform during quiet standing may result in an erroneous signal output for the two shear force components (F_x, F_y). This phenomenon is called crosstalk. It typically occurs between the orthogonal components of force or acceleration transducer signals. Crosstalk is expressed as the percentile error that occurs between two force components (F_z versus F_x, $F_y \leq 2\%$ for KIAG 9285). In some data specification sheets the expression 'transverse sensitivity' is used rather than crosstalk.

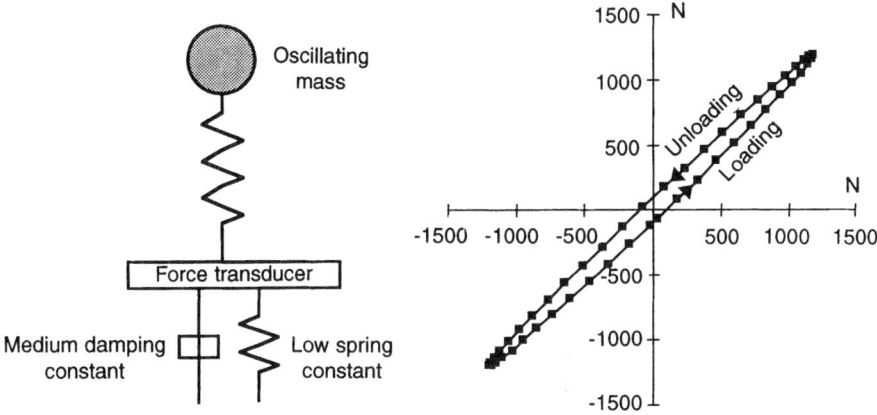

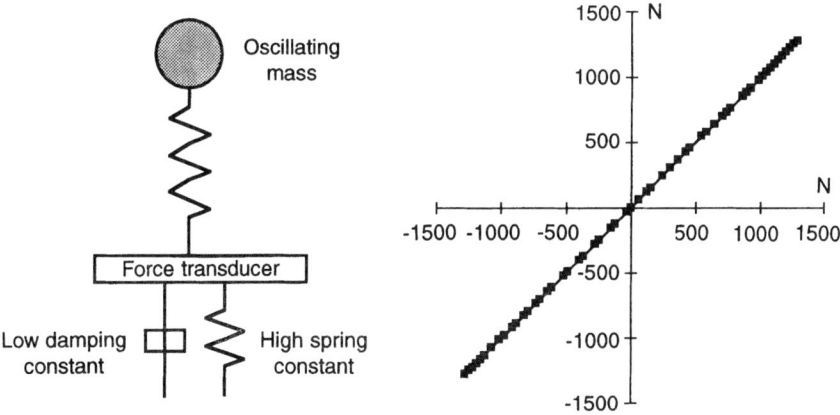

Figure 1. Mechanical model outputs for force platforms with different viscoelastic properties

TRANSDUCER TECHNOLOGIES

Because the majority of today's transducers create electrical output for convenient data acquisition by computers, only electromechanical technologies are described. Yet, to obtain profiles of pressure distribution under the foot and for some medical applications, optical methods are still being used. Elftman (1934) was one of the pioneers in visualizing pressure patterns under the foot. He used a rubber mat with several small pyramids resting on a glass plate. Depending on the local forces, the pyramids compress and create contact areas of different sizes between the glass plate and the rubber mat. Cinematographic methods can be employed to record the temporal pressure patterns. Fuji film (Fuji Photo Film USA Inc., 6200 Phyllis Drive, Cypress CA 90630) is often applied to determine surface pressure and congruence within articulations. However, Fuji film cannot be used to record temporal changes of forces, and it only allows a visual impression of the highest pressures that occurred during loading.

Electromechanical transducers produce a change in electrical properties when subjected to mechanical loads. Depending on the type of transducer, forces can create electrical charges, cause a change in capacitance, modify the electrical resistance or influence inductance. Because inductive transducers are based on relatively large displacements, they are rarely used in measuring instruments for biomechanical applications.

PIEZOELECTRIC TRANSDUCERS

Originating from the Greek *piezin*, meaning 'to press', piezoelectricity refers to a phenomenon whereby electrical charges are created on the surfaces of certain crystal materials by the application of forces. The brothers Pierre and Jacques Curie (Curie and Curie, 1880) were the first to report this effect for tourmaline. Quartz, another piezoelectric crystal structure found in nature, shows excellent transducer properties. It is very rigid and allows, even for short-duration impacts, accurate measurements with very low hysteresis (Figure 1b). Most high-precision force transducers use quartz as the sensor material. The electrical charge that is generated on the quartz surfaces is very low (2.30 pC/N for the longitudinal piezoelectric effect) and charge amplifiers have to be used for electronic processing (Tichy and Gautschi, 1980). Due to volume conduction within the material and because of the limited input impedance of charge amplifiers, quartz cannot show a purely static response. However, given the excellent insulation properties of synthetic quartz crystals (10^{+16} Ω) and the high input impedance values of modern FET charge amplifiers, measuring devices with quartz are 'semistatic'. For most biomechanical applications, the semistatic response provides sufficient accuracy for loading times lasting up to several minutes. Considering the paramount importance of insulation for proper transducer operation, even increased humidity in the air may lead to noticeable drifts of the output signal. It is important to protect all cable connections of a piezoelectric platform from moisture. Drying the connectors with a hair dryer may help to improve performance of piezoelectric transducers on a humid day or in a humid environment.

High-precision and inexpensive piezoceramic materials have been used for pressure distribution measurements (Hennig et al., 1982). Piezoceramic materials show very small and highly elastic deformations. As compared to quartz, piezoceramic materials generate approximately 100 times higher charges on their surfaces when identical forces are applied. The high charge generation allows the use of inexpensive charge amplifiers, thus allowing pressure distribution units with more than 1000 discrete sensors. For rapid loads, these transducers combine good linearity and a very low hysteresis. Whereas temperature has only a minor influence on the piezoelectric properties of quartz, piezoceramics also exhibit pyroelectric properties. Therefore, thermal insulation or a temperature equilibrium, as is normally present inside shoes, is necessary. Piezoceramic transducers have successfully been used for applications ranging from measuring the pressures under diabetic feet (Cavanagh et al., 1985) to analysing the foot–ground interaction during running and jumping (Milani and Hennig, 1994). Piezoelectric polymeric films (polyvinylidene fluoride, PVDF) and piezoelectric rubbers have also been developed and have been proposed for the measurements of forces. However, due to large hysteresis effects and unreliable reproducibility, these methods have been less successful. A more detailed description of different piezoelectric sensors and electronic processing procedures is given elsewhere (Hennig, 1988).

RESISTIVE TRANSDUCERS

A variety of different sensor types belong to the category of resistive transducers. The electrical resistance of the sensor material changes under tension or compression. Electrical strain gauges are thin and highly elastic metal wires, allowing the measurement of deformations of an object. With increasing strain, the length of a wire increases and its diameter decreases, causing higher electrical resistance. The gauge factor describes the electromechanical behaviour of strain gauge transducers. The gauge factor (GF) is determined by the quotient of the relative change in resistance ($\partial R/R$) towards the strain (relative change in length $\partial L/L$). Typically, gauge factors are approximately 2. The deformation of strain gauges should not exceed the elastic limits (typically $< 2\%$), to allow elastic recoil of the wire. Strain gauge transducers are bonded to elastic beams that primarily determine the characteristics and accuracy of the transducer assembly. Consequently, each measuring unit must be calibrated individually to relate the change in electrical resistance to force values. Because metal wires change their electrical resistance with varying temperature, strain gauges require temperature compensation techniques with the employment of bridge amplifiers. Piezoresistive transducers have a semiconductor material that changes its resistance with strain. Owing to typically large gauge factors (up to 200), they are very sensitive; unfortunately, they are also very sensitive to temperature changes. Moreover, their piezoresistivity may vary with strain, thus causing a nonlinear response in some ranges.

Volume conduction has also been used as a method for measuring forces. Silicone rubber sensors, filled with silver or other electrical conducting particles, have been manufactured. The measuring principle is similar to the effect of graphite microphones in older telephones. With increasing pressure the conducting particles are pressed closer together, increasing the surface contact between the conducting particles and thus lowering electrical resistance. These transducers show large hysteresis, especially for higher-frequency impact events. They can be used successfully as switches (e.g. to detect foot contact) but are not suitable as accurate transducers. Resistive contact sensors work on a similar principle to the volume conduction transducers. Exerting pressure on two intersecting conductive paint strips (Tekscan Inc., Boston, MA) causes increased contact between the conductive surfaces and leads to a reduction in electrical resistance. However, large hysteresis and low reliability have been reported for this type of sensor (McPoil, Cornwall and Yamada, 1995).

CAPACITIVE TRANSDUCERS

An electrical capacitor typically has two metal plates in parallel with each other with a dielectric material sandwiched in-between. A change in capacitance occurs when the distance between the two plates is varied. Applied forces can be determined through the compression of the elastic material (dielectric) between the plates. The change in capacitance can be transformed into an electric signal for registration by a data acquisition system. The simple construction and low material costs allow the manufacture of inexpensive pressure distribution mats with up to several thousand discrete capacitive transducers. Early attempts, using foam rubber between elastic conductive metal strips, were successful in producing inexpensive pressure distribution devices for biomechanical applications (Nicol and Hennig, 1976). More recently, silicone rubber mats with improved elastic characteristics have been used for the production of commercially available pressure distribution measuring platforms and insoles (Novel Inc., München, Germany). Capacitive transducers can be built in a flexible format to accommodate body curvatures and contours. Due to their soft and pliable nature, capacitive transducers can be employed for the measurements of seating pressures and the registration of body pressures on mattresses (Nicol and Rusteberg, 1993). As compared to the rigid nature of piezoelectric transducers with very low deformations of the transducer ($< 10^{-7}$%), large relative displacements (> 10%) are necessary for capacitive transducers. Large displacements and the viscoelastic nature of the rubber material cause, especially for short-duration impacts, noticeable hysteresis that limits measurement accuracy (Figure 1A). For many locomotor activities, with relatively low loading frequencies, the accuracy of capacitive transducers is sufficient.

SPECIFIC MEASUREMENT REQUIREMENTS

FORCE

Studies of human locomotion that involve measurement of force deal almost exclusively with the non-invasive recording of ground reaction forces. *In vivo* direct measurements of human achilles tendon tension are rare exceptions (Komi, 1990). Instrumented shoes (White et al., 1982) have been employed for a few investigations, but force platforms have been used in the majority of all locomotor studies. To facilitate comparisons between individuals with different body mass, body weight units (BW) are used for normalization purposes. During locomotion, ground reaction force (GRF) peaks have been shown to vary between one and six times body weight, with contact times from 100 ms to more than 1 s. It should be noted that considerably higher vertical (12 BW) and large shear forces (6 BW) have been measured during track and field events (Baumann, 1981) and court sports (McClay et al., 1994).

The resultant force pattern in walking is characterized by two peaks, similar in shape and magnitude (0.9–1.5 BW), that are separated by a trough. Prior to the first peak, there is an initial sharp spike that lasts only few milliseconds. The failure to register that spike indicates poor platform mounting, insufficient sampling frequency or unintended filtering by the processing software. The running pattern is characterized by an initial high-frequency force peak (1.5–2.5 BW) that is followed by a second larger (2.0–3.0 BW) but lower-frequency force peak (Miller, 1990). For fast running or sprinting, the first passive peak can be larger than the second one (6 BW versus 4 BW). Spectral analyses have revealed that most components of the force signal are concentrated below 100 Hz for both walking and running. The fast-rising GRF and its frequency components challenge force platform mountings. Mountings must consequently be rigid enough not to interfere with the physical characteristics of the platform. The usable frequency of the mounted force platform should be at least 300 Hz to capture the initial spike of walking and to adequately represent the spectral components of impact loading.

Targeting is a major problem in the measurement of GRF. It can be reduced effectively by larger platforms. However, larger platforms are characterized by lower natural frequency, and contralateral foot contact might occur during ipsilateral recording of walking gait. For most locomotion studies, however, larger platforms should be favoured. Nevertheless, care should be taken to limit analysis to recordings of single foot contacts when seeking centre of pressure information.

ACCELERATION

The rapidly increasing force at ground contact produces an impact shock. This impact shock can be measured by lightweight accelerometers mounted at different locations (shank, pelvis and head) along the locomotor system. In biomechanics, acceleration is commonly expressed in multiples of g (9.81 m/s^2). Through measurements with bone-mounted transducers, acceleration peaks of more than $5g$ and $11g$ have been recorded along the tibia during walking (Lafortune and Hennig, 1992) and running (Lafortune, Hennig and Valiant, 1995), respectively. Peaks of similar magnitude have been recorded perpendicular to the tibia (Lafortune, 1991). Spectral analyses have indicated that, for walking and running, almost 98% of the acceleration signal power is contained below 100 Hz, with the most powerful components concentrated below 25 Hz. Measurements obtained at the head with accelerometers mounted on bitebars reveal considerable attenuations of the shock both in magnitude and frequency.

The mounting of accelerometers in bone is an invasive technique to measure skeletal acceleration. For most locomotion studies, less traumatic superficial mounting is usually employed. The adequacy of this mounting is heavily dependent upon the technique used to secure the transducer to the bone (Schnabel and Hennig, 1995). The use of preloading can reduce but not eliminate the artefacts caused by the underlying soft tissues, with resonance between 20 and 60 Hz and attenuation of frequencies above 100 Hz (Lafortune, Hennig and Valiant, 1995).

Based upon the magnitude of the shock recorded at the tibia, accelerometers with a range of $\pm 20g$ should be adequate for most locomotion studies. The spectral analysis of bone signals indicates that accelerometers for use in locomotion should have useful frequencies from DC to a minimum of 100 Hz. The DC response is paramount in view of the low-frequency components of the shock, which are dominant. Also, only accelerometers that have a DC response can accurately measure the effect of gravity and segmental kinematics associated with locomotion.

PRESSURE DISTRIBUTION

Pressure is defined as the force divided by the area on which this force is acting. Pressures are measured in kilopascals (100 kPa = 10 N/cm^2). During locomotion the forces between the human body and the ground are distributed under the various supporting structures of the foot. Measurements of GRF with a force platform do not provide information about the loading of these individual structures. For more than 100 years researchers have been interested in the distribution of loads under the human foot during various activities. Early methods estimated plantar forces from impressions of the foot in plaster of paris and clay. Later techniques included optical methods with cinematographic recording (Elftman, 1934). During recent years, the availability of inexpensive force transducers and modern data acquisition systems has made the construction of various pressure distribution-measuring systems possible. Transducer technologies for pressure distribution devices are based on capacitive, piezoelectric and resistive principles. Typically, discrete pressure sensors are arranged in matrices to provide pressure distribution profiles. Pressure platform resolution is defined as the distance between the centres of adjacent discrete sensors. Resolutions of two sensors/cm^2 should be present to obtain detailed information under fine structures of the feet. Even for a relatively small pressure distribution platform of 45 cm $\times$ 23 cm (Novel SF-2), this resolution requires a large number of 2016 discrete sensors. Data acquisition, processing and storage are important issues for pressure distribution measurements. For instance, recording durations of 1 s at a sampling rate of 50 Hz for a pressure mat that comprises 2000 elements will result in 100 000 pressure data points. With this high volume of information, visual presentation and data reduction techniques become important.

Graphical representation of pressure distribution is commonly achieved through wire frame diagrams or isobarographs (Figure 2). These pressure maps can be obtained for each sampling interval or at specific instants during the foot–ground contact.

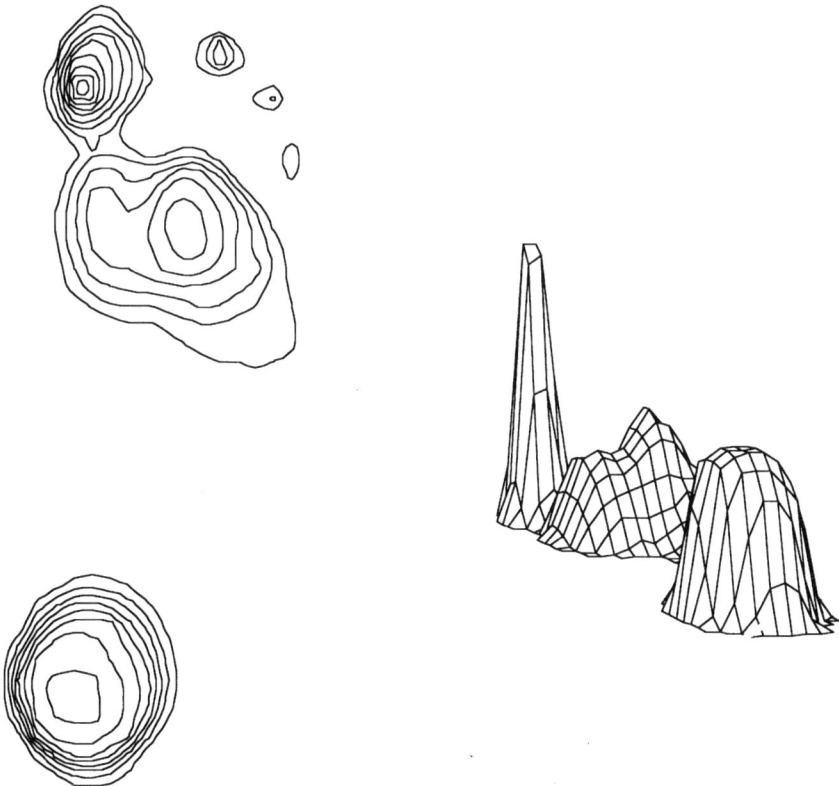

Figure 2. Two types of graphical representation of the peak pressure distribution pattern under a cavus foot during walking

A peak pressure analysis provides information about the highest pressures under the foot, as they occur throughout foot contact. An understanding of the load-bearing role of individual anatomical structures can also be obtained by the time integral of the pressure signals during foot contact with the ground. Relative impulses are calculated by dividing the local force–time integral (force = pressure × area) under a specific anatomical region by the sum of the force integrals under all anatomical regions of the foot surface. Relative loads are expressed as percentages of the total impulse. The division of the plantar surface into anatomical regions should be based on a geometric scheme, to compare results between subject groups and to allow a comparison of findings between different research groups. An example of a division of the foot into 10 anatomical regions, using a geometric algorithm, has been presented by Cavanagh et al. (1985). However, a major morphological foot study is still needed to identify meaningful anatomical foot regions.

Because footwear modifies the foot–ground interaction considerably, in-shoe pressure measurements are of special interest. Matrix sensor insoles can easily be placed inside shoes. Because most feet exhibit individual anatomical variations, the exact placement of the sensors under the anatomical sites of interest is not known. Moreover, depending upon shoe construction peculiarities, the relative positioning of the insole matrix under the foot may vary. To overcome these problems, pressure insoles with large numbers of small transducers are necessary. However, pressure insoles with a high density of accurate sensors are expensive, have restricted time resolution and can generate large amounts of data. It must also be recognized that a device depending on relatively large deformations of its viscoelastic sensing elements will act as an in-shoe cushioning element. Thus, it will modify the magnitude of the pressure that it attempts to measure. The use of a limited number of discrete rigid transducers (Figure 3) offers a viable alternative to gather in-shoe pressure information.

They are typically fixed with adhesive tape under anatomical foot structures that are manually palpated. The major advantage of this technique is an exact positioning of the sensors under the foot structures of interest, independent of individual foot shape. It also guarantees that the sensor locations remain independent from footware construction peculiarities. Their major disadvantage is an incomplete mapping of the foot–shoe interaction, as pressure knowledge is limited to the chosen sensor locations. The thickness of the individual sensors also causes a point-loading effect that will result in an overestimation of local pressures. However, it has been shown that the summed pressures of only eight sensors under the foot are well suited to reproduce the vertical GRF (Hennig and Milani, 1995a). The point-loading artefact of discrete sensors may also help in identifying areas of excessive pressure and should improve the contrast between different footwear constructions.

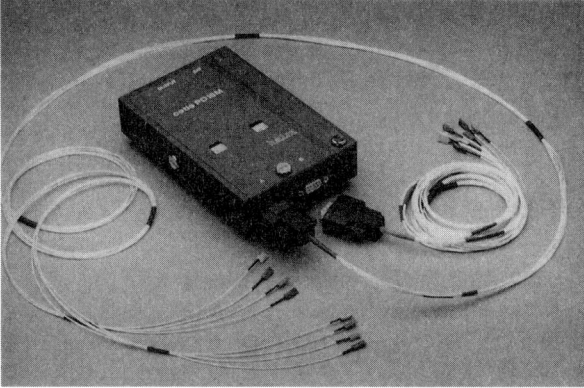

Figure 3. Pressure distribution unit with 16 single piezoelectric sensors (PD-16, Halm GmbH, Frankfurt, Germany)

TRANSDUCER CALIBRATION

Although manufacturers supply calibration information with their transducers, it is useful to perform calibration checks prior to major studies. For transducers with a static response this can easily be accomplished by the user. Positioning known weights at various locations on a force plate and comparing the output allows vertical sensitivity checks across the transducer surface and may reveal possible defects of sensing elements. Shear force calibration can be performed with a mechanical tensiometer. A weight-centreing pole with several barbell weights is positioned at various locations of the force plate. With a thin steel wire around the pole, the tensiometer can be used to pull in a perpendicular direction to the platform axes. The barbell weights on the platform should provide sufficient friction to apply shear forces. The steel string should be located close to the platform surface and the tensiometer should be pulled parallel to its surface.

Static or DC response accelerometers can easily be calibrated by taking advantage of gravity on earth. Readings can be obtained when the sensitive axis of the transducer is held perpendicular to the ground and again following a 180° rotation; the difference in transducer output corresponds to an acceleration change of $2g$. Transducers with non-static response are considerably more difficult to calibrate by the user. One needs a shaker table that can excite the accelerometer at various frequencies over a range of amplitudes. Most biomechanics laboratories do not possess such a calibration tool and users are therefore forced to rely upon the specification sheet of the manufacturer.

Calibration of pressure distribution transducers is also difficult to perform by users. Applying known weights onto the surface of a pressure matrix is not an appropriate procedure to calibrate arrays of pressure transducers. In practice, it is virtually impossible to create homogeneous pressures over the entire surface of the array. Air pressure calibration devices are necessary to apply homogeneous pressures. Typically, a very thin rubber membrane is vulcanized to the bottom of a steel frame with a pressure valve inlet. The pressure transducer matrix is placed underneath the rubber membrane surface and compressed air is used to vary pressure. The thin rubber membrane guarantees accommodation to unevenness of the pressure sensor matrix surface. The thinner and more flexible the membrane is, the more homogeneous the pressure will be across the surface of the transducer. Typically, natural rubber is used as the membrane material.

APPLICATION TO HUMAN LOCOMOTION: RUNNING

Sport shoe construction has been shown to have an influence on kinematic and kinetic variables during running (Nigg and Bahlsen, 1988). This section summarizes the results from a study that combined force, acceleration and pressure distribution measurements (Hennig, Milani and Lafortune, 1993; Hennig and Milani, 1995a). The information was used to evaluate the cushioning properties of running footwear. In addition, the relationships between the different measurements are discussed.

Twenty-two experienced runners performed five trials of rearfoot strike running at approximately 12 km/h (3.3 m/s) across a Kistler force platform (Type 9281 B) in each of 19 different sports shoe constructions (Figure 4). Simultaneously with the recording of GRFs, skin acceleration measurements at the tibia were performed with a low-mass accelerometer (Type EGAX-F-25; Entran Devices Inc., 10 Washington Avenue Fairfield, NJ 07004, USA). The

Figure 4. Subject running across a force platform for a comparative running shoe study

accelerometer was embedded in a small piece of balsa wood. It was fastened onto the skin covering the medial aspect of the tibia midway between the medial malleolus and medial tibial condyle. To improve mechanical coupling, the transducer assembly was glued to the skin and pressed against the shaft of the tibia with an elastic bandage. The mass of the transducer arrangement was less than 1.5 g, including the mass of the balsa wood. A pressure distribution unit Halm PD-16 (Halm GmbH, Frankfurt, Germany) with eight discrete piezo-ceramic transducers was used to record in-shoe pressure. The measuring elements were 4 by 4 mm with a thickness of less than 2 mm. Using adhesive tape, the transducers were positioned at the following anatomical locations underneath the foot: medial and lateral heel and midfoot, the metatarsal heads I, III and V, and the hallux (Figure 5). These locations were identified through visual examination and palpation of the foot.

The vertical component of GRF, the tibial axial acceleration and the output of the eight pressure sensors were sampled simultaneously at 1 kHz per channel. A vertical GRF threshold of 5 N was used to identify the onset of foot–ground contact. The variables used to describe force, axial tibial acceleration and pressure are given in Table 1.

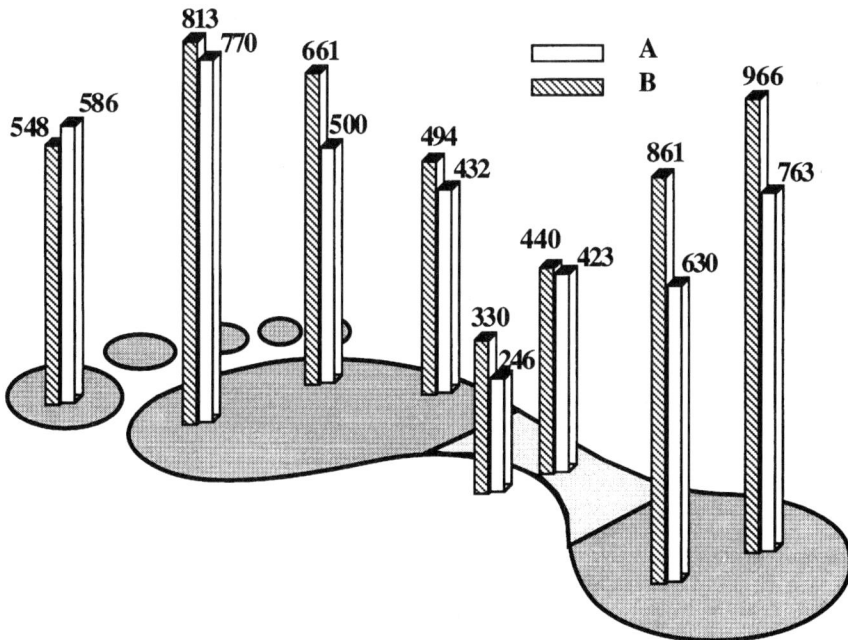

Figure 5. Peak pressure patterns (values in kPa) of 22 runners in two different sport shoe constructions (A and B)

Table 1. List of force, acceleration and pressure variables

Parameter	Code	Unit
Peak vertical force 1	PVF1	BW
Peak vertical force 2	PVF2	BW
Differential quotient to peak vertical force 1	DPVF	BW/s
Median power frequency of the vertical force	MPFZ	Hz
Peak positive axial acceleration	PPAC	g
Peak pressure (heel)	PP-heel	kPa
Peak pressure (midfoot)	PP-midf	kPa
Peak pressure (forefoot)	PP-foref	kPa

PVF1 represented the initial impact force peak and PVF2 the maximum force at midstance. The maximum force rate (differential quotient to peak vertical force DPVF) corresponded to the highest differential quotient of adjoining vertical GRF values divided by the time resolution of 1 ms. The median power frequency of the vertical force (MPFZ) signal was calculated from a 1024-point FFT power spectrum analysis. Spectral power components below 10 Hz were ignored to emphasize the impact aspect of the GRF.

Significant between-shoe differences ($p < 0.001$) were observed for all variables except the second force peak (PVF2). At foot strike, the lateral heel pressures were found to be higher than the medial values in all shoe constructions. Following this initial concentration of pressure under the lateral structures of the foot, a gradual shift towards medial structures took place in the midfoot and forefoot regions. During push-off, the load was almost exclusively concentrated under the first metatarsal head and the hallux. The peak pressure image (Figure 5) reveals large in-shoe pressure differences between footwear constructions. Shoe B shows substantially increased pressures in the heel area, whereas the forefoot pressures under the first metatarsal head and the hallux are almost identical to those in shoe A. Peak pressures have also been found to vary with shoe wear. After a distance of 220 km had been run in 19 different shoes, heel and forefoot peak pressures were found to have increased by almost 8% on average (Hennig and Milani, 1995b). For some shoe constructions, peak pressures increased by as much as 25% in the heel and forefoot regions.

High correlations were found between peak tibial axial acceleration and both maximum force rate ($r = +0.98$, Figure 6) and median power frequency ($r = +0.94$). Conversely, the correlation between peak tibial axial acceleration and initial peak GRF was only $r = +0.80$.

A good prediction of peak tibial accelerations could be obtained from force platform measurements. Using a stepwise regression analysis with two variables (DPVF, MPFZ), most of the variability in peak axial acceleration ($r^2 = 0.97$) could be explained by variations of the GRF variables. Using intracortical bone acceleration measurements with five subjects during running, good predictions

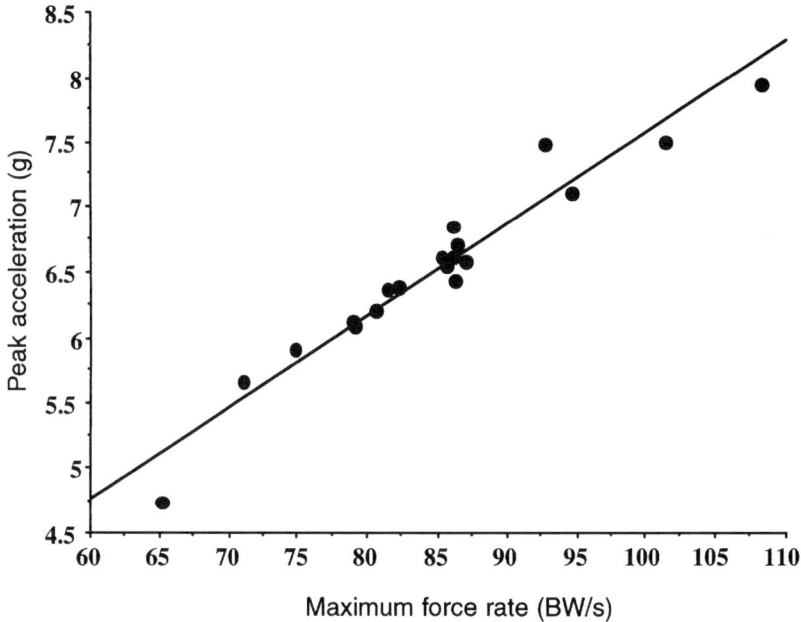

Figure 6. Relationship between peak tibial acceleration and maximum force rate. The values represent mean values from 22 subjects in each of 19 different footwear models

of peak tibial accelerations by ground reaction parameters ($r^2 = 0.99$) have also been reported (Hennig and Lafortune, 1991).

Surprisingly, a weak relationship existed between peak heel pressure and initial peak vertical force. The correlation between these two variables barely exceeded 0.5. This low correlation can be explained by the loading of midfoot and forefoot structures that began before the occurrence of PVF1 (Hennig and Milani, 1995a). Thus, depending on features of the shoes' midfoot and forefoot construction, the various foot structures will contribute differently to the magnitude of the first vertical GRF peak.

GRF, pressure and acceleration measurements have been shown to be valuable in evaluating the effects of footwear on the load and shock experienced by the human body during running. In spite of high correlations between some of the biomechanical variables, each measuring method provides unique information that contributes to the understanding of the interaction between the human body and the ground. Force platforms quantify the total load applied to the body, accelerometers measure the shock experienced by different structures of the body, and pressure transducers provide information about the local forces acting under discrete anatomical structures of the foot. The value of any one of these measurements is contingent upon the types of transducer used, their technical specifications, and adequate mounting for both force platforms and acceler-ometers.

REFERENCES

Baumann, W. (1981) On mechanical loads on the human body during sports activities. In: *Biomechanics VII-B* (ed. K. F. A. Morecki, K. Kedzior and A. Wit), Pwn-Polish Scientific Publishers, Warsaw, pp. 79–87.

Cavanagh, P. R., Hennig, E. M., Rodgers, M. M. and Sanderson, D. J. (1985) The measurement of pressure distribution on the plantar surface of diabetic feet. In: *Biomechanical Measurement in Orthopaedic Practice* (ed. M. Whittle and D. Harris), Clarendon Press, Oxford, pp. 159–166.

Curie, J. and Curie, P. (1880) Développement par compression de l'électricité polaire dans les cristeaux hémidèdres à faces inclinées. *Bull. Soc. Mineral*, **3**, 90–93.

Elftman, H. (1934) A cinematographic study of the distribution of pressure in the human foot. *Anat. Rec.*, **59**, 481–492.

Hennig, E. (1988) Piezoelectric sensors. In: *Encyclopedia of Medical Devices and Instrumentation* (ed. J. G. Webster), John Wiley and Sons, New York, pp. 2310–2319.

Hennig, E. M. and Lafortune, M. A. (1991) Relationships between ground reaction force and tibial bone acceleration parameters. *Int. J. Sport Biomech.*, **7**(3), 303–309.

Hennig, E. M. and Milani, T. L. (1995a) In-shoe pressure distribution for running in various types of footwear, *J. Appl. Biomech.*, **11**(3), 299–310.

Hennig, E. M. and Milani, T. L. (1995b) Biomechanical profiles of new against used running shoes. In: *Nineteenth Annual Meeting of the American Society of Biomechanics* (ed. K. R. Williams), American Society of Biomechanics, Palo Alto, pp. 34–44.

Hennig, E. M., Milani, T. L. and Lafortune, M. A (1993) The use of ground reaction force parameters for the prediction of peak tibial accelerations in running with different footwear. *J. Appl. Biomech.*, **9**(4), 306–314.

Hennig, E. M., Cavanagh, P. R., Albert, H. and Macmillan, N. H. (1982) A piezoelectric method of measuring the vertical contact stress beneath the human foot. *J. Biomed. Eng.*, **4**, 213–222.

Komi, P. V. (1990) Relevance of in vivo force measurements to human biomechanics. *J. Biomech.*, **23**(S1), 23–34.

Lafortune, M. A. (1991) Three-dimensional acceleration of the tibia during walking and running. *J. Biomech.*, **24**, 877–886.

Lafortune, M. A. and Hennig, E. M. (1992) Cushioning properties of footwear during walking: accelerometer and force platform measurements. *Clin. Biomech.*, **7**, 181–184.

Lafortune, M. A., Hennig, E. M. and Valiant, G. A. (1995) Tibial shock measured with bone and skin mounted transducers. *J. Biomech.*, **28**, 989–993.

McClay, I. S., Robinson, J. R., Andriacchi, T. P., Frederick, E. C., Gross, T., Martin, P., Valiant, G., Williams, K. R. and Cavanagh, P. R. (1994) A profile of ground reaction forces in professional basketball. *J. Appl. Biomech.*, **10**(3), 222–236.

McPoil, T. G., Cornwall, M. W. and Yamada, W. (1995) A comparison of two in-shoe plantar pressure measurement systems. *The Lower Extremity*, **2**, 95–103.

Milani, T. L. and Hennig, E. M. (1994) Druckverteilungsanalysen im Sportschuh beim Weitsprung unterschiedlicher Leistungsklassen. *Dtsch. Z. Sportmed.*, **45**(1), 4–8.

Miller, D. J. (1990) Ground reaction forces in distance running. In: *Biomechanics of Distance Running* (ed. P. R. Cavanagh), Human Kinetics, Champaign, Illinois, pp. 203–224.

Nicol, K. and Hennig, E. M. (1976) Time-dependent method for measuring force distribution using a flexible mat as a capacitor. In: *Biomechanics V-B* (ed. P. V. Komi), University Park Press, Baltimore, pp. 433–440.

Nicol, K. and Rusteberg, D. (1993) Pressure distribution on mattresses. In: *Biomechanics XIV* (ed. S. Bouisset, S. Métral and H. Monod), International Society for Biomechanics, Paris, pp. 942–943.

Nigg, B. M. and Bahlsen, H. A. (1988) Influence of heel flare and midsole construction on pronation, supination, and impact force for heel–toe running. *Int. J. Sport Biomech.*, **4**, 205–219.

Schnabel, G. and Hennig, E. M. (1995) The effect of skin mounting technique on tibial acceleration measurements during running. In: *Second Symposium on Footware Biomechanics* (ed. P. Brüggemann, M. Shorten, N. Frederick, A. Knicker, S. Luethi and G. Valiant), Deutsche Sporthochschule Köln, Köln, pp. 34–35.

Tichy, J. and Gautschi, G. (1980) *Piezoelektrische Meßtechnik*. Springer, Berlin.

White, S. C., Winter, D. A., King, A. I. and Nakhla, S. (1982) Kinetic analysis of gait using instrumental force-shoe. In: *Human Locomotion II* (ed. J. G. Reid, T. Bryant, S. Olney, B. Smith, J. Stevenson and R. Walmsley), Canadian Society for Biomechanics, Kingston, Ontario, pp. 92–93.

7

Gait Data: Terminology and Definitions

AURELIO CAPPOZZO[1], UGO DELLA CROCE[1]
AND LUIGI LUCCHETTI[2]

[1]Cattedra di Tecnologie Biomediche, Università degli Studi, Sassari, Italy
[2]Istituto di Fisiologia Umana, Università degli Studi 'La Sapienza', Rome, Italy

INTRODUCTION

Movement analysis, in order to assess musculoarticular function, both in physiological and clinical contexts, uses the quantitative description of joint kinematics, i.e. the relative motion between adjacent bones, and the prediction of forces transmitted by the tissues involved. Sometimes the potentially powerful concepts of work and energy are also used.

Basically, this requires two sets of data, a set of variables and a set of parameters:

1. The three-dimensional (3D) instantaneous position and orientation of the bones involved, and the external forces and couples acting on relevant body segments. These data must be available as a function of time during the execution of the physical exercise under analysis and are obtained using a stereophotogrammetric system and force plates.
2. Information concerning the inertial properties and anatomical data of body segments such as the musculoskeletal geometry and musculotendon parameters.

The effective use of movement analysis in professional contexts critically depends on a universal agreement on variable definitions, conventions and terminology. This is the case for the following reasons:

Three-dimensional Analysis of Human Locomotion. Edited by P. Allard, A. Cappozzo, A. Lundberg and C. Vaughan
© 1997 John Wiley & Sons Ltd. ISBN 0 471 96949 4

- to allow for a direct interpretation and comparison of data obtained from different laboratories;
- to allow for the establishment of databases;
- to allow for the application of anatomical data obtained in a sample population to a specific individual;
- to make the training of relevant operators easier, more effective, and laboratory-independent.

Following these considerations, this chapter deals with definitions, conventions and terminology with reference to the kinematic and force data. The present material is proposed as a basis for a standard in human movement biomechanics.

Quantities resulting from further processing of the above-mentioned variables will be dealt with in Chapters 8 and 9, where relevant mathematical procedures are discussed.

GLOBAL FRAMES: DEFINITIONS

Photogrammetric Axes

This is the frame with respect to which positional data of markers are provided by the stereophotogrammetric system; it is arbitrarily chosen and usually coincides with the photogrammetric calibration object system of axes.

Walk-path Axes

This is a frame which indicates the direction of progression and position and orientation of the floor with respect to the photogrammetric axes. It provides an overall description of the motor task assigned to the subject and is represented by the following right-handed set of axes:

X coincides with the walking direction assigned to the subject and points anteriorly;

Y is orthogonal to the floor and points upwards;

Z goes from the left to the right-hand side of the subject;

O the origin must lie on the floor and on the midsagittal plane assigned to the subject.

Information regarding the position and orientation of these axes must accompany photogrammetric data, unless these are provided, through a coordinate transformation, directly with respect to the walk-path axes.

We are aware that this convention may differ from that used by many people active in biomechanics, as demonstrated by a debate which is ongoing in the Biomechanics and Movement Science listserver. The vertical axis is, in fact, often labelled with Z. However, there are as many precedents in the literature exhibiting one convention as there are exhibiting the other. Indeed, there are no indisputable reasons to favour one over the other. The above convention is suggested as a standard for the very simple reason that two international organizations, i.e. the International Society of Biomechanics (Wu and Cavanagh, 1995) and the CAMARC programme, have recommended it (Benedetti et al., 1994; Cappozzo et al., 1994).

Plumb Line

This is the gravity line and points upwards. If the Y axis of the walk-path frame is not vertical, then relevant information must accompany kinematic data.

Force Plate Axes

This is the frame embedded in the force plate and with respect to which the relevant calibration matrix is provided. Relevant information is mandatory when ground reactions are reported. The positional relationship of these axes with respect to the walk-path axes may provide information on what foot hits what force plate.

Global Frames Calibration

This is the experimental procedure which allows for the determination of the position and orientation of the walk-path axes, the plumb line and the force plate axes relative to the photogrammetric axes. This may be achieved by using photogrammetric data of an adequate number of markers placed on the walk-path, on a plumb line and on the force plate(s) and the mathematical procedures illustrated in Chapter 8.

The mathematical description of the above-mentioned frames and of the transformations from one to the other is illustrated in a later section.

POINTS AND LOCAL FRAMES: DEFINITIONS

In the above-mentioned context, one basic objective is the reconstruction of the position and orientation of the bone under analysis, which entails the possibility of reconstructing the instantaneous spatial position in a global frame of any relevant point of the bone, and it is wished to do this irrespective of whether we are able to experimentally observe the trajectory of this point during movement or not. In this exercise, different types of points and local sets of axes are involved.

Anatomical Landmark

A point, or effectively a small area, that can be reliably identified within a biological structure (bone). Relevant examples are provided in the next section.

Marker Point

This indicates a location on the skin where a skin marker is positioned.

Anatomical Skin Marker

This is a marker positioned on the skin in a location approximating an anatomical landmark.

Technical Marker

This is a marker positioned in a location which has no anatomical relevance.

Cluster of Markers

This is a set of markers associated with a bone.

Bone-embedded Frame

This is a set of coordinate axes effectively rigid with the bone. It is evident, but worthy of emphasis, that in the very moment we talk about a bone-embedded frame we accept the hypothesis of rigidity of the bone, in the sense of the stereotype of classical mechanics. This frame is defined in a right-handed manner. Definitions related to the mathematical description of the position and orientation of such a frame are given in a later section.

After calibration of the stereophotogrammetric set-up and of the relevant global frames, the trajectories of the markers are recorded by the optoelectronic devices while the subject performs a given motor task. Using a 3D point reconstruction algorithm, the coordinates of the markers are thereafter estimated in the photo-grammetric frame at each sampled instant of time. From the reconstructed coordinates of the markers of a cluster, the position and orientation of the relevant bone is estimated versus time using the algorithms illustrated in Chapter 8. Relevant definitions are as follows.

Three-dimensional Point Reconstruction

This is the estimation of the instantaneous photogrammetric coordinates of a marker from its image coordinates.

Rigid-body Pose

This is the ensemble of position and orientation of a rigid body, i.e. of a rigid-body-embedded frame.

Rigid-body Pose Reconstruction

This is the procedure used to estimate the pose of a rigid-body-embedded frame using the coordinates of the markers of the relevent marker cluster.

Once the location in space of the bone-embedded frame is known, in order to solve the stated problem, information concerning the location of relevant anatomical landmarks with respect to this frame is needed. This information usually comes in the form of the local coordinates of the target points; these coordinates are, of course, time invariant and, as such, may be determined under the most favourable experimental circumstances. This is certainly not necessarily while the subject is performing his or her exercise. These coordinates are therefore to be looked upon as parameters of the skeletal model of a given individual. This leads to the following definition.

Anatomical Landmark Calibration

This is the *ad hoc* experiment used for the determination of the local coordinates of the anatomical landmarks.

If the anatomical landmark lies on the cutaneous surface, then its coordinates in the technical frame may be determined by placing a marker on it. The subject is asked to assume a posture which allows both the anatomical and the technical markers to be seen by two or more cameras. Then at least one frame is recorded. Obvious vector calculations follow. The procedure may have to be repeated for each anatomical landmark and the subject may be required to assume different postures in order to make both the anatomical and the technical markers visible to the cameras. Markers used for the identification of the anatomical landmarks are, of course, removed before the physical movement is performed. Anatomical landmark location may also be determined using a pointer on which a minimum of two markers has been mounted at an adequate distance. The experimenter points the tip of the pointer onto the anatomical landmark so that the markers on the pointer and the relevant body segment technical markers are visible to the cameras. Then at least one frame is recorded. This procedure is repeated for each anatomical landmark. Through geometrical calculations and using the reconstructed position of the pointer markers, the location of the anatomical landmarks may be determined. This method is usually more practical than the one presented previously, especially when the anatomical landmark is in an awkward position. When the anatomical landmark to be calibrated is internal to the body, i.e. at a distance from the surface greater than the skin and subcutaneous tissue thickness, the method of the pointer may still be used provided that the anatomical landmark is on the line defined by the two pointer markers and the position relative to them is known. If two landmarks are a short distance apart as seen by one or more cameras, and their positions are used to determine the orientation of a frame axis, then, in order to avoid macroscopic inaccuracies on this orientation, the following technique may be used. The two landmarks involved are calibrated simultaneously using a pointer which, besides indicating their locations, makes the orientation of the relevant axis visible to the operator (Gauffin, Areblad and Tropp, 1993). Typically, this technique may be applied to the femoral condyles and to the malleoli. Anatomical landmark calibration may also be carried out using roentgenographic and anatomical measurements (Andriacchi et al., 1980; Johnston, Brand and Crowninshield, 1979).

If no other practical way is available, other techniques entailing more or less important assumptions may be used. When an anatomical landmark may be approximated by a joint centre, as is the case with the centre of the acetabulum, the calibration may be carried out using a 'functional approach' based on the assumption that the joint centre coincides with the pivot point of the movement between the adjacent segments (Cappozzo, 1984; Bell, Pedersen and Brand, 1990; Cappozzo and Gazzani, 1990; Riley, Mann and Hodge, 1990). However, under pathological circumstances this method may be difficult to use and may be unreliable. The location of an anatomical landmark in the relevant frame may also be determined by assuming that, during a given posture, it coincides with another landmark in a different frame which can be calibrated more easily. For instance, the centre of the femoral head in the femoral technical frame may be determined by assuming that it coincides with the centre of the acetabulum while the subject assumes a predefined posture (e.g. a standing posture). Similarly, the intercondylar eminence in the tibia may be calibrated as coinciding with the midpoint between the femoral medial and lateral epicondyles while the subject assumes a predefined posture. It is clear that, even when using these approaches, no restriction is imposed on the six degrees of freedom of the bone involved; it may only be a matter of accuracy associated with the anatomical frame estimation.

In summary, in order to reconstruct the trajectory of anatomical landmarks in the global frame during movement, the following numerical information is required: the pose in space of a bone-embedded frame in time, and the local coordinates, relative to this latter frame, of the anatomical landmarks, which are time-invariant parameters.

A further problem in human movement analysis is the following. For analytical purposes, i.e. for the estimation of joint kinematics and dynamics, bone-embedded frames ought to meet the following requirements:

- Their determination from experimental data should be repeatable both inter- and intra-individually.
- In view of the quantitative description of the relevant joint kinematics, they should possibly incorporate or permit the determination of suitable axes with respect to which both rotations and translations of the joint may be defined.
- Since the analysis of the limb will be dynamic, they should permit an easy implementation of the estimation techniques aimed at the location of the body segment centre of mass and principal axes of inertia. In addition, sufficient information must be available to locate the reference system with respect to which the intersegmental loads are calculated.
- Requirements associated with the description of muscle and ligament lines of action and the location and orientation of the articulation surfaces must also be taken into careful consideration.

As will appear evident in the next chapter, bone-embedded frames, provided in a first instance by the experimental data, do not have the above-listed characteristics because their definition is based only on experimental requirements. For this reason these frames result in having thoroughly arbitrary and non-repeatable geometric relationships with respect to the bone. It is evident that the above-mentioned requirements are only met by frames rigidly associated with the anatomy of the bone. Their identification will therefore be based on the location of a number of anatomical landmarks.

Relevant definitions are as follows.

Technical Frame

This is a right-handed set of coordinate axes estimated from the positions of technical markers (cluster).

Anatomical Frame

This is a set of right-handed coordinate axes constructed from the positions of bone anatomical landmarks. The definitions of these frames deserve standardization. A proposal in this sense and with reference to the lower limb and the pelvis is given in Cappozzo et al. (1995) and summarized in the next section.

Joint Axes

As will be discussed in Chapter 9, for an effective application of kinematic data in the clinical practice, the six scalar quantities associated with the six degrees of freedom of the joint must be given with respect to an adequate set of axes (the joint axes) which does not necessarily coincide with the axes of the above-mentioned anatomical frames.

The joint axes are three axes, not necessarily orthogonal, with respect to which the joint six degrees of freedom are defined. The orientations of these axes are given in the relevant anatomical frames.

Details of possible solutions to this problem are provided in Chapter 9.

PELVIS AND LOWER LIMB ANATOMICAL LANDMARKS AND FRAMES

ANATOMICAL LANDMARKS

Anatomical landmarks in the pelvis, thigh, shank and foot which may be of relevance in the present context are shown in Figures 1, 2 and 3 and listed below, with reference to one lower limb.

Hip Bone (Figure 1)

ASIS   anterior superior iliac spine
PSIS   posterior superior iliac spine
AC   centre of the acetabulum

Femur (Figures 1 and 2)

FH   centre of the femoral head
GT   prominence of the greater trochanter external surface
ME   medial epicondyle
LE   lateral epicondyle
LP   anterolateral apex of the patellar surface ridge
MP   anteromedial apex of the patellar surface ridge
LC   most distal point of the lateral condyle
MC   most distal point of the medial condyle

Tibia and Fibula (Figures 2 and 3)

IE   intercondylar eminence
TT   prominence of the tibial tuberosity
HF   apex of head of the fibula
MM   distal apex of the medial malleolus
LM   distal apex of the lateral malleolus
MMP   most medial point of the ridge of the medial tibial plateau
MLP   most lateral point of the ridge of the lateral tibial plateau

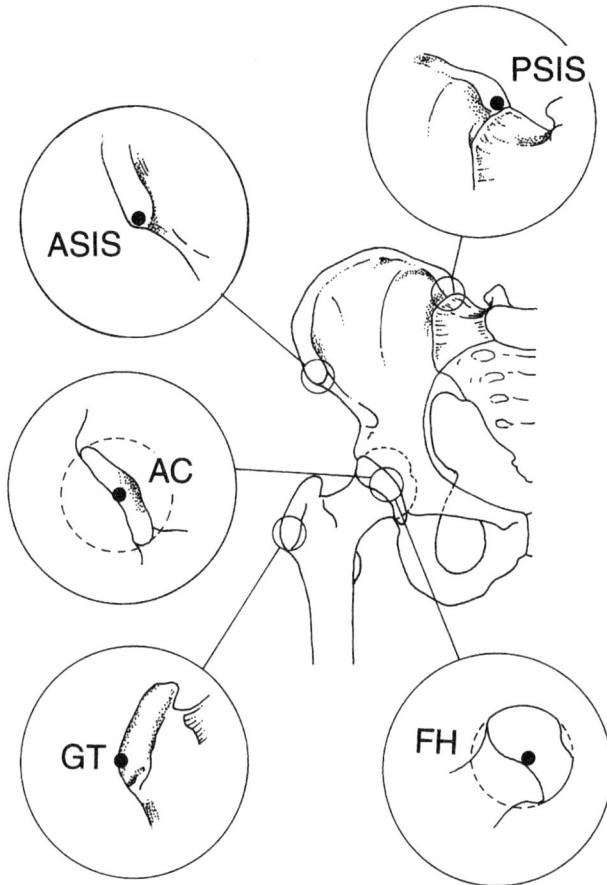

Figure 1. Anatomical landmarks of hip bone and of the proximal end of the femur

Foot (Figure 3)

CA   upper ridge of the calcaneus posterior surface
FM   dorsal aspect of first metatarsal head
SM   dorsal aspect of second metatarsal head
VM   dorsal aspect of fifth metatarsal head

Terminology used to describe these landmarks was derived from well-established anatomical literature (Davies and Coupland, 1969) and care was taken to describe the smallest possible area of the bone for each landmark.

Most of these anatomical landmarks are relatively easy to identify within small areas and can be located by palpation using the indications provided in

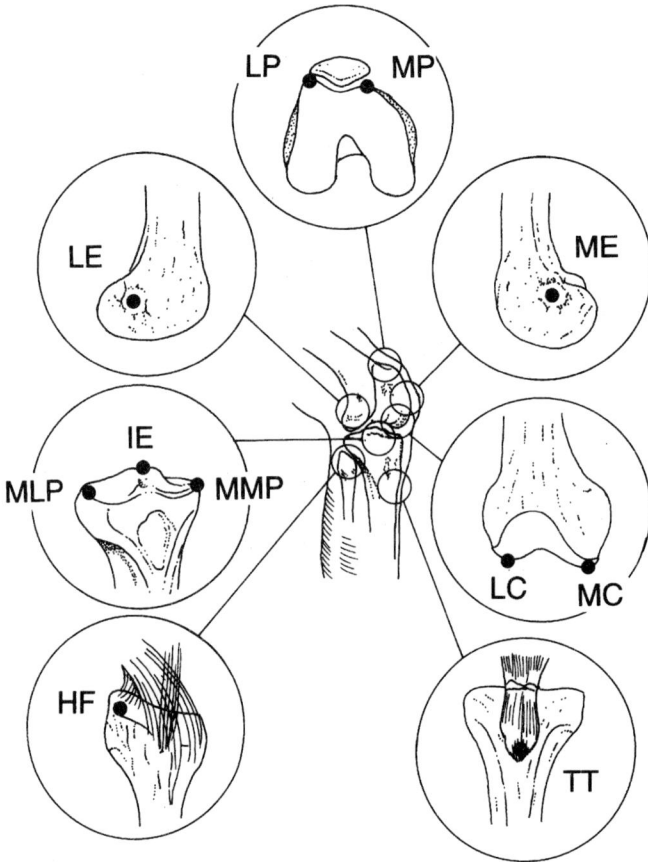

Figure 2. Anatomical landmarks of the distal end of femur and proximal end of tibia and fibula

Hoppenfield (1976) and Benedetti et al. (1994). Consequently, they can be calibrated using the techniques mentioned above. On the contrary, as previously briefly mentioned, landmarks AC, FH and IE are derived from these and/or other measurements.

As better seen below, the anatomical landmarks reported above for each body segment are redundant with respect to those strictly necessary to identify the anatomical frames. However, this redundancy may allow for best estimates of these anatomical frames, for identifying joint axes, and for a realistic graphical representation of the bones. In addition, some of the listed anatomical landmarks are good candidates as reduction points for the calculation of the intersegmental couples, seen as estimates of muscular and ligament moments, as well as for the estimation of joint linear displacements.

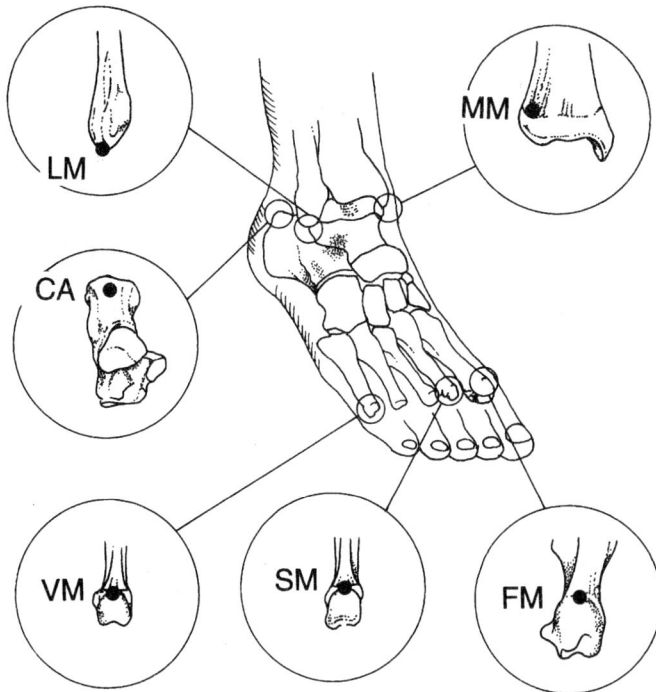

Figure 3. Anatomical landmarks of the distal end of tibia and fibula and of the foot

ANATOMICAL FRAMES

The pelvic girdle comprises three separate bones, the two hip bones and the sacrum. In this study the relative movement between these bones is negligible. Morphological sagittal asymmetries can, however, be taken into account.

The posterior and middle segments, and the metatarsal bones of the foot, are considered to be rigidly connected.

Based on the criteria mentioned above and the anatomical landmarks listed above, the following right-handed anatomical frames are proposed for pelvis and both right and left lower limb bones (Figure 4). In this description, anatomical planes are defined with respect to the standing subject in the 'anatomical position'.

Pelvis (Right and Left Hip Bones and Sacrum)

O_p The origin is at the midpoint between the anterior superior iliac spines (RASIS and LASIS).

z_p The z axis is oriented as the line passing through the ASISs with its positive direction from left to right.

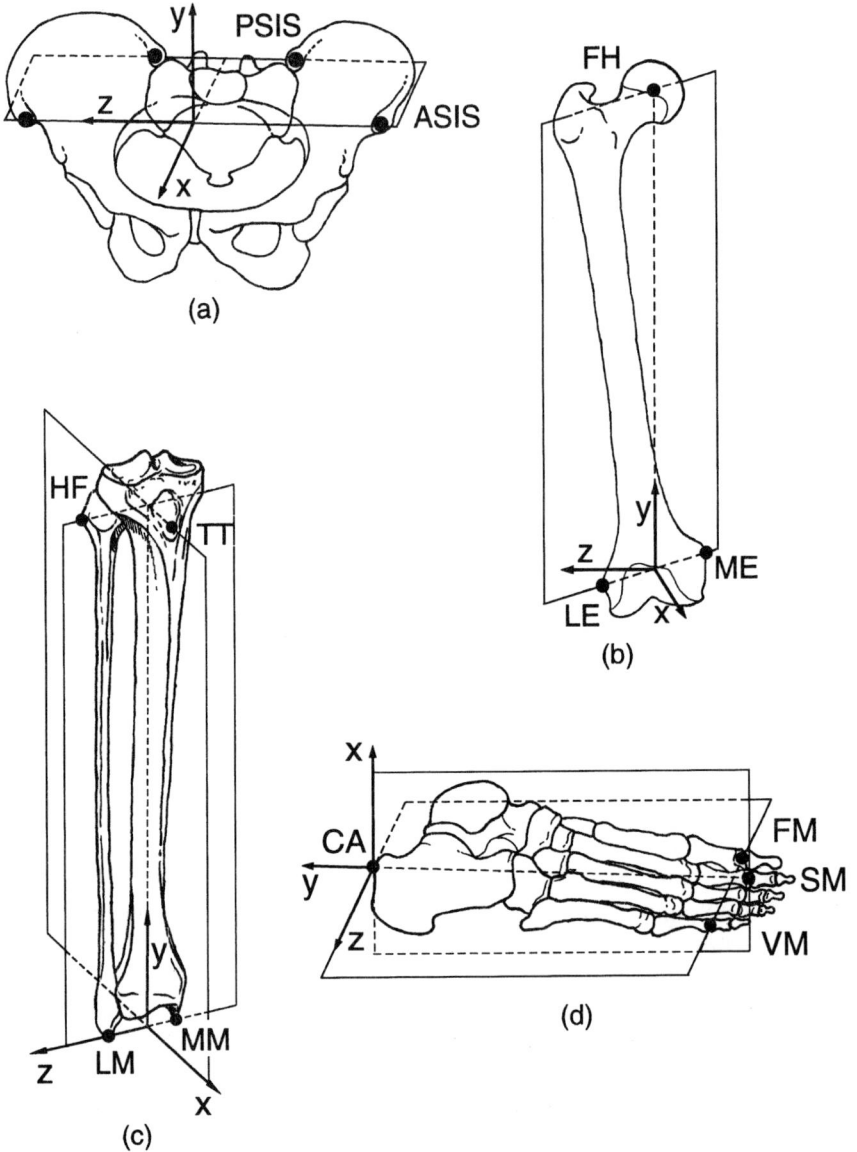

Figure 4. Anatomical frames

x_p The x axis lies in the quasi-transverse plane defined by the ASISs and the midpoint between the PSISs and with its positive direction forwards.

y_p The y axis is orthogonal to the xz plane and its positive direction is proximal.

Right and Left Thigh

O_t The origin is the midpoint between the lateral and medial epicondyles (LE and ME).

y_t The y axis joins the origin with the centre of the femoral head (FH) and its positive direction is proximal.

z_t The z axis lies in the quasi-frontal plane defined by the y axis and by the epicondyles with its positive direction from left to right.

x_t The x axis is orthogonal to the yz plane with its positive direction forwards.

Right and Left Shank

O_s The origin is located at the midpoint of the line joining the lower ends of the malleoli (MM and LM).

y_s The malleoli and the head of the fibula landmarks (HF) define a plane which is quasi-frontal. A quasi-sagittal plane, orthogonal to the quasi-frontal plane, is defined by the midpoint between the malleoli and the tibial tuberosity (TT). The y axis is defined by the intersection between the above-mentioned planes with its positive direction proximal.

z_s The z axis lies in the quasi-frontal plane with its positive direction from left to right.

x_s The x axis is orthogonal to the yz plane with its positive direction forwards.

Right and Left Foot (Talus + Calcaneus + Cuboid + Navicular + Lateral, Medial and Intermediate Cuneiforms + Metatarsals)

O_f The origin is located at the calcaneus landmark (CA).

y_f The calcaneus and the first and fifth metatarsal heads (FM and VM) define a plane which is quasi-transverse. A quasi-sagittal plane, othogonal to this latter plane, is defined by the calcaneus landmark and the second metatarsal head (SM). The y axis is defined by the intersection of these two planes and its positive direction is proximal.

z_f The z axis lies in the quasi-transverse plane and its positive direction is from left to right.

x_f The x axis is orthogonal to the yz plane and its positive direction is dorsal.

NUMERICAL DESCRIPTION OF A RIGID-BODY-EMBEDDED FRAME POSE

As appears evident from the above definitions, the most recurrent analytical problem in the present context is the description of the pose of a rigid body, be it a bone, a force plate or other, with respect to a reference observer. In the context of the present chapter and as a preamble to Chapters 8 and 9, it is therefore worth expounding on this topic with special reference to definitions.

The pose of a rigid-body-embedded frame, with respect to a reference observer, may be numerically described in a number of ways (Berme, Cappozzo and Meglan, 1990; Woltring, 1994; Nigg and Cole, 1994). We find it convenient to use two vectors: a position vector and an orientation vector. These two vectors represent, in fact, a translation and a rotation respectively, i.e. the displacements which take the frame from a reference location, e.g. coinciding with the global frame, to its actual location. We exploit a kinematic concept, or rather a theorem of the kinematics of rigid bodies which says that the most general displacement of a rigid body is a translation plus a rotation (Charles' theorem). Thus, the instantaneous position and orientation of a rigid-body-embedded frame (B—b_1, b_2, b_3) is defined and described with respect to the global frame (L—L_1, L_2, L_3) in each instant of time by (Spoor and Veldpaus, 1980; Woltring, 1991) a position vector of the origin

$$^L\mathbf{t}_B = [t_1 \quad t_2 \quad t_3]^T \tag{1}$$

and an orientation vector

$$^L\boldsymbol{\theta}_B = [\theta_1 \quad \theta_2 \quad \theta_3]^T \tag{2}$$

This latter vector can be calculated from the frame direction cosines (orientation matrix) as reported below and represents a compact form (three versus nine numbers) for representing the relevent information.

The orientation of a system of axes B with respect to a reference system of axes L can be represented by the orientation matrix

$$^L\mathbf{R}_B = \begin{bmatrix} R_{11} & R_{12} & R_{13} \\ R_{21} & R_{22} & R_{23} \\ R_{31} & R_{32} & R_{33} \end{bmatrix} \tag{3}$$

where R_{ij} is the direction cosine of the jth axis of B with respect to the ith axis of L, i.e. the cosine of the angle between the positive semi-axes i of L and j of B.

The orientation vector components can be calculated using the following relationships (Spoor and Veldpaus, 1980):

$$\sin\theta\,\mathbf{n} = \frac{1}{2} \begin{bmatrix} R_{32} - R_{23} \\ R_{13} - R_{31} \\ R_{21} - R_{12} \end{bmatrix} \tag{4}$$

$$\sin\theta = \tfrac{1}{2}\sqrt{(R_{32} - R_{23})^2 + (R_{13} - R_{31})^2 + (R_{21} - R_{12})^2}$$

$$\cos\theta = \tfrac{1}{2}(R_{11} + R_{22} + R_{33} - 1)$$

and therefore

$$\boldsymbol{\theta} = \operatorname{atan} 2\left(\frac{\sin \theta}{\cos \theta}\right) \tag{5}$$

where $\operatorname{atan} 2(y/x)$ is the arc whose tangent is y/x; it is evaluated taking into account the signs of both x and y and it is not defined for $y = x = 0$.

From θ and $\mathbf{n}$ the three components of the orientation vector $\boldsymbol{\theta}$ are obtained:

$$\boldsymbol{\theta} = \begin{bmatrix} \theta_1 \\ \theta_2 \\ \theta_3 \end{bmatrix} = \theta \begin{bmatrix} n_1 \\ n_2 \\ n_3 \end{bmatrix} = \frac{\theta}{2 \sin \theta} \begin{bmatrix} R_{32} - R_{23} \\ R_{13} - R_{31} \\ R_{21} - R_{12} \end{bmatrix} \tag{6}$$

In the singular case $\mathbf{R} = \mathbf{I}$ (identity matrix), it is assumed, by convention, that

$$\boldsymbol{\theta} = [0 \quad 0 \quad 0]^T$$

For the convenience of the user, we report here the equations which allow the calculation of the direction cosines, i.e. the orientation matrix elements, from the orientation vector components:

$$\mathbf{R}(\boldsymbol{\theta}) = \cos \theta \mathbf{I} + \frac{\sin \theta}{\theta} \mathbf{A}(\boldsymbol{\theta}) + \frac{1 - \cos \theta}{\theta^2} \boldsymbol{\theta}\boldsymbol{\theta}^T \tag{7}$$

where

$$\boldsymbol{\theta} = \begin{pmatrix} \theta_1 \\ \theta_2 \\ \theta_3 \end{pmatrix};$$

$$\theta = \sqrt{(\boldsymbol{\theta}^T \boldsymbol{\theta})}; \tag{8}$$

$$\mathbf{A}(\boldsymbol{\theta}) = \begin{pmatrix} 0 & -\theta_3 & \theta_2 \\ \theta_3 & 0 & -\theta_1 \\ -\theta_2 & \theta_1 & 0 \end{pmatrix}$$

If $\mathbf{R}$ is the orientation matrix of frame B with respect to frame L, then $\mathbf{R}^T$ is the orientation matrix of L with respect to B. This means that the columns of $\mathbf{R}$ contain the coordinates of vectors which are mutually perpendicular and that the same can be said about its rows. Hence $\mathbf{R}$ is an orthogonal matrix, which implies that

$$\mathbf{R}^{-1} = \mathbf{R}^T \text{ and } \det(\mathbf{R}) = 1$$

The mutual orthogonality of the columns of $\mathbf{R}$ and their being unity vectors introduce six constraints which allow the nine elements of $\mathbf{R}$ to describe only three degrees of freedom (DOF) of the system. Such rotational DOF must be added to the three translational DOF provided by the position vector $\mathbf{t}$ (origin of B in frame L), resulting in the six DOF associated with a general orthogonal frame, and consequently with a rigid body, in 3D space.

Besides their use for representing the relative pose of two orthogonal reference frames or, which is the same, the pose of a rigid body with respect to a global frame, $\mathbf{R}$ and $\mathbf{t}$ can be employed to 'switch' from a reference frame to the other, i.e. to transform the coordinates of any point in a frame into its coordinates in the other frame. In this case $\mathbf{R}$ is normally indicated with the more expressive name of rotation matrix. If $\mathbf{R}$ and $\mathbf{t}$ represent the pose of frame B with respect to L, then for each point P

$$^L\mathbf{p} = \mathbf{R} \cdot {}^B\mathbf{p} + \mathbf{t} \tag{9}$$

where $^L\mathbf{p}$ includes the coordinates of point P in frame L and $^B\mathbf{p}$ the same in frame B. If we are interested in the inverse transformation, we just need to recall that $\mathbf{R}^{-1} = \mathbf{R}^T$, so that

$$^B\mathbf{p} = \mathbf{R}^T \cdot ({}^L\mathbf{p} - \mathbf{t}) \tag{10}$$

The equation above shows that $\mathbf{R}^T$ and $-\mathbf{R}^T\mathbf{t}$ are, respectively, the orientation matrix and position vector of frame L with respect to frame B.

RIGID-BODY LANDMARK POSITION ESTIMATION

To the data describing the position and orientation of the rigid-body-embedded frame, the time-invariant local coordinates of n relevant landmarks may be added. Position vector of the n landmarks:

$$^B\mathbf{a}_k = [a_{1k} \quad a_{2k} \quad a_{3k}]^T, \ k = 1, 2, \ldots, n \tag{11}$$

Using the definitions given above, the global position of the landmarks may be calculated as follows:

$$^L\mathbf{a} = \mathbf{R} \cdot {}^B\mathbf{a} + \mathbf{t} \tag{12}$$

CONCLUSIONS

This chapter presents a proposal for standardization of nomenclature in movement analysis in general and gait analysis in particular. However, it should be pointed out that the structured language illustrated above also carries a message concerning the relevant methodological approach. We consider this latter contribution to be as worthy of consideration as the former.

The material illustrated herein is also relevant to the problem associated with gait data archiving and sharing. This is thoroughly dealt with in Chapter 19. Here we simply remark that the methodological approach which allows for bone movement description through relevant technical frame pose and anatomical landmark calibration parameters helps in the solution of this problem. In fact, it allows for marker configuration/protocol flexibility and for future updating of joint kinematics and dynamics estimation procedures. Consistent with this approach and specifically aimed at gait data archiving and sharing, Cappozzo and Della Croce (1994) have proposed a data format based on ASCII code and on a defined syntax and lexicon.

ACKNOWLEDGEMENTS

This work was carried out within the CEC programme AIM—project A-2002 CAMARC-II. The constructive discussions which the authors have had with the project partners about the problems addressed in this paper are gratefully acknowledged. Copies of the CAMARC-II Internal Reports and Deliverables quoted in this paper may be requested from the Project Coordinator, Professor Tommaso Leo, Università degli Studi di Ancona, Dipartimento di Elettronica ed Automatica, Via Brecce Bianche, I-60131 Ancona, Italy.

REFERENCES

Andriacchi, T. P., Andersson, J. B. J., Fermier, R. W., Stern, D. and Galante, J. O. (1980) A study of lower limb mechanics during stair climbing. *J. Bone Joint Surg.*, **62A**, 749–757.

Bell, A. L., Pedersen, D. R. and Brand, R. A. (1990) A comparison of the accuracy of several hip center location prediction methods. *J. Biomech.*, **23**, 617–621.

Benedetti, M. G., Cappozzo, A., Catani, F. and Leardini, A. (1994) *Anatomical Landmark Definition and Identification.* CAMARC II Internal Report, 15 March.

Berme, N., Cappozzo, A. and Meglan, J. (1990) Kinematics. In: *Biomechanics of Human Movement: Applications in Rehabilitation, Sports and Ergonomics*, (ed. N. Berme and A. Cappozzo), Bertec Corporation, Worthington, Ohio, pp. 89–102.

Cappozzo, A. (1984) Gait analysis methodology. *Hum. Movem. Sci.*, **3**, 27–54.

Cappozzo, A. and Della Croce, U. (1994) *The PGD Lexicon.* CAMARC II Internal Report, 15 May.

Cappozzo, A. and Gazzani, F. (1990) Joint kinematics. In: *Biomechanics of Human Movement—Applications in Rehabilitation, Sports and Ergonomics* (ed. N. Berme and A. Cappozzo), Bertec Corporation, Worthington, Ohio, pp. 263–274.

Cappozzo, A., Catani, F., Della Croce, U. and Leardini, A. (1994) *Anatomical Frame Definition and Determination.* CAMARC II Internal Report, 2 May.

Cappozzo, A., Catani, F., Della Croce, U. and Leardini, A. (1995) Position and orientation of bones during movement: anatomical frame definition and determination. *Cl. Biomech.*, **4**, 171–178.

Davies, D. V. and Coupland, R. E. (1969) *Gray's Anatomy*, 34th edn. Longmans, Green and Co. Ltd, London and Harlow.

Gauffin, H., Areblad, M. and Tropp, H. (1993) Three-dimensional analysis of the talocrural and subtalar joints in single-limb stance. *Cl. Biomech.*, **8**, 307–314.

Hoppenfeld, S. (1976) *Physical Examination of the Spine and Extremities.* Appleton Century Crofts, East Norwalk, CT.

Johnston, R. C., Brand, R. A. and Crowninshield, R. D. (1979) Reconstruction of the hip. *J. Bone Joint Surg.*, **61A**, 639–652.

Nigg, B. M. and Cole, G. K. (1994) *Biomechanics of the Musculo-skeletal System.* John Wiley & Sons, Chichester, pp. 254–286.

Riley, P. O., Mann, R. W. and Hodge, W. A. (1990) Modelling of the biomechanics of posture and balance. *J. Biomech.*, **23**, 503–506.

Spoor, C. W. and Veldpaus, F. E. (1980) Rigid body motion calculated from spatial co-ordinates of markers. *J. Biomech.*, **13**, 391–393.

Woltring, H. J. (1991) Definition and calculus of attitude angles, instantaneous helical axes and instantaneous centres of rotation from noisy position and attitude data. In: *Proceedings of the International Symposium on Three-dimensional Analysis of Human Movement* (ed. P. Allard), Montréal, Canada, 28–31 July, pp. 59–62.

Woltring, H. J. (1994) 3-D attitude representation of human joints: a standardization proposal. *J. Biomech.*, **27**, 1399–1414.

Wu, G. and Cavanagh, P. (1995) ISB recommendations for standardization in the reporting of kinematic data. *J. Biomech.*, **28**, 1257–1261.

8

Bone Position and Orientation Reconstruction Using External Markers

ANGELO CAPPELLO[1], AURELIO CAPPOZZO[2], UGO DELLA CROCE[2]
AND ALBERTO LEARDINI[3]

[1]Dipartimento di Elettronica, Informatica e Sistemistica,
Università degli Studi di Bologna, Bologna, Italy
[2]Cattedra di Tecnologie Biomediche, Università degli Studi, Sassari, Italy
[3]Laboratorio di Analisi del Movimento, Istituti Ortopedici Rizzoli, Bologna, Italy

INTRODUCTION

As already mentioned in previous chapters, a basic objective in movement analysis is the reconstruction of the bone motion in the laboratory reference system. This is usually achieved using photogrammetry. From the reconstructed marker trajectories, the pose of the relevant bone, represented by the bone-embedded frame position and orientation vectors, is estimated versus time using a suitable mathematical procedure referred to as rigid-body pose estimator.

The most critical aspect of this exercise resides in the fact that reconstructed marker trajectories are not stationary with respect to the underlying bone. This is obviously in contradiction with the objective of reconstructing a bone-embedded frame and represents the experimental problem which has prevented movement analysis from giving the scientific and professional contributions that it could, in principle, provide.

These undesired movements belong to two classes:

1. One movement is apparent and depends on the error with which marker

Three-dimensional Analysis of Human Locomotion. Edited by P. Allard, A. Cappozzo, A. Lundberg and C. Vaughan
© 1997 John Wiley & Sons Ltd. ISBN 0 471 96949 4

coordinates are reconstructed in the laboratory frame (photogrammetric error);
2. Another, which is real, is due to the relative movement between the marker and the underlying bone, mostly associated with the interposition of both passive and active soft tissues (skin movement artefacts).

Because of these displacements, which, as will be shown, may be of a substantial magnitude, the cluster of markers associated with a body segment undergoes deformation, i.e. intermarker distances change, and/or move, with respect to the bone. Thus the estimated time histories of the pose of the cluster of markers will tend to differ from those which apply to the bone.

Often, rigid-body pose estimators are used that do not accommodate these errors and artefacts and, in each instant of time, assume the cluster to be rigid with the bone and not deformable. In these algorithms the bone-embedded frame axes are constructed using geometric rules and may be alluded to as non-optimal pose (NOP) estimators (Johnston, Brand and Crowninshield, 1979; Andriacchi et al., 1980; Cappozzo, 1984; Kadaba, Ramakrishnan and Wootten, 1990; Riley, Mann and Hodge, 1990; Davis et al., 1991; Vaughan, Davis and O'Connor, 1992). High-frequency photogrammetric errors are dealt with by smoothing either raw data or end results. Pose estimators, which embed the minimization of the effect of the above-mentioned experimental errors on the results, started to be described in the literature in the early 1980s and have gained popularity in the last few years (Spoor and Veldpaus, 1980; Arun, Huang and Blostein, 1987; Veldpaus, Woltring and Dortmans, 1988; Magnani et al., 1993; Söderkvist and Wedin, 1993; Wang, Rezgui and Verriest, 1993; Cappello et al., 1994; Challis, 1994, 1995; La Palombara, Cappello and Leardini, 1995; Cheze, Fregly and Dimnet, 1995; Cappello, La Palombara and Leardini, 1996). The suggested methods accommodate for the deformability of the marker cluster and exploit the following circumstances. Since the bone-embedded frame has six degrees of freedom and the marker cluster carries a number of degrees of freedom equal to three times the number of markers, i.e. equal or greater than nine, the pose estimator may exploit this redundancy in the data and allow for an optimal identification, in each instant of time, of the pose of the bone-embedded frame. In this case it is alluded to as the least squares pose (LSP) estimator.

This chapter aims at critically illustrating the above-mentioned rigid-body pose estimators, providing some evidence for the major unresolved problems. First, the properties of the marker clusters used in movement analysis will be illustrated. Then, the basic characteristics of the skin movement artefacts will be presented. A review of the above-mentioned optimization techniques will follow. Finally, the accuracy with which anatomical landmarks and bone position and orientation vectors can be estimated during movement will be analysed, making reference to actual data. This accuracy will also be looked upon as a function of the shape and size characteristics of the marker cluster.

CLUSTERS OF EXTERNAL MARKERS

The marker points need to be selected according to the following experimental requirements.

1. Sufficient measurements (three image coordinates) should be available on each marker from the available cameras at any given time; this means that, for a given experiment, the light emitted or reflected from markers should be visible to a sufficient number of cameras.
2. The number of markers associated with each bone (technical cluster) must be equal to three or more. Both the distance between the markers associated with each cluster and the offset of each marker from the line joining any other two should be sufficiently large so that error propagation from reconstructed marker coordinates to the bone orientation in space is minimal (see considerations about cluster size and shape in a later section).
3. The relative movement between markers and underlying bone should be minimal (see next section).
4. Mounting the markers on the experimental subject should be a quick and easy operation. It should be possible to place markers despite the presence of appliances such as orthoses, prostheses or external fracture fixators.

Markers are associated with external and not bony (internal) points. This means that invasive experimental approaches are not taken into consideration.

A cluster of markers will be referred to as a rigid cluster if its behaviour is that of a rigid body, i.e. if the distance between any couple of markers does not vary significantly during the motion of the body segment. If this condition is not complied with, the cluster will be called deformable. Examples of rigid clusters (Figure 1) are arrays of markers conveniently arranged on rigid plates, which can be fastened to the segment under analysis, or on metal frameworks directly fixed to the bone. The latter structures imply the use of invasive implant techniques and are recommended only when their presence is required for the patient's clinical treatment (e.g. external fixation devices in fracture-healing therapy). Common examples of deformable clusters are collections of markers attached either directly to the skin or onto an elastic band wrapped around the segment. As opposed to skin markers, the use of fixtures has the following advantages:

1. Marker mounting on patients is facilitated (especially when active markers are dealt with, because one cable per fixture may be used);
2. Sufficiently wide elastic bands help to reduce soft-tissue movements;
3. Marker light emission may be suitably orientated.

When using plates or other rigid structures not directly fixed to the bone, the whole cluster will move rigidly with respect to the bone during the segment

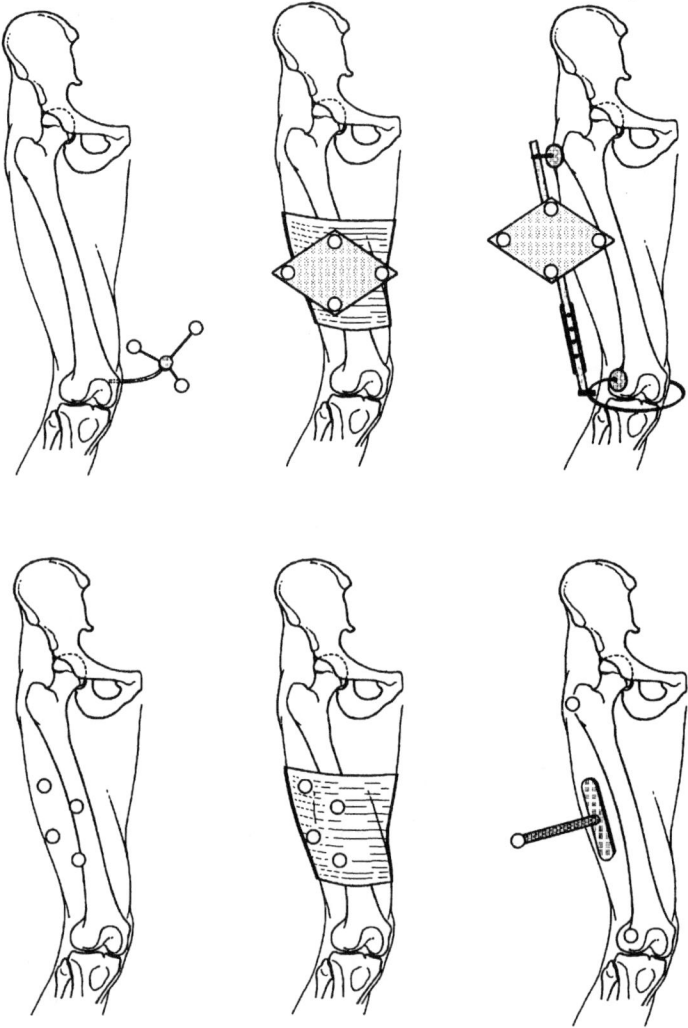

Figure 1. Examples of rigid (top) and deformable (bottom) technical clusters

motion because of the presence of interposed soft tissues. Deformable clusters will show, instead, both rigid movements and deformations. A major objective in bone motion reconstruction is to minimize the above-mentioned inaccuracies by improving: (1) placement and fixation techniques (before data acquisition); and (2) data conditioning and enhancement through smoothing and optimization routines (after data acquisition).

Since the estimation of velocity and acceleration is not relevant in this chapter and the objective is the assessment of the above-mentioned optimization routines, smoothing procedures are not used. The combined use of optimization and smoothing procedures is dealt with in La Palombara, Cappello and Leardini (1995) and Cappello, La Palombara and Leardini (1996).

SKIN MOVEMENT ARTEFACTS

This section deals with the experimental artefacts which affect the reconstruction of a bone-embedded system of axes during the execution of a motor task through the use of a stereophotogrammetric system. The term experimental artefact refers to an error which originates at the interface between the measurement instrument and the substrate which is the object of the measurement. This definition makes it different from the errors which originate within the instrument and makes it clear that it depends on the particular method and the conditions under which the measurement is conducted. The most important artefact, and source of error in absolute terms, in the present context, is associated, as already mentioned, with the skin movement artefacts. Information about the magnitude and the direction of these artefacts is presented below with special reference to the thigh and shank during various motor tasks. This information was obtained by Cappozzo et al. (1996) by exploiting experimental situations in which an observer, i.e. a set of axes rigidly coupled to the bone, could be made available. In particular, experiments were conducted using patients wearing external fracture fixation devices. Markers were mounted on the device and ensured that the observer was rigidly coupled to the bone. Roentgen photogrammetry was also used in order to be able to have information on subjects with more trophic musculature than the patients wearing fixation devices.

   The instantaneous positions of skin markers, located above anatomical land-marks or in other regions of the segments involved, could be reconstructed with respect to truly bone-embedded anatomical frames during different movements and will be referred to as position artefact. The dispersion of these positions, i.e. the standard deviation of the relevant coordinates, was taken as a measure of the skin movement artefact. Figure 2 shows the position artefact standard deviations of skin markers located above the GT (greater trochanter external prominence), LE (lateral epicondyle), HF (head of the fibula) and LM (lateral malleolus) as a function of joint rotation ranges (Cappozzo et al., 1996). Shaded areas summarize data from four subjects performing different movements. In Figure 3 the coordinates in the relevant anatomical frame of the same anatomical markers as well as other skin markers located on the shank are shown versus time during a walking cycle. Similar results with reference to other skin markers placed on the thigh are also depicted as recorded during a flexion–extension exercise of the entire limb.

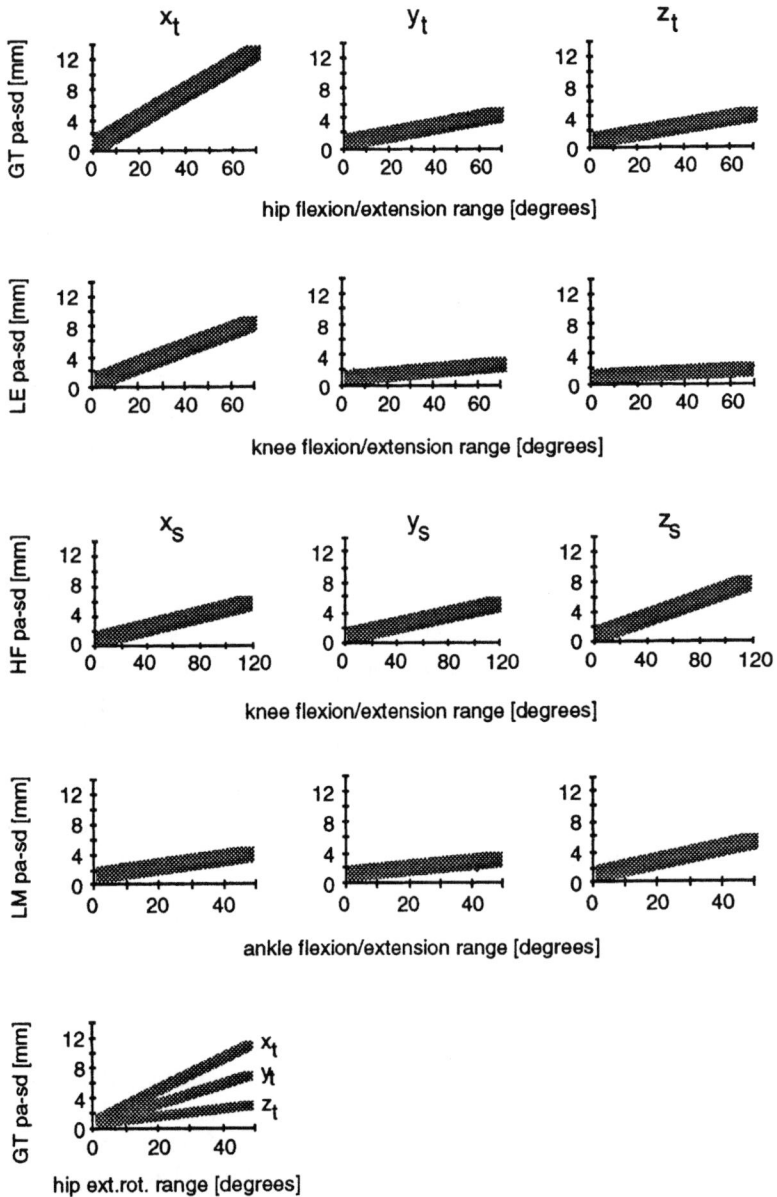

Figure 2. Position artefact standard deviation (pa-sd) of the indicated anatomical marker local coordinates as a function of joint rotation range. Data relative to four subjects performing different movements are summarized with the shaded areas. The pa-sd describes the dispersion of these coordinates with respect to their mean values during movement (walking, cycling, flexion–extension exercises, and hip internal–external rotation)

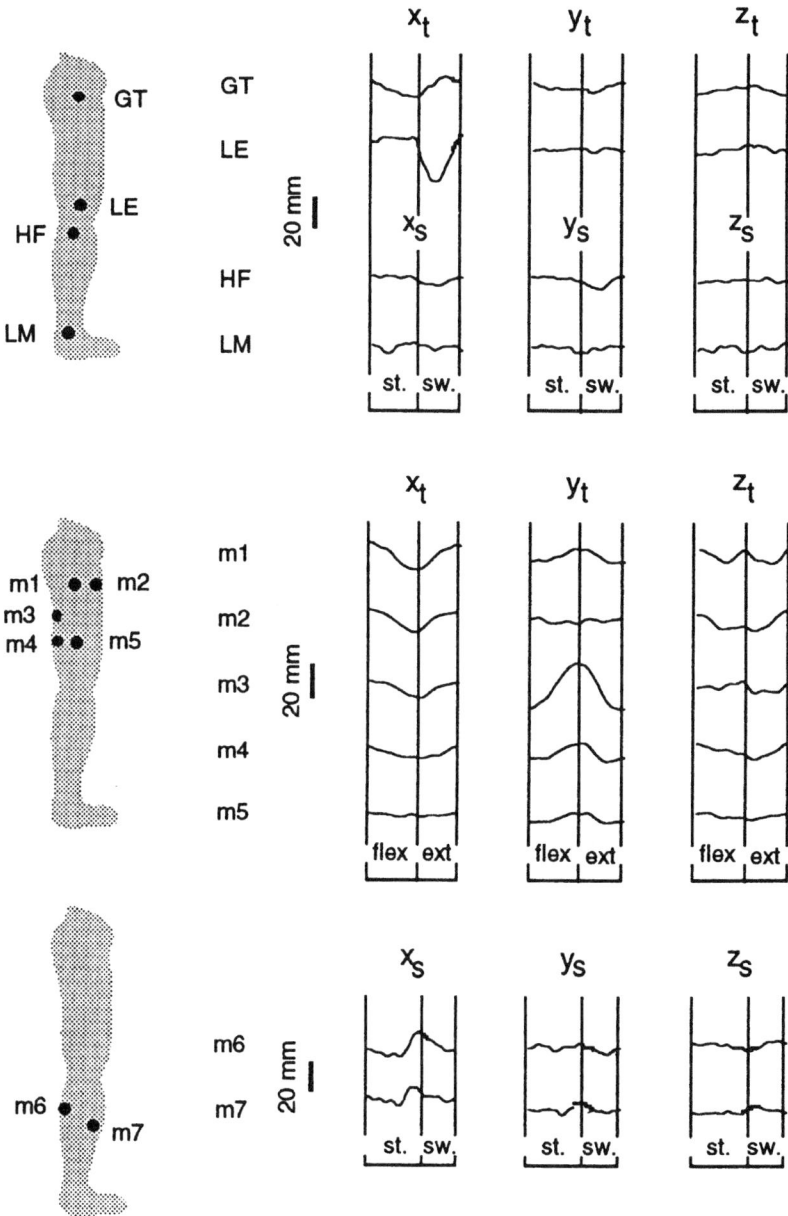

Figure 3. Position artefact coordinates in the relevant anatomical frames of the indicated skin markers. Top and bottom plots refer to a walking cycle. The plot in the middle refers to a hip and knee flexion–extension exercise

THIGH SKIN MARKER ARTEFACTS

These results indicate that during hip flexion–extension a marker located on the greater trochanter (GT) undergoes displacements with respect to the underlying bone, mostly in the anteroposterior direction. This artefact has a magnitude which varies approximately linearly with respect to the joint flexion angle and reaches a peak value of 35 mm when the hip joint moves in the sagittal plane by 80°. Artefact components in the mediolateral and longitudinal directions are usually within 15 mm. During level walking, the GT skin marker moves forwards and upwards during the first two-thirds of the swing phase (hip flexion), and backwards and downwards during the rest of the cycle (hip extension). External rotation of the hip causes a relative displacement in the posteroanterior direction which increases with the rotation angle and may reach 30 mm when this angular excursion equals 45°. A marker located on the lateral epicondyle (LE) of the femur is also very mobile. It moves backwards as the knee flexes. In the anteroposterior direction, the artefact may be of approximately 25 mm for a flexion of 70°. Artefacts in the longitudinal and mediolateral directions are relatively small, i.e. within 10 mm. During walking, this skin marker moves mostly along the anteroposterior axis, backwards during the first part of the swing phase (knee flexion) and forwards during the second half of the swing phase (knee extension). During stance phase, it undergoes relatively minor displacement. Experiments carried out using skin markers located on the thigh in positions other than the above anatomical landmarks show that skin portions in the proximal area of the thigh are subject to larger movements than distal portions. Proximal skin markers also introduce remarkably large artefacts while the hip rotates.

SHANK SKIN MARKER ARTEFACTS

During knee flexion–extension, a marker located on the head of the fibula (HF) undergoes displacements with about the same magnitude in all three directions. This artefact is smaller than that observed for the thigh markers. For a knee flexion angle within 70° it may reach 15 mm. However, it may have a magnitude around 25 mm in the anteroposterior and longitudinal directions for a flexion angle of 110°. During walking, and therefore for a knee flexion in the range 0–60°, the HF skin marker moves backwards and downwards during the first half of the swing phase (knee flexion) and forwards and upwards during the second half (knee extension). During stance, movements are small. A skin marker located on the lateral malleolus (LM) undergoes smaller displacements than the above-mentioned skin markers. The relevant artefact was found, having a maximal amplitude within 10 mm in all directions and for all movements analysed. No obvious relationship between this artefact and joint angle amplitudes was found. During walking, the LM skin marker moves roughly backwards as the ankle plantarflexes and forwards during dorsiflexion. The other two artefact components are too small to reveal a trend. During walking, the marker located on the lateral aspect of the gastrocnemius shows a remarkable artefact, whereas the marker placed on the tibialis anterior is affected by smaller artefact. This is clearly correlated with this muscle's contraction.

SUMMARY

The results described above confirm that skin marker artefacts are usually overwhelming with respect to both systematic and random photogrammetric errors. During movement, markers located directly on the skin above anatomical landmarks such as the greater trochanter, lateral epicondyle of the femur, head of the fibula and lateral malleolus undergo displacements relative to the underlying bone which are roughly proportional to the angular displacement of the closest joint. During walking, these are in the range 10–30 mm. Skin markers located on the lateral portion of the thigh and of the tibia and far from joint areas may exhibit significantly smaller movements.

These results clearly indicate that if the thigh marker cluster includes markers located on the proximal portion of the segment, rotations of hip and knee will not be correctly represented.

The estimation of knee joint kinematics during walking, using NOP estimators and clusters made of three skin markers, have been shown by Cappozzo et al. (1996) to be affected by inaccuracies which, for flexion–extension, adduction–abduction and internal–external rotation, may roughly amount to 10%, 50% and 100% of the respective movement range angle. The use of optimization techniques can be effective in order to reduce the propagation of the above-mentioned experimental errors to these results.

POSE ESTIMATION FROM CLUSTERS OF EXTERNAL MARKERS

A number of techniques have been proposed to determine the orientation matrix, **R**, and the position vector, **t**, of the local-to-laboratory frame transformation (see Chapter 7), each having different computational requirements and making different assumptions about the data (Challis, 1994). These techniques can be classified into non-optimal pose estimators and least square pose estimators.

NON-OPTIMAL POSE ESTIMATOR FROM THREE-MARKER CLUSTERS

A local reference frame (Figure 4) can be readily and non-ambiguously defined starting from the coordinates of three non-collinear landmarks (either anatomical

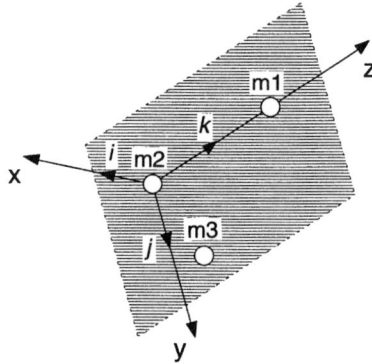

Figure 4. Definition of a reference frame based on three markers

or technical) (Chao and Morrey, 1978). If $\mathbf{m}_1$, $\mathbf{m}_2$ and $\mathbf{m}_3$ are the three-dimensional vectors containing the laboratory coordinates of three markers, we can take $\mathbf{m}_2$ as the origin of the local frame, set the z axis along the line oriented as $\mathbf{m}_1 - \mathbf{m}_2$, the x axis perpendicular to the plane defined by the three markers and directed as the vector $\mathbf{v} = (\mathbf{m}_3 - \mathbf{m}_2) \times (\mathbf{m}_1 - \mathbf{m}_2)$ and the y axis perpendicular to both x and z, i.e. directed as $(\mathbf{m}_1 - \mathbf{m}_2) \times \mathbf{v}$. Position and orientation vectors ($\mathbf{t}$ and $\mathbf{R}$) of the local frame with respect to the laboratory frame can thereafter be calculated.

This procedure assumes that the position of the three points is error-free. When measurement errors are not negligible, as in the present case, the errors on marker positions propagate to the body-embedded frame pose and consequently to the $\mathbf{t}$ and $\mathbf{R}$ estimates. Furthermore, the redundancy of information due to the availability of nine numbers associated with the marker coordinates versus the six degrees of freedom of the rigid body is not exploited. This means that other methods may be used to optimally solve the problem even if only three markers are available.

LEAST SQUARES POSE ESTIMATORS FROM CLUSTERS OF MARKERS

A number of estimation techniques have been proposed for the determination of rigid-body transformation parameters in the presence of noise. Laub and Shiflett (1982) proposed a method based on linear algebra and matrix perturbation theory to account for the lack of precision in the data. This technique is limited to three non-collinear landmarks and gives biased estimates (Challis, 1994). These are strong limitations which suggest the use of least squares methods also in the case of there being three points available.

Spoor and Veldpaus (1980) propose a least squares method requiring the calculation of the eigenvectors of a three-by-three data matrix. Veldpaus,

Woltring and Dortmans (1988) suggest slight improvements to this algorithm by which computation of the eigenvectors may be avoided. This method has been since refined by the use of the better-conditioned singular value decomposition (SVD) algorithm. This improves the stability of the method in the presence of noise when the cluster is composed of almost aligned markers (Söderkvist and Wedin, 1993).

To overcome the difficulties related to error anisotropy and to provide evaluation of the accuracy of the estimated degrees of freedom (DOF), Magnani et al. (1993) propose an iterative weighted least squares (WLS) optimization algorithm based on the linearization of the model-predicted laboratory coordinates with respect to the unknown orientation and position components.

The Mathematical Problem

The problem is to estimate the orientation matrix $\mathbf{R}$ (or a related orientation vector) and the position vector $\mathbf{t}$ of a rigid body (the bone), on a frame-by-frame basis, starting from the reconstructed coordinates of three or more non-collinear markers coupled to the body segment. These markers are organized in a rigid or deformable cluster, the motion of which can be decomposed in a rigid movement with respect to the bone and a cluster deformation. The position of the markers in the laboratory frame can be regarded as affected by additive noise due mainly to: (1) stretching of the skin; (2) muscle contraction; (3) inertial phenomena; and (4) stereophotogrammetric noise.

Assume that we have a cluster of $n \geq 3$ landmarks, at least three of which are not collinear, coupled to a rigid body in a more or less rigid connection, and let $\mathbf{p}_k$ and $\mathbf{l}_k$, $k = 1, 2, \ldots, n$ be the three-dimensional positions of these landmarks in the local and laboratory frame, respectively.

The cluster $\mathbf{p}_k$, $k = 1, 2, \ldots, n$ assumes here the meaning of a rigid local model to be used for body pose assessment. This rigid model can be the cluster configuration in a given instant of time during the experiment, or an average configuration of the cluster during the analysed motion (Cappello et al. 1994; Cheze, Fregly and Dimnet, 1995). The problem of identifying the optimal rigid model is not dealt with here. This is justified by the fact that results of an empirical investigation, conducted by the authors, have shown that this is not a critical issue when bone position and orientation are to be estimated.

Let us define the following quantities which depend only on technical frame definition and measured data:

$$\bar{\mathbf{p}} = \frac{1}{n}\sum_{k=1}^{n}\mathbf{p}_k, \quad \bar{\mathbf{l}} = \frac{1}{n}\sum_{k=1}^{n}\mathbf{l}_k \qquad 3 \times 1 \qquad \text{Centroids of the local and laboratory clusters}$$

$$\mathbf{P} = [\mathbf{p}_1 - \bar{\mathbf{p}}, \mathbf{p}_2 - \bar{\mathbf{p}}, \ldots, \mathbf{p}_n - \bar{\mathbf{p}}] \quad 3 \times n \qquad \text{Local cluster with centroid at the origin}$$

$\mathbf{L} = [\mathbf{l}_1 - \bar{\mathbf{l}}, \mathbf{l}_2 - \bar{\mathbf{l}}, \ldots, \mathbf{l}_n - -\bar{\mathbf{l}}]$ $3 \times n$ Laboratory cluster with centroid at the origin

$\mathbf{K} = \mathbf{L}\mathbf{L}^T/n$ 3×3 Cluster dispersion matrix

$\mathbf{G} = \mathbf{L}\mathbf{P}^T$ 3×3 Cross-dispersion matrix (Woltring et al., 1994)

We want to determine a rotation matrix $\mathbf{R}$ and a translation vector $\mathbf{t}$ that map optimally the points $\mathbf{p}_k$ to the points $\mathbf{l}_k$, $k = 1, 2, \ldots, n$. The operator which optimally maps the rigid-cluster model to the relevant noisy configuration will be indicated by the term rigid model fitting (RMF) operator. Because of the measurement errors, the mapping is not exact:

$$\mathbf{l}_k = \mathbf{R}\mathbf{p}_k + \mathbf{t} + \mathbf{e}_k, \quad k = 1, 2, \ldots, n \tag{1}$$

where $\mathbf{e}$ represents an additive noise. The problem is now reduced to the application of some estimation scheme to compute the optimal orientation matrix $\hat{\mathbf{R}}$ and position vector $\hat{\mathbf{t}}$.

Due to the lack of information about the statistical properties of noise, particularly when systematically distributed errors are present, an unweighted least squares estimator is generally adopted. In this case, the following minimization problem must be solved:

$$\min_{\mathbf{R},\mathbf{t}} \Sigma^2(\mathbf{R}, \mathbf{t}) = \sum_{k=1}^n \|\mathbf{e}_k\|^2 = \sum_{k=1}^n \|\mathbf{R}\mathbf{p}_k + \mathbf{t} - \mathbf{l}_k\|^2$$

$$\text{with } \mathbf{R}\mathbf{R}^T = \mathbf{R}^T\mathbf{R} = \mathbf{I}, \det(\mathbf{R}) = 1 \tag{2}$$

Figure 5 shows a simple interpretation of this problem with reference to the pose estimation of a deformable cluster. Minimizing the sum of the squared distances between ordered pairs of markers corresponds also to finding the minimum elastic energy stored in the springs, assumed to have a common stiffness value. This figure suggests a possible way to weight the reconstruction error of each marker or along each direction differently. In fact, it is sufficient to use springs with higher stiffness for the most reliable markers and/or directions. From a mathematical point of view, this means the adoption of a WLS scheme where a weighting matrix $\mathbf{W}$ must be specified to account for the *a priori* knowledge about noise anisotropy and/or measurement reliability (Magnani et al., 1993; Cappello, La Palombara and Leardini, 1996). The WLS estimator is unbiased only if the observation noise belongs to a zero-mean uncorrelated gaussian distribution in each direction. Furthermore, the absence of correlation between noise affecting the different markers and the different directions is assumed.

A generalization of this approach, taking into account both anisotropic noise and correlated marker displacements, is possible by using a Markov estimator,

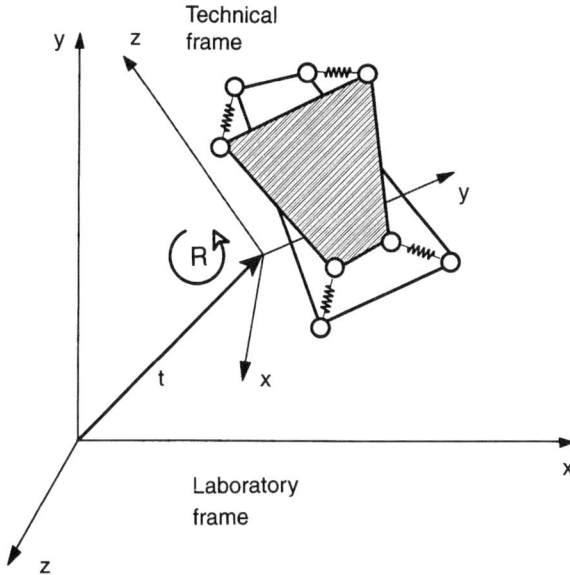

Figure 5. Physical interpretation of the pose estimation of a deformable cluster though a least squares method. The rigid-cluster model is indicated by the filled area

but this requires that the noise correlation matrix be completely known. Details about SVD and WLS algorithms are given below.

Singular Value Decomposition Algorithm

The SVD method was first used in the present context by Hanson and Norris (1981). The method uses the SVD (Golub and van Loan, 1989) of the 3×3 cross-dispersion matrix **G**, derived from the local and laboratory marker positions. The main advantages with this method, compared to those previously mentioned, consist of: (1) an improved stability, essential for ill-conditioned problems; and (2) the ease of implementation and analysis. Since the method has been extensively published elsewhere (Hanson and Norris, 1981; Arun, Huang and Blostein, 1987), we will only briefly present it here.

The algorithm develops in three stages:

1. SVD of the cross-dispersion matrix, $\mathbf{G} = \mathbf{U}\mathbf{D}\mathbf{V}^T$.
   U and **V** are 3×3 orthogonal matrices $(\det(\mathbf{U}) = \pm 1, \det(\mathbf{V}) = \pm 1)$ and $\mathbf{D} = \mathrm{diag}(\sigma_1, \sigma_2, \sigma_3)$ where $\sigma_1 \geqslant \sigma_2 \geqslant \sigma_3 \geqslant 0$ are the singular values of **G**.
2. Orientation matrix estimation.
   $\hat{\mathbf{R}} = \mathbf{U}\,\mathrm{diag}(1, 1, \det(\mathbf{U}\mathbf{V}^T))\mathbf{V}^T$ where the term $\det(\mathbf{U}\mathbf{V}^T)$ has been added to avoid the estimation of a reflection matrix $(\det(\hat{\mathbf{R}}) = -1)$. The orientation vector estimate $\hat{\boldsymbol{\theta}}$ can thus be obtained as described in the previous chapter.

3. Position vector estimation, $\hat{\mathbf{t}} = \bar{\mathbf{l}} - \hat{\mathbf{R}}\bar{\mathbf{p}}$.

   In order to have uncorrelated position and orientation vector estimates, it is necessary to assume $\bar{\mathbf{p}} = \mathbf{0}$.

The SVD algorithm is available in most matrix computation packages and does not require an initial estimate for the unknown transformation parameters, and is thus particularly efficient from a computational point of view. The main difficulty is that it does not provide a measure of DOF accuracy and cannot explicitly take into account *a priori* information on noise statistics such as, for example, anisotropy.

Weighted Least Squares Algorithm

To overcome these problems, an iterative minimization procedure based on the classical linearization technique has been proposed (Magnani et al., 1993; Cappello, La Palombara and Leardini, 1996) where a local rigid model of the cluster is used to assess the body segment position and orientation during motion. The model is known *a priori* when a rigid cluster is manufactured by accurate processes, or can be estimated by some calibration procedure. Body position and orientation are estimated by calculating the translation and rotation which match, in the WLS sense, the cluster model to the actual cluster as it has been observed by the measurement system. The above-mentioned position and orientation can be denoted by a pair of three-dimensional (3D) vectors that we will call $\mathbf{t}$ and $\boldsymbol{\theta}$, respectively (Woltring, 1994).

For this purpose a WLS procedure was used to estimate the 6×1 DOF column vector, $\hat{\mathbf{D}}_f = [\hat{\boldsymbol{\theta}}; \hat{\mathbf{t}}]$, by minimizing the weighted sum of the squared distances between the measured marker coordinates, $\mathbf{l}_k$, $k = 1, 2, \ldots, n$, and the calculated ones, $\mathbf{q}_k = \mathbf{R}(\hat{\boldsymbol{\theta}})\mathbf{p}_k + \hat{\mathbf{t}}$, where $\mathbf{R}(\hat{\boldsymbol{\theta}})$ is the orientation matrix corresponding to the estimated orientation vector, $\hat{\boldsymbol{\theta}}$, and $\hat{\mathbf{t}}$ is the estimated position vector. Each of the error components along the three axes can be weighted differently and, in principle, cross-correlation between noise components can also be taken into account. If the noise introduced by the measurement system is additive, we can assume

$$\mathbf{l}_k = \mathbf{R}(\boldsymbol{\theta})\mathbf{p}_k + \mathbf{t} + \mathbf{e}_k, \quad k = 1, 2, \ldots, n \tag{3}$$

where the error mean is $E\{\mathbf{e}_k\} = \mathbf{0}$, and the covariance matrix, $E\{\mathbf{e}_k\mathbf{e}_k^T\} = \mathbf{C}$, is supposed to be known and invariant for the different markers. In the absence of cross-correlation between noise components, $\mathbf{C}$ is a diagonal matrix whose elements represent the variances along the three directions. If we define the $(3n \times 1)$ column vectors $\mathbf{L} = [\mathbf{l}_1; \ldots; \mathbf{l}_n]$, $\mathbf{Q} = [\mathbf{q}_1; \ldots; \mathbf{q}_n]$ and $\mathbf{E} = [\mathbf{e}_1; \ldots; \mathbf{e}_n]$, we can write $\mathbf{L} = \mathbf{Q}(\mathbf{D}_f) + \mathbf{E}$ and the cost function to be minimized can be expressed as follows:

$$F(\mathbf{D}_f) = [\mathbf{L} - \mathbf{Q}(\mathbf{D}_f)]^T \mathbf{W}[\mathbf{L} - \mathbf{Q}(\mathbf{D}_f)] \tag{4}$$

where $\mathbf{W}$ is a block diagonal $(3n \times 3n)$ positive definite weighting matrix which allows us to cope with anisotropic noise. $\mathbf{W}$ is made up by n identical (3×3) matrices, $\mathbf{W}_s$, lined along its principal diagonal and by zero elements elsewhere. If noise is normally distributed and we set $\mathbf{W}_s = \mathbf{C}^{-1}$, the WLS estimator, $\hat{\mathbf{D}}_f$, coincides with the Gauss–Markov estimator and the procedure provides unbiased DOF estimates.

Owing to the nonlinear relationship between the model output $\mathbf{Q}$ and the parameter vector $\mathbf{D}_f$, the WLS minimization algorithm must be iterative. Based on model output linearization, it starts from a reasonable initial guess of $\mathbf{D}_f$. A suitable guess may be, for instance, the optimal value estimated at the previous frame. Details about the iterative algorithm and the DOF covariance matrix estimation are given in Cappello, La Palombara and Leardini (1996).

The WLS procedure is less efficient than other least squares procedures, such as the SVD algorithm, as far as computing time requirements are concerned. It offers, nevertheless, some positive features, such as the capability to adjust to noise conditions and, above all, the possibility of providing useful measures of the accuracy of the estimated DOFs.

CLUSTER DESIGN AND ERROR PROPAGATION TO DOFS

The effective use of the mathematical algorithms illustrated above strongly depends on the size and shape of the marker cluster.

In order to give a quantitative description of the size and shape of a cluster used for body segment movement reconstruction, three indices can be used:

- *Size index*, given by the root mean square distance of the markers from the cluster centroid and calculated as $\sqrt{(\text{trace}\,(\mathbf{K}))}$, where $\mathbf{K}$ is the cluster dispersion matrix as defined above. This index should be maximized to reduce the effect of the noise superimposed on the marker coordinates.
- *3D distribution index*, represented by the sum of the squared distances of the markers from the best approximating plane in a least squares sense (Söderkvist and Wedin, 1993). This index is maximum for isotropically distributed clusters and zero for two-dimensional (2D) clusters. In the last case, a 2D distribution index can be used.
- *2D distribution index*, represented by the sum of the squared distances from the best approximating straight line. This index should be maximized to avoid ill-determined rotation around this axis.

The number of markers n of the cluster also plays an important role, since redundancy may provide more accurate results. Obviously, the use of markers with highly systematic movement with respect to the bone, as occurs in proximity to the bone prominences, should be avoided.

The noise affecting the marker coordinates causes both position and orientation errors with respect to the original pose. The accuracy of the DOF estimates, for small errors with respect to a given position and orientation, can be obtained by the following error propagation model:

$$\hat{\mathbf{t}} = \Delta\mathbf{t}, \ \hat{\mathbf{R}} = \mathbf{I} + \Delta\mathbf{R} \tag{5}$$

where $\Delta\mathbf{R} = \mathbf{S}(\Delta\boldsymbol{\theta})$ and $\mathbf{S}$ indicates the skew-symmetric matrix (Woltring et al., 1985). For simplicity, but without loss of generality, local and laboratory reference frames are assumed here to coincide. Consequently, the covariance matrices of the DOF errors are

$$\text{Cov}\,(\Delta\mathbf{t}) = \frac{\hat{\sigma}^2}{n}\mathbf{I}, \ \text{Cov}\,(\Delta\boldsymbol{\theta}) = \frac{\hat{\sigma}^2}{n}\mathbf{A}^{-1} \tag{6}$$

where $\mathbf{A} = \text{trace}\,(\mathbf{K})\mathbf{I} - \mathbf{K}$, and $\hat{\sigma}^2$ is the estimated noise variance along each direction.

Figure 6 shows the root mean square value of the norm of the orientation error as a function of the size index/noise standard deviation ratio for both a 2D and 3D cluster configuration. These results suggest that the size index should be at least 10 times the standard deviation of the marker position error. Furthermore, there is no particular advantage in adopting 3D configurations as opposed to 2D configurations: planar or quasi-planar clusters composed of four or five well-distributed markers (with high 2D distribution index) represent a good compromise between practical realizability and expected estimation accuracy.

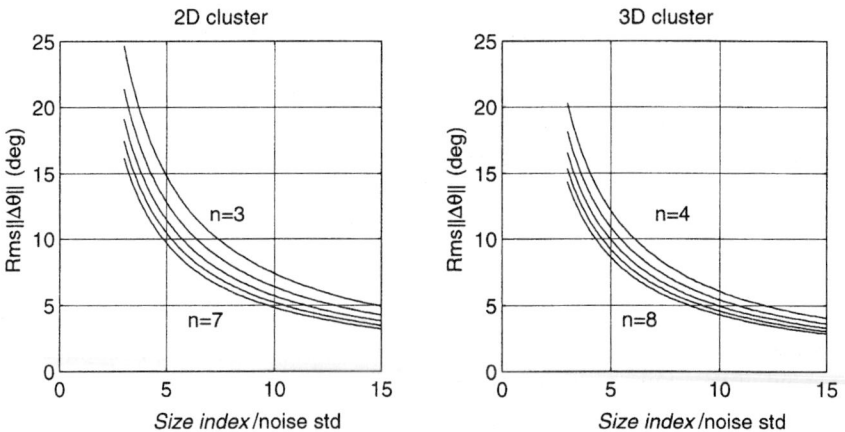

Figure 6. Root mean square value of the norm of the orientation error as a function of the size index/noise standard deviation ratio for 2D and 3D isotropic clusters, respectively. n = number of markers

A CASE STUDY: ESTIMATION OF FEMUR POSE DURING CYCLING

METHODS

The reconstruction of bone pose and anatomical landmark laboratory trajectory during the execution of a motion task is illustrated below with reference to actual experimental data.

The experiments were carried out in the Laboratory of Motion Analysis, Istituti Ortopedici Rizzoli, Bologna, Italy. The laboratory was based on a passive-marker optoelectronic stereophotogrammetric system run at a rate of 100 frames/s. The test subject was a young male wearing an external fixation device for the treatment of a fracture in the left femur (Figure 7). He performed a cycling exercise at a cadence of about 0.6 cycles/s. Eight markers (numbered from 8 to 15 in Figure 7) were stuck on the skin of the subject's left thigh. Subsets of three or more of these skin markers provided skin clusters. These were, of course, affected by both instrumental errors and skin movement artefacts. Four additional markers were located on the external fixator. This latter set of markers, numbered from 4 to 7 in Figure 7, could be considered to be affected by instrumental errors only and provided a cluster which could be considered rigid and truly embedded with the bone (reference cluster). The trajectories of all of these markers were recorded during movement. A further data set concerned the calibration parameters, i.e. the local coordinates of the GT, LE and ME anatomical landmarks, and was acquired using the methodology illustrated in Chapter 7.

Using these data, the poses of the reference and skin clusters, i.e. of the reference technical frame and the skin technical frames respectively, were estimated using rigid-body pose estimators. As illustrated in previous sections, these estimators could include or not include optimization criteria. From any technical frame pose, together with the relevant local coordinates of the anatomical landmarks, the laboratory trajectories of these anatomical landmarks, as well as the associated anatomical frame pose, could be estimated. Depending on the technical frame used, these latter anatomical frames will be referred to as reference anatomical frame or skin anatomical frame.

Some improvement in the reconstruction of both anatomical landmark laboratory trajectories and anatomical frame pose could be obtained by performing multiple calibrations, i.e. applying the calibration procedure as described by Cappozzo et al. (1995) while the body segment assumed different poses in the range of the movement under analysis. In this case, the calibration parameters were not constant and were considered to be a function of the joint angle. In this way they embedded the skin movement artefact.

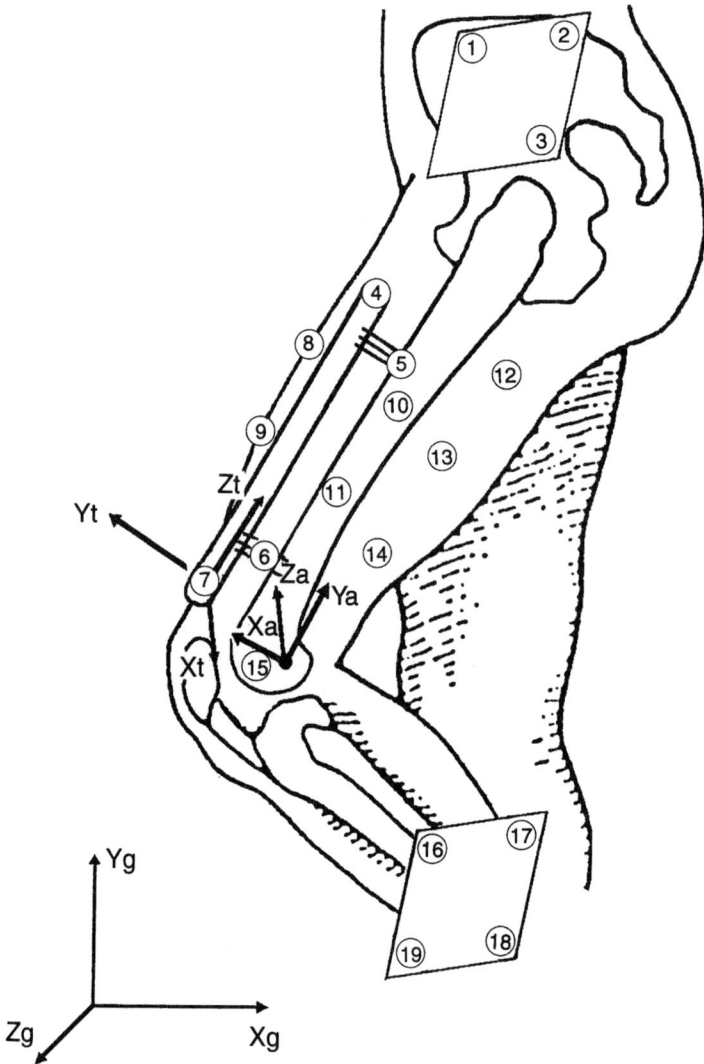

Figure 7. Skin and fixator marker locations. The reference anatomical and technical frames are also shown, together with the laboratory set of axes

RESULTS

Cluster Selection

Figure 8 shows how markers move in the reference anatomical frame during movement. The pseudo-motion of markers 4 to 7, attached to the fixation device, is due to stereophotogrammetric measurement inaccuracies. The displacement of

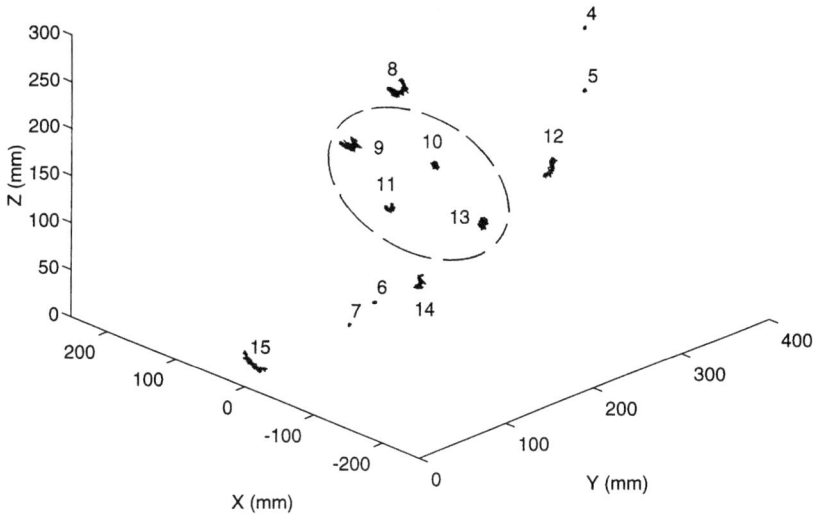

Figure 8. Marker trajectories during cycling in the reference anatomical frame. The selected cluster markers 9–10–11–13 are also indicated (see Figure 7)

skin markers is mostly associated with the skin movement artefacts; it is one order of magnitude higher than that of the fixator markers and shows a systematic behaviour, i.e. the components along the three axes are strongly cross-correlated. The root mean square displacements are 6.7, 5.5, 4.1, 3.9, 8.2, 5.6, 6.2 and 9.4 mm for skin markers 8–15, respectively.

Among the many different skin clusters that could be defined using the eight available skin markers, in agreement with the criteria for cluster design and taking into consideration the above-mentioned experimental results, the cluster 9–10–11–13 was chosen for the present analysis.

Figure 9 shows the trajectories (*XY* projection) of the selected cluster markers in the skin technical frame reconstructed using the SVD algorithm. The GT anatomical landmark trajectory in the same frame is also shown. This erroneous GT movement is the superposition of a random component and of a cyclic deterministic trajectory due to skin movement artefacts. This latter trajectory may be estimated and relevant knowledge used to minimize the skin artefact effects on the end results. This was done through a double anatomical landmark calibration, in correspondence to the two extreme positions of the relevant joint and by assuming linear behaviour of the artefact between these two values. The linear interpolation is justified by the considerations given on pp. 151–155 and in Figure 2 in particular. In the present cycling motion, the two frames used corresponded approximately to maximum flexion (*F*) and maximum extension (*E*) of the knee joint.

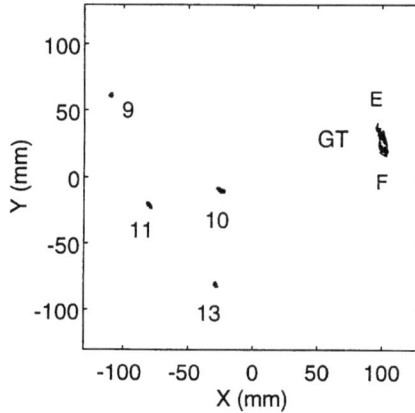

Figure 9. Trajectories of skin markers 9–10–11–13 in the cluster technical frame (XY projection). The position of the anatomical landmark GT has been reconstructed from the reference technical frame. F and E represent the frame corresponding to maximum flexion (pedal up) and extension (pedal down), respectively

Anatomical Landmark Reconstruction

Figure 10a presents what we assume to be the 'true' components of the GT trajectory in the laboratory frame during one cycle of the analysed movement. They have been reconstructed applying the optimized estimation procedure to the reference cluster 4–5–6–7. The trajectory of the GT in the laboratory frame was also estimated using the skin cluster selected previously. Both estimations were carried out using one set of calibration parameters (F = maximum knee flexion, pedal up, or E = maximum knee extension, pedal down). Figure 10b displays the difference between the two trajectories, i.e. the error which affects the estimation through the skin cluster.

The GT laboratory coordinates were also estimated using the double calibration procedure alluded to above ($E + F$). The difference between this trajectory and the reference trajectory is shown in Figure 10b. The most significant difference occurs along the X and Y laboratory coordinates, where the reconstruction errors are very small when the body pose is maximally similar to the one in the calibration frame considered. The Z component, on the contrary, appears almost independent of the choice of the calibration frame.

With the double calibration, visible improvements are obtained. Table 1 shows the reconstruction errors (root mean square error) obtained with one calibration frame (F or E) or two calibrations. The use of a double calibration frame has a strong positive effect, particularly in the sagittal plane.

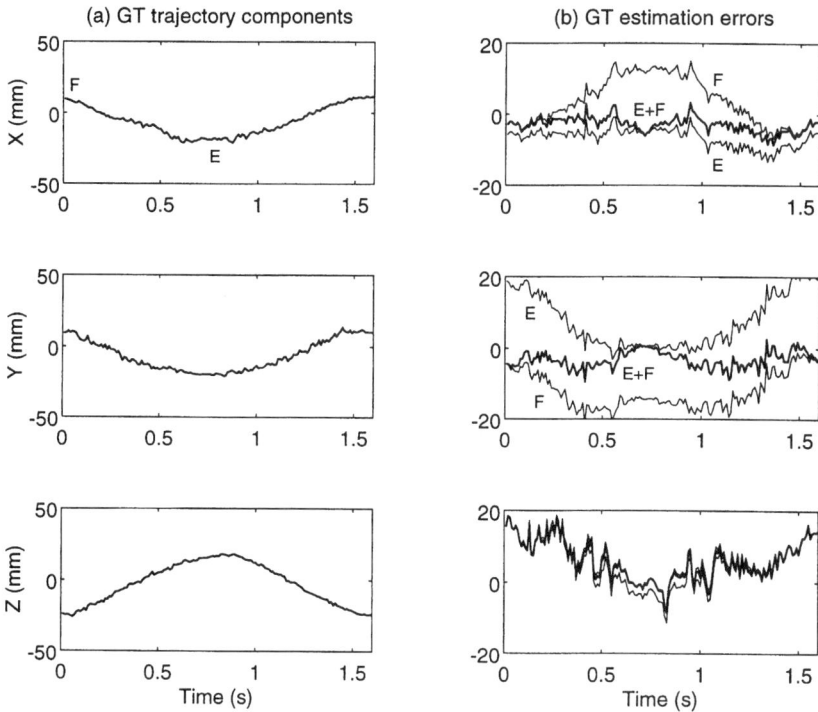

Figure 10. (a) Trajectory of GT in the laboratory frame during one cycle of the analysed cycling movement as obtained using external fixator data, i.e. reference data. (b) Reconstruction errors of the GT position by single (F = flexion, E = extension) or double ($F + E$) calibration

Table 1. GT position reconstruction errors with single (F = flexion, E = extension) and double ($F + E$) calibration

	F	E	$F + E$
RMS X (mm)	7.5	7.0	3.5
RMS Y (mm)	13.3	10.7	3.9
RMS Z (mm)	8.0	8.8	8.3

Bone Pose Estimation

A similar comparison was carried out on the orientation and position components, relative to the initial pose of the skin anatomical frame as calculated using single and double calibration data (Figure 11). These results were obtained by repeating the above-described steps for each anatomical landmark and then estimating the anatomical frame position and orientation.

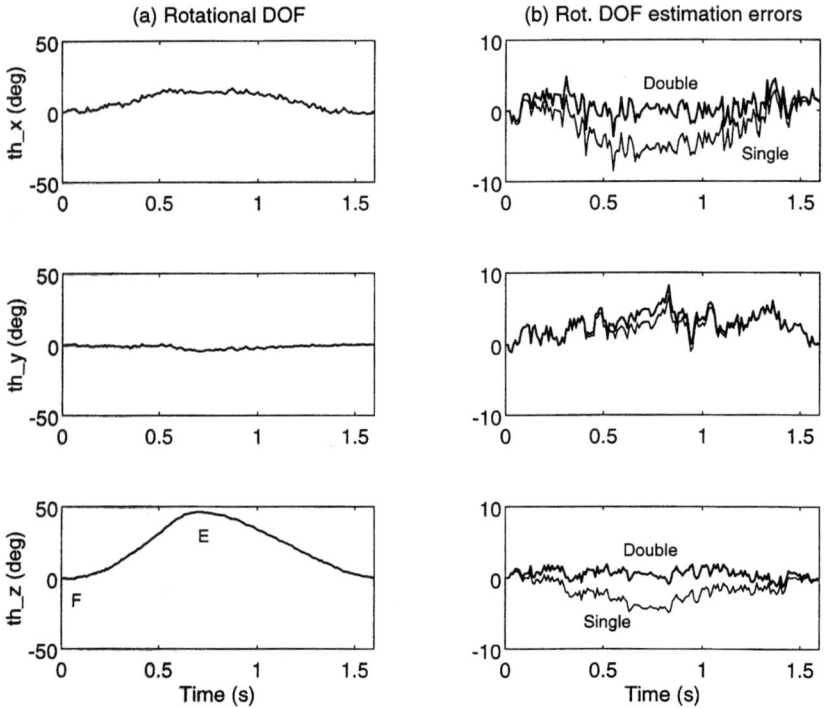

Figure 11. (a) Reference rotational DOFs of the reference anatomical frame in the laboratory frame. (b) Bone orientation errors between reference values and those calculated from the cluster technical frame using single (F = flexion, E = extension) or double ($F + E$) calibration

Table 2 shows the position and orientation reconstruction errors (root mean square error) obtained with one calibration frame (F or E) or two calibrations ($F + E$). Results are scarcely sensitive to the calibration frame adopted, while the double calibration procedure improves the estimates of both position and

Table 2. Bone position and orientation reconstruction errors with single (F = flexion, E = extension) and double ($F + E$) calibration

	F	E	$F + E$
RMS θ_X (deg)	3.7	3.5	1.7
RMS θ_Y (deg)	2.8	2.8	3.3
RMS θ_Z (deg)	2.3	2.3	0.8
RMS t_X (mm)	2.4	2.4	2.4
RMS t_Y (mm)	2.8	2.9	2.9
RMS t_Z (mm)	5.9	5.8	2.3

orientation components. The advantages are clearly evident in the orientation components θ_X and θ_Z and the position component t_Z.

In order to underline the usefulness of cluster design to the minimization of propagated errors, the cluster size and noise variance have been evaluated, giving size index $= 67$ and $\hat{\sigma}^2 = 25$ (size index/noise standard deviation $= 13.4$). The RMS orientation errors in Table 2 are of the same order of magnitude as those predicted in Figure 6 for an isotropic planar cluster of four markers with a size index/noise standard deviation $= 14$.

SUMMARY

The analysis of this case study has demonstrated that:

- optimization techniques should be used systematically to estimate bone position and orientation from technical clusters of markers;
- because of the lack of information about measurement error statistics, the SVD algorithm represents the best choice for computational efficiency;
- even when four well-distributed markers are selected to form the technical cluster, errors in bone orientation and position have a root mean square value in the order of $3°$ and 3 mm, respectively, and this requires attention and further analysis to avoid excessive influence on joint mechanics description.

ACKNOWLEDGEMENTS

The authors wish to express their gratitude to Dr Pier Francesco La Palombara and Dr Francesco Pensalfini for their invaluable contributions to this study. This word was supported in part by the CEC programme AIM—project A-2002 CAMARC II (Computer Aided Movement Analysis in a Rehabilitation Context II).

REFERENCES

Andriacchi, T. P., Andersson, J. B. J., Fermier, R. W., Stern, D. and Galante, J. O. (1980) A study of lower limb mechanics during stair climbing. *J. Bone Joint Surg.*, **62A** 749–757.

Arun, K., Huang, T. and Blostein, S. (1987) Least-squares fitting of two 3-D point sets. *IEEE Trans. Patt. Anal. Machine. Intell. Pami*, **9**(5), 698–700.

Cappello, A., La Palombara, P. F. and Leardini, A. (1996) Optimization and smoothing techniques in movement analysis. *Int. J. Biomed. Comput.*, **41**(3), 137–151.

Cappello, A., Leardini, A., Catani, F. and La Palombara, P. F. (1994) Selection and validation of skin array technical references based on optimal rigid model estimation. In: *Proceedings of the III International Symposium on 3-D Analysis of Hum Movenent*, Stockholm, 5–8 July, pp. 15–18.

Cappozzo, A. (1984) Gait analysis methodology. *Hum. Movem. Sci.*, **3**, 25–54.

Cappozzo, A., Catani, F., Leardini, A., Benedetti, M. G. and Della Croce, U. (1996)

Position and orientation in space of bones during movement: experimental artefacts. *Clin. Biomech.*, **11**(2), 90–100.

Cappozzo, A., Catani, F., Della Croce, U. and Leardini, A. (1995) Position and orientation of bones during movement: anatomical frame definition and determination. *Clin. Biomech.*, **10**(4), 171–178.

Challis, J. H. (1994) An examination of procedures for determining body segment attitude from noisy biomechanical data. *Med. Eng. Phys.*, **17**, 83–90.

Challis, J. H. (1995) A procedure for determining rigid body transformation parameters. *J. Biomech.*, **6**, 733–737.

Chao, E. Y. and Morrey, B. F. (1978) Three-dimensional rotation of the elbow. *J. Biomech.*, **11**, 57–73.

Cheze, L., Fregly, B. J. and Dimnet, J. (1995) A solidification procedure to facilitate kinematic analyses based on video system data. *J. Biomech.*, **28**, 879–884.

Davis, R. B., Ounpuu, S., Tyburski, D. J. and Deluca, P. A. (1991) A comparison of two-dimensional and three-dimensional techniques for the determination of joint rotation angles. In: *Proceedings of the International Symposium on 3-D Analysis of Human Movement*, Montréal, Canada, 28–31 July, pp. 67–70.

Golub, G. H. and van Loan, C. F. (1989) *Matrix Computation*, 2nd edn. Johns Hopkins University Press, Baltimore.

Hanson, R. and Norris, M. (1981) Analysis of measurements based on the singular value decomposition. *SIAM J. Sci. Stat. Comput.*, **2**, 363–373.

Johnston, R. C., Brand, R. A. and Crowninshield, R. D. (1979) Reconstruction of the hip. *J. Bone Joint Surg.*, **61A**, 639–652.

Kadaba, M. P., Ramakrishnan, H. K. and Wootten, M. E. (1990) Measurement of lower extremity kinematics during level walking. *J. Orthop. Res*, **8**, 383–392.

La Palombara, P. F., Cappello, A. and Leardini, A. (1995) Combining optimization and smoothing techniques in human motion analysis. In: *Computer Simulations in Biomedicine* (ed. H. Power and R. T. Hart), Computational Mechanics Publications, Boston, pp. 401–408.

Laub, A. J. and Shiflett, G. R. (1982) A linear algebra approach to the analysis of rigid body displacement from initial and final position data. *J. Appl. Mech.*, **49**, 213–216.

Magnani, G., Angeloni, C., Leardini, A. and Cappello, A. (1993) Optimal estimation of rigid body position and attitude from noisy marker coordinates. In: *Proceedings of the XIV Congress of ISB*, Paris, 4–8 July, pp. 820–821.

Riley, P. O., Mann, R. W. and Hodge, W. A. (1990) Modelling of the biomechanics of posture and balance. *J. Biomech.*, **23**, 503–506.

Söderkvist, I. and Wedin, P. A. (1993) Determining the movements of the skeleton using well-configured markers. *J. Biomech.*, **12**, 1473–1477.

Spoor, C. W. and Veldpaus, F. E. (1980) Rigid body motion calculated from spatial co-ordinates of markers. *J. Biomech.*, **13**, 391–393.

Vaughan, C. L., Davis, B. L. and O'Connor, J. C. (1992) *Dynamics of Human Gait*. Human Kinetics Publishers, Champaign, Illinois.

Veldpaus, F. E., Woltring, H. J. and Dortmans, L. J. M. G. (1988) A least-squares algorithm for the equiform transformation from spatial marker coordinates. *J. Biomech.*, **21** 45–54.

Wang, X., Rezgui, M. A. and Verriest, J. P. (1993) Using the polar decomposition theorem to determine the rotation matrix from noisy landmark measurements in the study of human joint kinematics. In: *Proceedings of the 2nd International Symposium on 3D Analysis of Human Movement*, Poitiers, 30 June to 3 July, pp. 53–56.

Woltring, H. J. (1994) 3-D attitude representation of human joints: a standardization proposal. *J. Biomech.*, **27**, 1399–1414.

Woltring, H. J., Huiskes, R., de Lange, A. and Veldpaus, F. E. (1985) Finite centroid and helical axis estimation from noisy landmark measurements in the study of human joint kinematics. *J. Biomech.*, **18**, 379–389.

Woltring, H. J., Long, K., Osterbauer, P. J. and Fuhr, A. W. (1994) Instantaneous helical axis estimation from 3-D data in neck kinematics for whiplash diagnostics. *J. Biomech.*, **27**(12), 1415–1432.

9

Joint Kinematics

SANDRO FIORETTI[1], AURELIO CAPPOZZO[2]
AND LUIGI LUCCHETTI[3]

[1] Dipartimento di Elettronica ed Automatica, Università degli Studi di Ancona, Ancona, Italy
[2] Cattedra di Tecnologie Biomediche, Università degli Studi, Sassari, Italy
[3] Istituto di Fisiologia Umana, Università degli Studi 'La Sapienza', Rome, Italy

INTRODUCTION

The expression joint kinematics alludes to the description of the relative movement between two adjacent bones. Here we deal with such a description while a human subject is performing a physical exercise in general and locomotion in particular. The mechanics of rigid bodies provide a number of ways to numerically represent these relative movements. Whether the objective of a biomechanical investigation is to contribute to knowledge or to supply information that may be useful for the treatment of a patient, the method used must satisfy the fundamental rule of reliability. This allows for intraindividual and interindividual comparison and for the generalization of observations. If relevant results are to be used in a clinical context, then it is of crucial importance that joint movements be represented by making reference to definitions that tend to be consistent with the standard terminology used by the practitioner.

In summary, the experimental and analytical methods to be used for the description of articular movements during the execution of a physical exercise should comply with the following requirements:

- The methods should supply results that are repeatable and therefore based on unambiguous definitions and consistently identifiable measurable quantities.
- The methods should supply results that can be expressed in the established anatomical and physiological terminology.

Three-dimensional Analysis of Human Locomotion. Edited by P. Allard, A. Cappozzo, A. Lundberg and C. Vaughan
© 1997 John Wiley & Sons Ltd. ISBN 0 471 96949 4

- The description of joint movements obtained with these methods should be consistent with the qualitative descriptions already provided in the anatomical and physiological literature.

The objective of this chapter is to review the possible descriptions of joint kinematics, giving special emphasis to the above-listed requirements. The laboratory positions and orientations of relevant bones during movement are assumed to have been made available through the experimental and analytical methods illustrated in Chapters 7 and 8.

MECHANICAL CONSIDERATIONS

POINT COORDINATE TRANSFORMATION

As mentioned in Chapter 7, it is possible to describe, in any sampled instant of time during movement, the position and orientation, in the laboratory frame (LF), of an orthogonal system of reference considered rigid with a bone (bone-embedded frame, BEF) through an orientation matrix $\mathbf{R}$ and a position vector $\mathbf{t}$. In addition, the anatomical landmark calibration procedure provides the position vector $^B\mathbf{p}$ in the BEF of any relevant point of the bone.

Therefore, for any point P of the bone, the following relationship holds:

$$^L\mathbf{p} = \mathbf{R}^B\mathbf{p} + \mathbf{t} \tag{1}$$

where $^L\mathbf{p}$ is the position vector of point P in the LF.

Equation (1) solves the basic problem in movement biomechanics, i.e. the determination of the laboratory trajectory of any relevant point of the bone, whether this is directly observable or not.

RIGID-BODY POSITION AND ORIENTATION

Given the orientation matrices $\mathbf{R}_p$ and $\mathbf{R}_d$ and the position vectors $\mathbf{t}_p$ and $\mathbf{t}_d$ in the LF for two adjacent bones, proximal and distal respectively, from equation (1), written for each bone and through elimination of the global position vector, it follows that:

$$\mathbf{R}_j = \mathbf{R}_p^T\mathbf{R}_d \qquad \mathbf{t}_j = \mathbf{R}_p^T(\mathbf{t}_d - \mathbf{t}_p) \tag{2}$$

where $\mathbf{R}_j$ (joint orientation matrix) and $\mathbf{t}_j$ (joint position vector) are the orientation matrix and position vector of the distal BEF relative to the proximal BEF, which, from now on, will be considered the reference frame (Figure 1). This matrix and this vector describe joint kinematics completely.

Let $(\mathbf{i}_X \, \mathbf{j}_Y \, \mathbf{k}_Z)$ and $(\mathbf{i}_x \, \mathbf{j}_y \, \mathbf{k}_z)$ be the unit vectors of the proximal BEF $\{XYZ\}$ and of the distal BEF axes $\{xyz\}$, respectively. Then matrix $\mathbf{R}_j$ can be expressed in terms of scalar products of unit as follows:

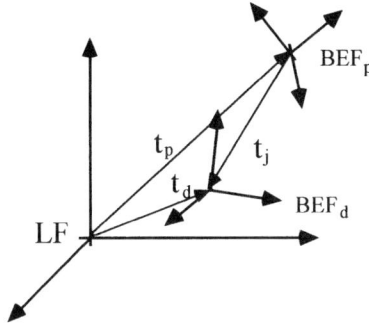

Figure 1. Schematic representation of laboratory frame (LF) and proximal and distal bone-embedded frame (BEF)

$$\mathbf{R}_j = \begin{bmatrix} \mathbf{i}_X \cdot \mathbf{i}_x & \mathbf{i}_X \cdot \mathbf{j}_y & \mathbf{i}_X \cdot \mathbf{k}_z \\ \mathbf{j}_Y \cdot \mathbf{i}_x & \mathbf{j}_Y \cdot \mathbf{j}_y & \mathbf{j}_Y \cdot \mathbf{k}_z \\ \mathbf{k}_Z \cdot \mathbf{i}_x & \mathbf{k}_Z \cdot \mathbf{j}_y & \mathbf{k}_Z \cdot \mathbf{k}_z \end{bmatrix} \tag{3}$$

which shows that it is orthonormal. This circumstance introduces six constraints which allow the nine elements of $\mathbf{R}_j$ to be represented by only three independent coordinates which, when added to the three components of vector $\mathbf{t}_j$, result in six independent coordinates. These are the six degrees of freedom (DOF), three associated with the position and three with the orientation of a rigid body relative to another.

The three scalar coordinates associated with orientation may be represented in the different ways discussed below.

If we suppose that the two frames are equi-oriented and that the $\{xyz\}$ BEF is then rotated by an angle α about the X (or x) axis, then the resulting orientation matrix, as derived from equation (3), is

$$\mathbf{R}_{X,\alpha} = \mathbf{R}_{x,\alpha} = \mathbf{R}_\alpha = \begin{bmatrix} 1 & 0 & 0 \\ 0 & \cos\alpha & -\sin\alpha \\ 0 & \sin\alpha & \cos\alpha \end{bmatrix} \tag{4}$$

Similarly, the orientation matrices resulting from rotations about the Y (or y) axis (β) or about the Z (or z) axis (γ) are given respectively by

$$\mathbf{R}_{Y,\beta} = \mathbf{R}_{y,\beta} = \mathbf{R}_\beta = \begin{bmatrix} \cos\beta & 0 & \sin\beta \\ 0 & 1 & 0 \\ -\sin\beta & 0 & \cos\beta \end{bmatrix} \tag{5}$$

and

$$\mathbf{R}_{Z,\gamma} = \mathbf{R}_{z,\gamma} = \mathbf{R}_{\gamma} = \begin{bmatrix} \cos\gamma & -\sin\gamma & 0 \\ \sin\gamma & \cos\gamma & 0 \\ 0 & 0 & 1 \end{bmatrix} \tag{6}$$

These matrices are referred to as basic rotation matrices (Fu, Gonzalez and Lee, 1988; Kane, Likins and Levinson, 1983).

Any orientation of the distal BEF with respect to the proximal BEF can be thought of as the result of three successive and ordered basic rotations about two or three different axes (belonging either to the reference or to the moving system). The relevant orientation matrix may be obtained using the following rules (Fu, Gonzalez and Lee, 1988).

- Initially, both $\{XYZ\}$ and $\{xyz\}$ BEFs are supposed to be coincident, and hence the orientation matrix is the 3×3 identity matrix $\mathbf{I}$.
- If the $\{xyz\}$ BEF rotates about an axis of the $\{XYZ\}$ BEF, then one has to pre-multiply its present (resultant) orientation matrix with the appropriate basic rotation matrix.
- If the $\{xyz\}$ BEF rotates about one of its own axes, then one has to post-multiply its previous (resultant) orientation matrix with the appropriate basic rotation matrix.

For example, the transformation matrix that describes the orientation of the $\{xyz\}$ frame with respect to the $\{XYZ\}$ reference system obtained through a rotation α about the X axis, followed by a rotation β about the y axis (about the direction it assumes in space after the first rotation is performed), followed by a rotation γ about the z axis (about the direction it assumes in space after the first two rotations are performed), is given by

$$\mathbf{R}_j = [(\mathbf{R}_{X,\alpha}\mathbf{I})\mathbf{R}_{y,\beta}]\mathbf{R}_{z,\gamma} = \mathbf{R}_{\alpha}\mathbf{R}_{\beta}\mathbf{R}_{\gamma} \tag{7}$$

It should be noted that this latter expression can also be interpreted as representing other sequences, such as a first rotation γ about the Z axis, followed by a second rotation β about the Y axis, followed by a rotation α about the X axis, and so on.

Another point worthy of mention is the following: since matrix multiplication is not commutative (except for very particular cases), careful attention must be paid to the order in which these matrices are multiplied and therefore the order in which basic rotations are assumed to be performed.

Thus joint kinematics may be described giving in each sampled instant of time six scalar coordinates, three components of vector $\mathbf{t}_j$ and the three above-mentioned independent angles derived, given $\mathbf{R}_j$ and an ordered sequence, from an equation of the type given in equation (7) and calculated with respect to a given set of three axes.

Instead of defining the orientation matrix in terms of a sequence of three ordered single-axis rotations as seen above, the orientation of the distal BEF with respect to the proximal BEF can be thought as a single rotation by an angle θ about an axis. This axis is described through a unit vector $\mathbf{n}$ (Kane, Likins and Levinson, 1983; Fu, Gonzalez and Lee, 1988). The angle θ is the amount of rotation about $\mathbf{n}$, in the counter-clockwise direction, required by the distal BEF to reach its current orientation from the reference orientation (superimposed on the proximal BEF). Thus the relative orientation of the BEFs may be described in terms of an orientation vector $\boldsymbol{\theta}_j = \theta\mathbf{n}$ (Woltring et al., 1987; Woltring, 1991a), the scalar components of which may be represented in either BEFs which, apart from a sign inversion, are identical (Woltring and Fioretti, 1989), or in any set of axes, whether they are orthogonal or not.

As indicated in Chapter 7, the orientation vector may be calculated from the relevant orientation matrix. It is an axial vector and as such it belongs to the class of pseudovectors. The reader is referred to Woltring (1994) for further details.

It should be noted that the components of the orientation vector $\boldsymbol{\theta}_j$ do not correspond to any physical rotation about the axes used to represent them numerically but merely represent an algebraic method to express a vector in a given coordinate system. Moreover, unlike position vectors, the orientation vectors are not additive (Woltring, 1991b; Kane, Likins and Levinson, 1983). Additivity is valid only under very particular conditions such as successive rotations about parallel axes (planar movements) or infinitesimal rotations about the reference orientation (Kane, Likins and Levinson, 1983; Bottema and Roth, 1979; Wittenburg, 1977; Goldstein, 1970).

Thus joint kinematics may also be described using three components of vector $\mathbf{t}_j$ and three components of the orientation vector $\boldsymbol{\theta}_j$ calculated with respect to a given set of three axes.

BIOMECHANICAL CONSIDERATIONS

The problem to be solved, for an effective application of the above-mentioned kinematic data in any application field, and in order to be consistent with the requirements listed in the Introduction, resides in an adequate choice of the six independent scalar quantities which describe the relevant six DOF. In biomechanics and in order to be easily interpreted, these six numbers must lend themselves to be interpreted as three separate rotations and three separate translations of one BEF with respect to the other. In this context there are two levels of conventions that should be adopted. One is anatomical and is associated with the definition of the axes with respect to which the six DOF are calculated. These axes are referred to as joint axes and must be defined using anatomical landmarks as reference points. The second convention relates to the mathematical algorithm used to calculate the six scalar components involved among the several ones illustrated in the previous section.

ANATOMICAL CONVENTION

An anatomical convention for each joint should thus be identified. As will be shown below, small variations in the location of these axes can have a significant effect on the description of joint kinematics. This is especially true with reference to those DOF that undergo small displacements. In addition, these axes should be chosen so that when basic movements, as described by functional anatomy, e.g. pure flexion–extension, adduction–abduction or internal–external rotation, are performed, only the relevant variable is significantly different from zero. In connection with this, it is noted that conventions used in the literature are never validated during the execution of these basic movements; on the contrary, only data obtained during walking or other complex movements are presented which do not effectively allow for such a control.

One possible criterion for the determination of the orientation of the joint axes may be based on using the mean functional axes about which basic rotations occur. These axes must, of course, be defined with respect to the anatomy of the bone, i.e. in standardized anatomical bone-embedded frames (ABEF) (see Chapter 7). However, they should not be subject dependent, but, on the contrary, they should be the result of a statistical survey within a sample of an able-bodied population and associated with it. The use of axes determined on a given individual would hide his or her differences with respect to the reference population.

The joint axes which meet the above-mentioned requirements do not in general coincide with the axes of the ABEFs, as opposed to what is often seen in the literature. Consequently, the direction cosines of the joint axes in the respective ABEF should be provided. This, to the authors' knowledge, has not been done yet. Consequently, despite the above-mentioned observation, in the examples reported in a subsequent section, the joint axes will be made to coincide with axes of the ABEFs.

MATHEMATICAL CONVENTION

Given the proximal and distal bones which define the joint under analysis, it is assumed here that their position and orientation are defined with reference to the respective ABEFs (Figure 2).

Rotational Degrees of Freedom

As seen previously, the orientation of the distal bone with respect to the proximal bone at any sampled instant of time may be defined either by a conventional sequence of three basic rotations, or by a single rotation about an axis, that take the distal segment from the reference orientation to its actual orientation. The reference orientation of the joint is generally defined by the coincidence of the

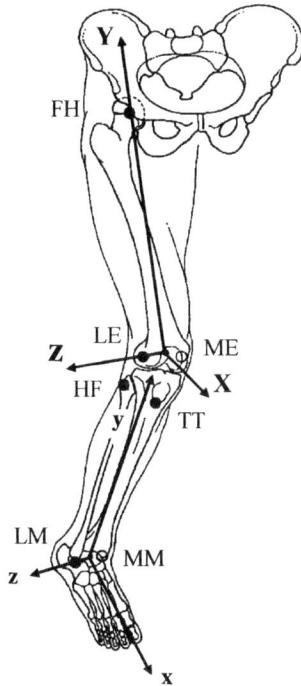

Figure 2. Proximal $\{XYZ\}$ and distal $\{xyz\}$ ABEFs for thigh and shank respectively. The ABEFs are defined on the base of the following anatomical landmarks: FH, centre of the femoral head; ME, medial epicondyle; LE, lateral epicondyle; TT, prominence of the tibial tuberosity; HF, apex of head of the fibula; MM, distal apex of the medial malleolus; LM, distal apex of the lateral malleolus

orientation of the two ABEFs. This definition allows for both intra- and interindividual repeatability. It should, however, be noted that the end results depend on the choice of the reference orientation, which is therefore not to be taken for granted and be subjected to standardization.

Different mathematical conventions which may be used in the present context are illustrated below.

Cardan Angles

These are the most widely used angles in biomechanics. By definition, they are obtained as an ordered sequence of three basic rotations thought to occur about three separate axes. Of all the possible different conventions that can be derived, attention is focused here on a particular one, which was proposed in Grood and Suntay (1983) and Chao (1980). The first rotation (γ) is about the Z axis of the proximal ABEF (or, equivalently, about the z axis of the distal ABEF), the second rotation (α) is about the x axis of the distal ABEF in the orientation it assumes after the first rotation is performed (referred to as floating axis), and the third rotation (β) is about the y axis of the distal ABEF in the new orientation after the first and second rotations. The orientation matrix is therefore given by the following product of basic rotation matrices:

$$\mathbf{R}_j = R_{Z,\gamma} R_{x,\alpha} R_{y,\beta} = R_\gamma R_\alpha R_\beta =$$

$$\begin{bmatrix} \cos\gamma\cos\beta - \sin\gamma\sin\alpha\sin\beta & -\sin\gamma\cos\alpha & \cos\gamma\sin\beta + \sin\gamma\sin\alpha\cos\beta \\ \sin\gamma\cos\beta + \cos\gamma\sin\alpha\sin\beta & \cos\gamma\cos\alpha & \sin\gamma\sin\beta - \cos\gamma\sin\alpha\cos\beta \\ -\cos\alpha\sin\beta & \sin\alpha & \cos\alpha\cos\beta \end{bmatrix}$$

$$(8)$$

The Cardan angles α, β and γ can be obtained by the following trigonometric equations:

$$\alpha = \sin^{-1}[\mathbf{R}_j(3, 2)]$$

$$\beta = \sin^{-1}[-\mathbf{R}_j(3, 1)/\cos\alpha] \qquad (9)$$

$$\gamma = \sin^{-1}[-\mathbf{R}_j(1, 2)/\cos\alpha]$$

The axes about which rotations are assumed to occur (joint axes), Z, y and the floating axes, define a set of axes that are not mutually orthogonal. The Z and y axes are embedded in the proximal and distal bones respectively, and the floating axis is the axis orthogonal to both Z and y in any given instant of time. Thus, if Z can be thought to represent the flexion–extension axis, y the internal–external rotation axis and the floating axis the abduction–adduction axis of the distal segment with respect to the proximal one, the functional meaning of γ, β and α is that of flexion–extension, internal–external rotation and abduction–adduction respectively, and the cardanic convention expressed in equation (8) is equivalent to that suggested by Grood and Suntay (1983).

As is evident in equation (9) a singularity condition (gimbal lock) occurs when rotation α about the floating axis is equal to $k\pi/2$ ($k = 1, 2, \ldots$). If the floating axis is chosen so that during functional movements it corresponds to small rotations (such as abduction–adduction or endorotation–exorotation in the case of knee joint), gimbal lock can be avoided.

It is worth noting that Cardan angles are different from Eulerian angles used in classical mechanics. In fact, these latter are obtained as an ordered sequence of three basic rotations about two different axes (one axis is used twice). In this case it is difficult to associate a functional meaning with the three angular variables, because two of the angles correspond to rotations about the same axis, though in different positions in space. Moreover, in the case of Eulerian angles, the gimbal-lock phenomenon occurs when $\mathbf{R}_j = \mathbf{I}$, i.e. just in the neutral position. Thus, given the ABEFs indicated for femur and tibia in Figure 2, one has to expect gimbal lock just when the two coordinate systems are equi-oriented, which may well occur during walking, for instance.

Scalar Components of the Orientation Vector

From a geometric point of view, the components of the orientation vector can be expressed in any coordinate system (not necessarily orthogonal).

Meglan et al. (1990) suggested expressing the orientation vector in a non-orthogonal base, i.e. projecting the orientation vector onto the non-orthogonal frame constituted by the joint axes as proposed by Grood and Suntay (1983) and illustrated above. The orientation vector components, in this case, can be obtained as oblique projections as follows. Let $\mathbf{u}_1$, $\mathbf{u}_2$ and $\mathbf{u}_3$ be the unit vectors describing the directions of the joint axes expressed in the proximal ABEF; the projection of $\boldsymbol{\theta}_j = \theta\mathbf{n}$ on these base vectors is the set of coefficients e_1, e_2, e_3 satisfying the following relationship:

$$\boldsymbol{\theta}_j = e_1\mathbf{u}_1 + e_2\mathbf{u}_2 + e_3\mathbf{u}_3$$

i.e.

$$\mathbf{e} = \mathbf{U}^{-1}\boldsymbol{\theta}_j \tag{10}$$

where $\mathbf{e} = (e_1 e_2 e_3)^T$, and $\mathbf{U}$ is the 3×3 matrix whose columns coincide with the vectors of the non-orthogonal base $\mathbf{u}_1$, $\mathbf{u}_2$ and $\mathbf{u}_3$, i.e. $\mathbf{U} = [\mathbf{u}_1 \ \mathbf{u}_2 \ \mathbf{u}_3]$.

In the case of joint axes defined according to Grood and Suntay's convention as described by equation (8), the matrix $\mathbf{U}$ is given by

$$\mathbf{U} = \begin{bmatrix} 0 & \cos\gamma & -\sin\gamma\cos\alpha \\ 0 & \sin\gamma & \cos\gamma\cos\alpha \\ 1 & 0 & \sin\alpha \end{bmatrix} \tag{11}$$

where α and γ are the same angles as in equation (8).

The use of the orientation vector for angle representation is a 'robust' procedure. In fact, it has been shown (Woltring et al. 1985; Woltring, 1990) that while $\mathbf{n}$ becomes undefined when θ approaches zero, both θ and the product $\theta\mathbf{n}$ are always well defined and the gimbal-lock phenomenon does not occur.

A Geometric Approach to the Joint Angle Calculation

An approach which is certainly worthy of mention consists of the projection in a given instant of time of an axis of one ABEF onto a plane of the other ABEF, and in the determination of the orientation of the projected axis in this latter plane (Paul, 1992).

For instance, with reference to the knee joint:

- Flexion–extension angle—the angle formed by the y axis of the shank and the projection of the Y axis of the thigh onto the xy plane of the shank.

- Adduction–abduction angle—the angle formed by the y axis of the shank and the projection of the Y axis of the thigh onto the yz plane of the shank.
- Internal–external rotation angle (inward–outward)—the angle formed by the x axis of the shank and the projection of the X axis of the thigh onto the xz plane of the shank.

This approach has the advantage of being very intuitive and similar to that already in use in functional anatomy, although only qualitatively. In fact, angles are assessed following a completely arbitrary description with no correspondence between the sequence of rotations characteristic of the Cardan angles and the components of the orientation vector.

Translatory Degrees of Freedom

The translatory movement of the joint is described, making reference to the position of a point, rigid with the distal bone, with respect to the proximal bone. Relevant points can be, for example, the centre of the head of the femur, the intercondylar eminence of the tibia and the centre of the body of the talus. At any given instant in time, the position vector of these points in the proximal frame is projected onto a properly chosen system of axes. For this purpose, different mathematical conventions can be used. For example, the system of axes could be chosen equal to the anatomical frame of the proximal body segment or equal to the joint axes used to describe rotations. However, it should be strongly emphasized that:

- translatory movements are usually very small and prone to be hidden by the experimental artefacts and errors;
- no commonly accepted convention exists that defines functionally significant translational directions when the joint is in an arbitrary configuration (Andrews, 1984; Grood and Suntay, 1983).

In the following, attention will be focused on rotational movements only.

EXPERIMENTAL ASSESSMENT

In the preceding sections the nonlinear character of the orientation matrix $\mathbf{R}_j$ has been demonstrated. It has been shown that the orientation matrix can be obtained through the product of basic rotation matrices (see, for instance, equation (8)). Since matrix multiplication is not in general commutative, careful attention must be paid to the order in which these matrices are multiplied. From a numerial point of view, this implies that, given an orientation matrix, the three angular values that can be computed from it are not unique but depend on the chosen order. As an example, Table 1 reports the results yielded by various cardanic sequences applied to two different orientation matrices.

From these data it can be seen that the largest rotation (γ) is least sensitive to the chosen sequence, while the other two angles (α and β) heavily depend on it. This effect becomes more evident as γ increases.

A comparison of different methods for joint angle computation was carried out. Results were relative to a gait analysis test performed on a healthy subject. The joint taken into account was the right knee joint. Four methods were compared:

1. the cardanic convention suggested by Grood and Suntay (1983) (zxy sequence);
2. the orthogonal projections of the orientation vector on the proximal (thigh) ABEF;
3. the non-orthogonal projections of the orientation vector on the joint axes defined by Grood and Suntay's cardanic convention (as in (1));
4. joint angles obtained following the geometric approach described in a previous section.

Data were collected with an active-marker optoelectronic system (CoSTEL) and the BEFs were determined as described in Chapter 7. Cardanic angles and the orientation vector were calculated using equations (9) and as indicated by equation (6) in Chapter 7, respectively. For the determination of non-orthogonal projections of the orientation vector, equation (10) was used.

Resulting joint angles are shown in Figure 3. These curves have similar characteristics in the sense that the correspondent angles obtained using the four methods show almost the same shape in time. This is particularly evident for the flexion–extension angle, which is the largest angle in terms of range of motion. Method (3) gives a reduced flexion–extension angle and an enhanced abduction in correspondence with the maximum knee flexion. The rotation of the shank

Table 1. Angle values (degree) obtained using different cardanic sequences (indicated by the sequence of the relevant axes) for two different orientation matrices (upper and lower rows, respectively)

yxz	xyz	zyx	xyz	yzx	zxy
$\alpha = 0.6$	$\alpha = 19.4$	$\alpha = 10.0$	$\alpha = 0.6$	$\alpha = 1.3$	$\alpha = 10.0$
$\beta = 11.2$	$\beta = 21.7$	$\beta = 4.9$	$\beta = 11.2$	$\beta = 10.0$	$\beta = 5.0$
$\gamma = 60.5$	$\gamma = 58.5$	$\gamma = 60.9$	$\gamma = 60.4$	$\gamma = 60.5$	$\gamma = 60.0$
$\alpha = 6.1$	$\alpha = 11.5$	$\alpha = 10.0$	$\alpha = 6.2$	$\alpha = 7.1$	$\alpha = 10.0$
$\beta = 9.4$	$\beta = 10.7$	$\beta = 4.9$	$\beta = 9.3$	$\beta = 5.7$	$\beta = 5.0$
$\gamma = 30.9$	$\gamma = 29.5$	$\gamma = 30.9$	$\gamma = 29.9$	$\gamma = 30.7$	$\gamma = 30.0$

α = rotation about x axis; β = rotation about y axis; γ = rotation about z axis.

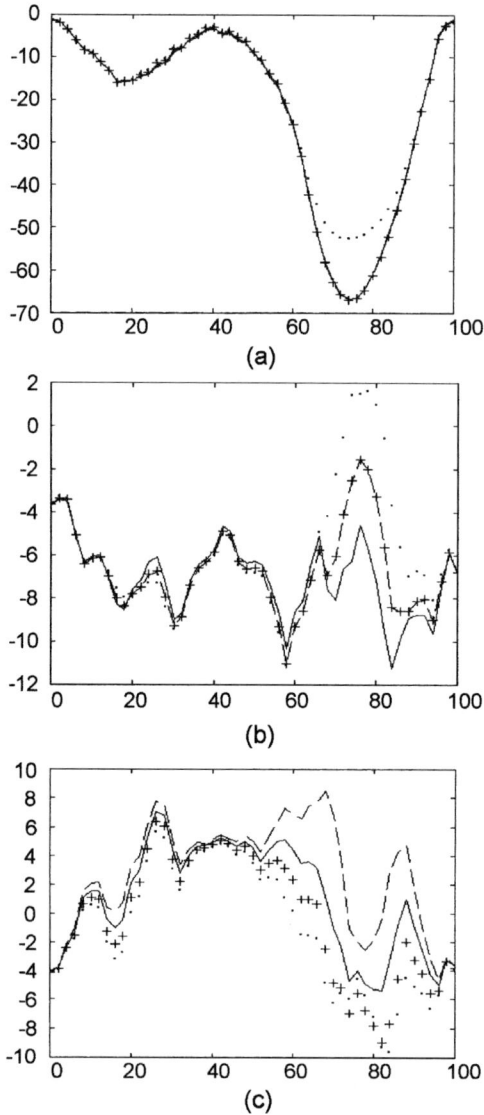

Figure 3. Flexion–extension (a), abduction–adduction (b) and internal–external (c) angles at the knee joint for a normal subject as a function of the percentage of the gait cycle. Swing phase is from 60% to 100% of the gait cycle. Increasing angles are associated with extension, adduction and internal rotation, respectively. Cardanic angles according to Grood and Suntay's convention are marked with (+). The projections of the orientation vector on the thigh anatomical orthogonal axes are represented by a solid line. The projections of the orientation vector on the non-orthogonal joint axes of the chosen cardanic convention are presented by the dotted line. Angles calculated according to the geometric approach are shown by the dashed line

with respect to the thigh is the angle that exhibits the most significant differences among the four methods.

The ranges of variation for the four angles are given in Table 2. This confirms that the various methods do not give rise to excessively different patterns, though method (2) exhibits the least ranges of variation for the abduction–adduction and the internal–external rotations.

A further comparison was carried out in order to test the sensitivity of the four methods to an incorrect determination of the orientation of the flexion–extension axis. This results in errors that affect the orientation matrix and, consequently, the angles. In the case of the knee joint, the flexion–extension axis is the most critical to be estimated because, in general, this is defined as the line passing through the medial and lateral femoral condyles. The condylar surfaces are fairly large, making it difficult to identify the two relevant anatomical landmarks, and, in addition, these are relatively close. Consequently, the direction of the flexion–extension axis may suffer from lack of reliability and its orientation may vary by $\pm 10°$ (Ramakrishnan and Kadaba, 1991).

Using the experimental data referred to above, in order to quantify the effects of the above-mentioned uncertainty, the flexion–extension axis of the knee was analytically rotated in the transverse plane from internal 15° to external 15°, by steps of 5° (positive internal, negative external). This was done applying the following similarity transformation to matrix $\mathbf{R}_j$:

$$\mathbf{R}_j^* = \mathbf{T}\mathbf{R}_j\mathbf{T}^{-1}$$

where $\mathbf{T}$ is the transformation matrix of a rotation about the Y axis.

Results with the four different methods here analysed are shown in Figures 4–7 and the maximum variations of the joint angles with respect to the nominal trajectory are shown in Table 3.

As can be seen from Figures 4–7, the largest differences occur in correspondence of the largest value of flexion. For all the four tested methods, the most affected angle is abduction–adduction. Internal–external rotation was unaffected for method 2 and only slightly for method 3. The slope of the abduction–adduction curve may be either positive or negative during the swing phase as a function of the direction of the flexion–extension axis, i.e. adduction

Table 2. Range of variation of the knee joint angles (degree) as function of four different methods used for their computation

Method	Flexion–extension	Abduction–adduction	Internal–external
1	65.4	9.5	15.3
2	65.3	7.9	12.4
3	51.1	12.8	15.3
4	65.6	9.5	13.2

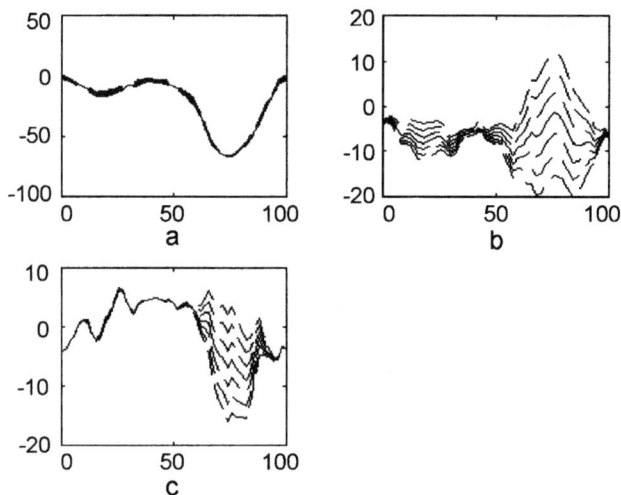

Figure 4. Joint angles obtained according to Grood and Suntay's cardanic convention results of the sensitivity analysis performed on the knee joint angles of a normal subject during level walking and normalized in percentage of gait cycle. (a) Flexion–extension angle. (b) Abduction–adduction angle. (c) Internal–external rotation. Dashed lines represent perturbations of the knee flexion–extension axis by steps of 5° from −15° to +15°. Reference data are represented by a solid line

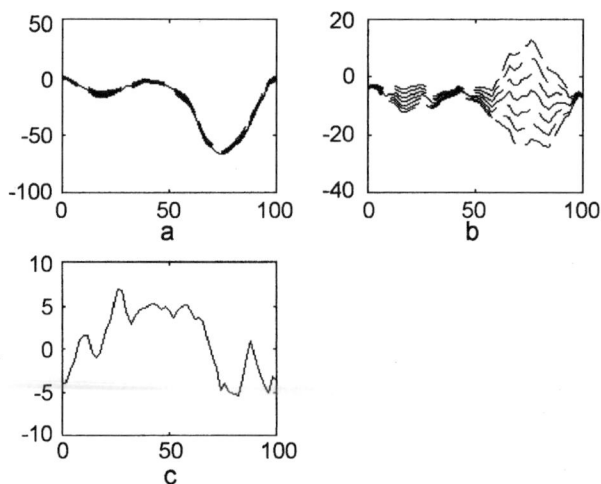

Figure 5. Joint angles obtained as components of the orientation vector in the thigh-embedded coordinate system (see caption of Figure 4)

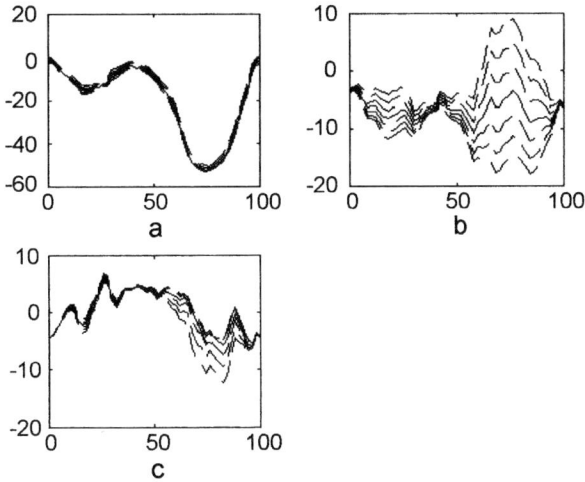

Figure 6. Joint angles obtained as components of the orientation vector in the non-orthogonal joint axes of Grood and Suntay's cardanic convention (see caption of Figure 4)

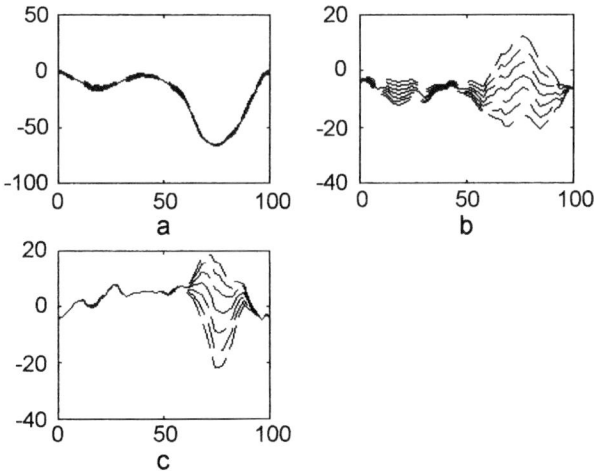

Figure 7. Joint angles obtained according to the geometric approach (see caption of Figure 4)

can be interpreted as abduction, or vice versa. Thus, the dispersion of the abduction–adduction and of the internal rotation angles during the swing phase is so large that their values during this phase of gait cannot be considered reliable.

Table 3. Maximum variation of the knee joint angles (degree) with respect to the nominal trajectory as a result of the sensitivity test. The direction of the flexion–extension axis is supposed to vary over a range of 30° from −15° to +15°. See also text and Figures 4–7

Method	Flexion–extension	Abduction–adduction	Internal–external
1	1.0	13.8	9.0
2	3.9	17.5	0.0
3	2.5	12.4	6.5
4	1.6	14.3	21.0

CONCLUSIONS

It has sometimes been suggested that the interpretation of the cardanic representation as a physically meaningful, ordered sequence of rotations is effective from a didactic point of view. On the contrary, the fact that the projections of the orientation vector on designated reference systems do not correspond to physically realizable rotations has been advocated as a difficulty for the interpretation of the results. The same doubt may be cast on the geometric approach.

The experimental comparison illustrated above shows that all four tested methods are not significantly different as far as the values of angles are concerned. During walking, the major, though limited, differences occur in abduction–adduction and internal–external rotations when the flexion angle becomes large, i.e. during swing phase. In addition, a proper interpretation of the angular values during the swing phase is critical in all the four tested methods because these values depend largely on the accuracy with which the ABEF axes have been determined.

In summary, these results show that it is difficult to claim the superiority of one approach with respect to another and also indicate that this is not the real problem. The major difficulties reside in the experimental artefacts and errors, as illustrated in Chapter 8.

REFERENCES

Andrews, J. G. (1984) On the specification of joint configurations and motions. *J. Biomech.*, **17**, 155–158.

Bottema, O. and Roth, B. (1979) *Theoretical Kinematics*. North Holland, Amsterdam.

Chao, E. Y. S. (1980) Justification of tri-axial goniometer for the measurement of joint rotation. *J. Biomech.*, **13**(12), 998–1006.

Fu, K. S., Gonzalez, R. C. and Lee, C. S. G. (1988) *Robotics: Control, Sensing, Vision and Intelligence*. McGraw-Hill, New York.

Goldstein, H. (1970) *Classical Mechanics.*. Addison-Wesley, Reading, Massachusetts.

Grood, E. S. and Suntay, W. J. (1983) A joint coordinate system for the clinical description of three-dimensional motions: application to the knee. *J. Biomech. Eng., Trans. ASME*, **105**, 136–144.

Kane, T. R., Likins, P. W. and Levinson, D. A. (1983) *Spacecraft Dynamics*. McGraw-Hill, New York.

Meglan, D. A., Pisciotta, J., Berme, N. and Simon, S. R. (1990) Effective use of non-sagittal plane joint angles in clinical gait analysis. In: *Proceedings of the 30th Annual Meeting, Orthopaedic Research Society*, 5–8 February, New Orleans, Louisiana, pp. 76–77.

Paul, J. P. (1992) *Terminology and Units*. Deliverable no. 4, CEC Programme AIM, Project A-2002: CAMARC-II.

Ramakrishnan, H. K. and Kadaba, M. P. (1991) On the estimation of joint kinematics during gait. *J. Biomech.*, **24**(10), 969–977.

Wittenburg, J. (1977) *Dynamics of Systems of Rigid Bodies*. B. G. Teubner, Stuttgart.

Woltring, H. J. (1990) Model and measurement error influences in data processing. In: *Biomechanics of Human Movement: Applications in Rehabilitation, Sports and Ergonomics* (ed. A. Cappozzo and N. Berme), Bertec Corporation, Worthington, Ohio, pp. 203–230.

Woltring, H. J. (1991a) Estimation and calculation of 3-D joint movement. *Hum. Movem. Sci.*, **10**(5), 603–616.

Woltring, H. J. (1991b) Definition and calculus of attitude angles, instantaneous helical axes and instantaneous centres of rotation from noisy position and attitude data. In: *Proceedings of the International Symposium on 3-D Analysis of Human Movement* (ed. P. Allard and J. P. Blanchi), 28–31 July, Montreal, Quebec, pp. 59–62.

Woltring, H. J. (1994) 3-D attitude representation of human joints: a standardization proposal. *J. Biomech.*, **27**(12), 1399–1414.

Woltring, H. J. and Fioretti, S. (1989) Representation and photogrammetric calculation of 3-D joint movement. In: *Proceedings of the First IOC World Congress on Sport Science* (ed. C. J. Dillman, R. C. Nelson, B. M. Nigg, R. O. Voy and M. M. Newsom), US Olympic Committee, Colorado Spring, CO, pp. 350–351.

Woltring, H. J., Huiskes, R., de Lange, A. and Veldpaus, F. E. (1985) Finite centroid and helical axis estimation from noisy landmark measurement in the study of human joint kinematics. *J. Biomech.*, **18**(5), 379–389.

Woltring, H. J., Lange, A. de, Kauer, J. G. M. and Huiskes, R. (1987) Instantaneous helical axis estimation via natural, cross validated splines. In: *Biomechanics: Basic and Applied Research* (ed. G. Bergmann, A. Kolber and A. Rohlmann), Martinus Nijhoff Publishers, Dordrecht, pp. 121–128.

10

Inverse Dynamics in Human Locomotion

SORIN SIEGLER AND WEN LIU

Department of Mechanical Engineering and Institute of Biomedical Engineering,
Drexel University, Philadelphia, PA, USA

INTRODUCTION

Human locomotion is governed by the general rules of physics as formulated by Newton's laws. To study the dynamics of human locomotion, the governing equations of motion can be derived according to the Newton–Euler formulation (Cappozzo, Leo and Pedotti, 1975; Hardt and Mann, 1980; Vaughan, Davis and O'Connor, 1992), Lagrange technique or Kane's method (Zajac and Winters, 1990; Yamaguchi and Zajac, 1990; Kuo, 1995). Regardless of which method is used, the result is a set of differential equations which include the driving forces (forces and moments), kinematic data (accelerations and velocities) and inertial properties (mass, centre of mass and moment of inertia) of body segments. These equations can be solved for the kinematics (direct dynamics approach) or for the driving forces (inverse dynamics approach). In most artificial mechanical systems, such as robots, the driving forces are known and the task of the dynamic study is to solve the equations of motion for the resulting kinematics. In contrast, there are situations when the system's kinematics are specified and the equations of motion have to be solved for the driving forces of the system. An example of this approach is the control of a robot to follow a desired trajectory. In this situation, the desired trajectory of the end effector is specified and the equations of motion of the robot are solved for the required actuator torques.

Three-dimensional Analysis of Human Locomotion. Edited by P. Allard, A. Cappozzo, A. Lundberg and C. Vaughan
© 1997 John Wiley & Sons Ltd. ISBN 0 471 96949 4

Both the direct dynamics approach and the inverse dynamics approach have been used in the study of human locomotion. The direct dynamics approach, shown schematically in Figure 1, was used by a number of investigators such as Hatze (1976) to gain insight into the control strategies used by the central nervous system in controlling movement. Hatze (1976) developed a model of the musculoskeletal system to study various activities such as a long jump. The input to the model was the neural excitation rate to the muscles estimated from surface muscle electromyographic signals. Muscles were modelled using Hill's muscle model (Hill, 1938), which allowed the conversion from neural excitation input to muscle force output. Hatze then proceeded with the mechanical musculoskeletal model and the solution of the governing equations of motion to obtain the resulting kinematics. Other investigators utilized simplified models of the human body and used the direct dynamic approach to perform simulations of human locomotion. The comparison of measured ground reaction forces with ground reaction force data obtained through the simulation was often used to validate the model (Beckett and Chang, 1968; Chow and Jackobson, 1971; Siegler, Seliktar and Hyman, 1982; Pandy and Berme, 1989a,b). The complexity, inaccuracies and time-consuming nature of the direct dynamics approach prevent it from being used routinely in the gait laboratory.

The inverse dynamic approach (Figure 2) is being used routinely to analyse both healthy and pathological gait. This analysis is directed at improving our understanding of the mechanisms involved in motor control of human locomotion, assisting in identifying criteria for design of artificial joints and prosthetic devices, and assisting in the diagnosis of the underlying pathologies which result in an observed abnormal gait pattern. In this approach, measured kinematic data are combined with estimated segmental inertial properties and ground reaction force data to estimate the resultant moment and resultant force acting at each major joint in a model of the subject's body. These data are often further processed to derive the muscle forces and joint reaction forces. The validity of these predictions is often determined by comparing the muscle activation

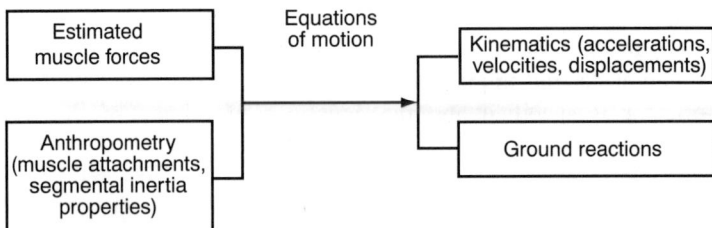

Figure 1. Flow chart diagram for the direct dynamics approach

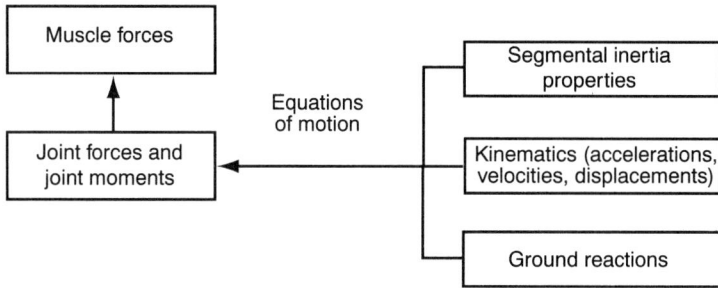

Figure 2. Flow chart diagram for the inverse dynamics approach

patterns as determined through the dynamic analysis to those determined through direct measurement of muscle electromyographic activity.

The necessary kinematic data are obtained by measuring the displacements of body segments using a motion analysis system (video-based, optoelectronic, etc.). The displacement data are then differentiated to derive the velocities and accelerations of the segments. Alternatively, data from accelerometers mounted on body segments can be used. The acceleration data must then be integrated to obtain the required velocities and displacements. Details of the kinematic analysis are provided in Chapter 9. Segmental inertial properties can be obtained either through direct measurements on the subject or through regression equations based on a number of anthropometric measurements, including body weight, stature and specific geometric dimensions of individual segments. The ground reaction forces can be measured by force plates or alternatively estimated from the kinematic data.

Once a solution for the joint moments and joint forces is obtained, these data can be used to estimate the muscle forces and joint reaction forces. This estimation problem represents one of the most difficult problems in the analysis of human locomotion. It arises due to the large number of muscles spanning each joint in the human body. Since more than one of these muscles can be contracting at a time, the estimation of forces developed by individual muscles is an indeterminate problem. Techniques to solve this problem were based either on grouping muscles with similar function and thus eliminating redundancy (Paul, 1966; Morrison, 1970; Procter and Paul, 1982; Nissan, 1981; Collins, 1994, 1995) or applying optimization criteria to resolve the muscle force distribution (Seireg and Arvikar, 1973; Hardt, 1978; Crownin-shield and Brand, 1981; Vaughan, Hay and Andrews, 1982b; Pederson et al., 1987; Davy and Audo, 1987; Yamaguchi and Zajac, 1990). None of these techniques was found to be fully satisfactory. The first approach leads to oversimplification, while the second suffers from lack of reliable validation procedures.

MODELS OF THE HUMAN BODY FOR THE INVERSE DYNAMICS APPROACH

Selecting a mathematical model is fundamental to the analysis of human locomotion. In all previous studies, the body was modelled as a system of interconnected rigid links. The rigidity assumption, although an obvious idealization (consider, for example, the deformation of the foot during the stance phase of gait), is necessary for the dynamic analysis which assumes rigid bodies. Further characterization of the model requires specifying whether the model is two- or three-dimensional, the number of interconnecting links representing the body, and the number and type of kinematic constraints present at each joint. In general, the model selected should have the simplest possible structure which will properly address the questions of the study. In practice this is often a difficult task requiring careful consideration of issues such as experimental, numerical and analytical complexity and realistic representation of the phenomenon to be investigated. As our knowledge of the properties of the musculoskeletal system expands, the models used to represent this expanded knowledge tend to be more complex and more realistic.

Human locomotion is a three-dimensional (3D) phenomenon occurring simultaneously in the sagittal, coronal and transverse planes. Earlier studies of human locomotion relied on 2D sagittal plane analysis (Cappozzo, Leo and Pedotti, 1975; Davy and Audo, 1987). A fundamental hypothesis in this 2D approach is that the important mechanical events in gait occur in the sagittal plane and that events out of the sagittal plane do not significantly affect the dynamics of the body in the sagittal plane.

Many pathologies which affect gait performance are primarily out-of-sagittal-plane phenomena. Examples include pathologies which affect balance and stability, soft-tissue injuries such as ligamentous injuries which occur in athletic activities and central nervous system damage which may affect the symmetry of gait performance. Therefore, in order to provide the clinical community with a useful diagnostic tool based on gait analysis, the analysis needed to be expanded to become truly 3D. This in turn necessitated the development of 3D models of the human body (Seireg and Arvikar, 1975; Crowninshield and Brand, 1981; Zarrugh, 1981; Yamaguchi and Zajac, 1990; Vaughan, Hay and Andrews, 1982a) and the improvement of the experimental techniques to measure the kinematics and kinetics of the body in 3D space.

Three-dimensional models of various complexity were used in the past. For example, Hatze (1977) used a 17-segment model to analyse the dynamics of the long jump and other activities. Vaughan, Hay and Andrews (1982a) used a 14-segment model to analyse the dynamics of various athletic activities. However, most of the studies of the 3D dynamics of human locomotion relied on a seven-segment model (Figure 3). In this model the head, neck, arms and torso were combined into one rigid segment. Several studies have shown that such a

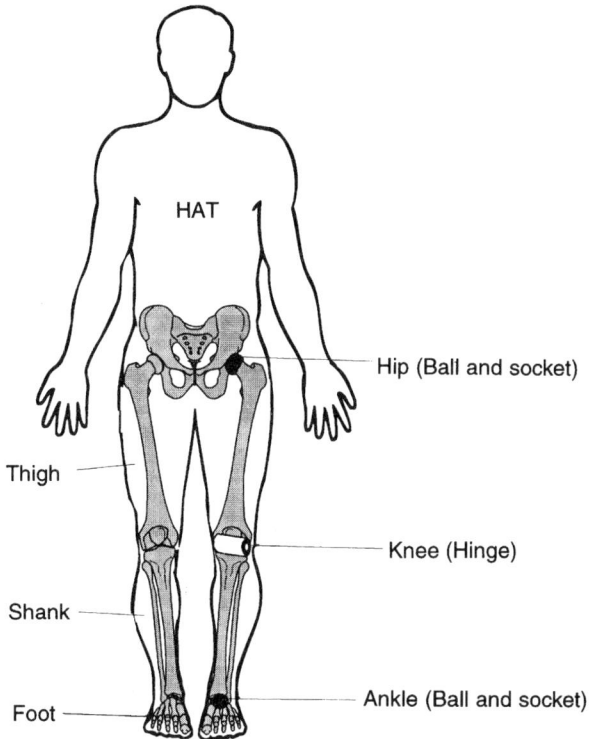

Figure 3. Seven-segment model of the human body commonly used for inverse dynamic solutions. The head, neck, arms and torso were combined into one rigid segment (HAT). The thighs, shanks and feet were represented by six other rigid segments

combination results in negligible errors when calculating parameters such as joint moments and mechanical energy of body segments (Cappozzo, 1981; Townsend, 1981; Stokes, Andersson and Forssberg, 1989). The thighs, shanks and feet were represented by six other rigid links (Seireg and Arvikar, 1975; Zarrugh, 1981; Yamaguchi and Zajac, 1990; Vaughan, Hay and Andrews, 1992).

In the seven-link model a total of six joints must be specified. These include the hip, knee and ankle of each lower limb. The number of degrees of freedom of the model depends on the number of kinematic constraints imposed on each joint. When using the Newton–Euler formulation based on successive free body diagrams of individual segments, it is not necessary to impose *a priori* constraints on the joints (Cappozzo, Leo and Pedotti, 1975; Hardt and Mann, 1980; Zarrugh, 1981; Crowninshield and Brand, 1981; Vaughan, Davis and O'Connor, 1992). All components of the intersegmental loading were included in the motion equations. However, when using other dynamic formulations, such as the Lagrange formulation, the joint kinematic constraints must be specified.

The hip joint was almost always considered to be a ball and socket joint with three rotational degrees of freedom. However, different kinematic constraints were used to model the knee. It was modelled as a hinge joint (Cappozzo, Leo and Pedotti, 1975; Hardt, 1978; Pederson et al., 1987), a ball and socket joint (Apkarian, Naumann and Cairns, 1989), or as a four-bar cruciate linkage in which the anterior cruciate ligament and the posterior cruciate ligaments were represented as rigid links in a four-bar mechanism (Collins, 1995). At the present time it is difficult to recommend a best knee model for the 3D analysis. However, it is recommended to choose the knee model based on the specific problems addressed in the study.

The ankle joint was modelled as having a single rotational degree of freedom (Crowninshield and Brand, 1981; Pandy and Berme, 1988a; Yamaguchi and Zajac, 1990), two rotational degrees of freedom (Procter and Paul, 1982; Rohrle et al., 1984; Pederson et al., 1987) and three rotational degrees of freedom: a ball and socket joint (Hardt, 1978; Apkarian, Naumann and Cairns, 1989). It is generally agreed that the single rotational degree of freedom may be adequate for sagittal plane motion analysis but is inadequate for a full 3D analysis. There is still controversy over whether the ankle joint complex can be represented by two rotational degrees of freedom or whether three rotational degrees of freedom are required (Siegler, Chem and Schneck, 1988; van den Bogert, Smith and Nigg, 1994). Until the controversy is resolved, we recommend the use of the three-degrees-of-freedom model, which is not only more general but also avoids the complexity of determining the location and orientation of the two axes in the two-degrees-of-freedom model (van den Bogert, Smith and Nigg, 1994).

The most common seven-link 3D model, used for the inverse dynamic analysis, had a ball and socket joint for the hip, a hinge joint for the knee and a ball and socket joint for the ankle. This reduces the total number of degrees of freedom of the seven-link model to a total of 14. The joint constraints can be used to reduce the number of equations of motion (Davy and Audo, 1987; Pandy and Berme, 1989a,b). Alternatively, the equations of motion of the model can be developed using n generalized coordinates, where n represents the total number of degrees of freedom of the system. This approach is routinely used in the Lagrange or Kane approach (Yamaguchi and Zajac, 1990; Kuo, 1995).

BODY SEGMENT INERTIAL PARAMETERS

After establishing a mathematical model to perform a dynamic analysis of human movement, and after the segmentation process has been completed, the inertial properties of each segment must be estimated. These inertial properties include the mass of the segment, its length, the location of its centre of mass with respect to a local coordinate frame embedded in the segment, the orientation of the principal moments of inertia and the values of the principal moments of inertia passing through the segment's centre of mass.

A number of methods were used in the past to obtain direct estimates of the inertial properties of body segments. These methods are briefly described below.

DIRECT MEASUREMENTS ON CADAVERS

All the inertial parameters of body segments can be measured directly from dismembered cadavers (Braune and Fischer, 1889; Dempster, 1955; Clauser, McConville and Young, 1969; Chandler et al., 1975). Segmental volume can be assessed using the water immersion technique. Average density can then be calculated by dividing the segment's mass by its volume. Specific densities of tissues such as bone, skin and muscle in a given segment can be determined with great accuracy (Dempster, 1955). Centre of mass can be determined using either suspension, balancing or reaction board techniques. Moments of inertia about the centre of mass were measured using oscillation techniques. In all but the work of Chandler et al. (1975), symmetry of the body segments about their long axes was assumed. This simplified the determination of the principal moments of inertia, since the principal directions then coincide with the long axis of the segment and the medial–lateral and anterior–posterior axes.

There are a number of severe limitations when one attempts to use these data in order to estimate the inertial parameters of a specific individual. These include limited sample size, age discrepancy (most of the cadaveric studies were conducted on old individuals) and morphological discrepancy. Despite these limitations, the use of these data is widespread. As will be discussed later, some investigators have used density values obtained from cadavers to supplement their geometrical measurements, while others used statistical regression equations to adopt the data to a specific subject.

DIRECT MEASUREMENTS ON LIVING SUBJECTS

The limitations imposed by the cadaveric measurement motivated the development of techniques to measure the inertial properties of body segments on living subjects.

Mass and density of individual segments can be determined directly using a number of techniques, including the reaction board technique (Bernstein, 1967) and various imaging techniques such as magnetic resonance imaging (Martin et al., 1989; Mungiole and Martin, 1990) or computerized tomography (Brooks and Jacobs, 1975; Huang and Wu, 1976; Zatsiorsky and Seluyanov, 1983, 1985). Likewise, the location of a segment's centre of mass can be determined using the techniques indicated above. It should be pointed out that the reaction board technique can be used to estimate either the location of the centre of mass or the segment's mass but cannot be used to simultaneously estimate both parameters on one individual. Principal moments of inertia can be determined using either one of the imaging techniques described above or quick release experiments (Bresler and Frankel, 1950; Cavanagh and Gregor, 1974; Hatze, 1975). This latest technique is subject to inaccuracies related to muscle activation during the test.

The various imaging techniques described above can provide very accurate *in vivo* estimates of the complete set of segmental inertial properties. However, exposure to radiation, high cost and lengthy procedures preclude, at least at present, these techniques from being used on a routine subject-specific basis. However, such techniques can provide reference data based on which specific inertial properties can be derived through statistical regression equations. The other direct techniques indicated in this section are time-consuming, incomplete and of questionable accuracy.

INDIRECT ASSESSMENT OF INERTIAL PROPERTIES

The major difficulties in assessing all the inertial properties on individual subjects motivated the development of the indirect approach. This approach is based on a statistical regression analysis which correlates anthropometric measurement from the individual subject with values of inertial properties which were previously obtained from direct measurements performed on populations of cadavers or living subjects. The earlier regression equations were based on one or two independent anthropometric variables such as body mass and stature (Braune and Fischer, 1889; Dempster, 1955). However, these do not take into account segment variations between individuals leading to gross inaccuracies. To improve the estimates of segmental inertial properties and account for variations in segmental properties, multi-variable linear regression equations were introduced (Clauser, McConville and Young, 1969; Hinrich, 1985). Hinrich (1985) based the regression equations on the cadaveric data of Chandler et al. (1975) obtained from six cadavers. Although this technique provided significant improvement over the ones based on weight and stature only, it had one severe limitation. This limitation was related to the fact that linear equations expressing relationships between dimensionally distinct properties are theoretically questionable.

Nonlinear, dimensionally consistent regression equations based on geometrical simplifications were proposed by Yeadon (1989), Vaughan, Davis and O'Connor (1992) and Zatsiorsky and Seluyanov (1985). The regression equations of both Yeadon (1989) and Vaughan, Davis and O'Connor (1992) were based on the cadaveric data of Chandler et al. (1975) and suffered from low reliability due to very limited sample size. Furthermore, extensive verification of these regression equations was not conducted. Nevertheless, they appear to make the best use of the limited cadaveric data presently available.

A similar nonlinear technique based on data obtained from 100 living subjects was presented by Zatsiorsky, Seluyanov and Chugunova (1990). In his method the body was divided into 16 segments. For the nonlinear regression all segments were considered to be cylindrical. The regression equations proposed were:

$$m_i = k_i L_i C_i^2$$

$$I_{si} = K_{si} m_i L_i^2$$

$$I_{fi} = K_{fi} m_i L_i^2$$

$$I_{li} = K_{li} m_i C_i^2$$

where m_i = segment's mass, L_i = length of segment, C_i = circumference of segment, and s, f, and l are sagittal, frontal and longitudinal directions, respectively.

These nonlinear relationships were similar to the ones proposed by Yeadon (1989). It was indicated that the above regression equations can be used for subjects whose statures are different from those of the population from which the regression equations were developed. However, no such statistical validation results were presented and the level of accuracy for the extrapolated value was not included. Nevertheless, these regression equations and the data on which they are based provide the most comprehensive and reliable data set available today. Recently, an adjustment to Zatsiorsky's data has been proposed by de Leva (1997) to reference them to the joint centre.

MATHEMATICAL MODELLING

In this approach (Hanavan, 1964; Jensen, 1978; Hatze, 1980) body segments were modelled by simple geometric shapes such as circular ellipsoids and elliptical cylinders. Each shape was assumed to have a uniform density with values adopted from cadaver data. The latest can be improved by using density values obtained from a large population of living subjects (Zatsiorsky and Seluyanov, 1983, 1985).

Jensen (1978) used the method developed earlier by Weinbach (1938) and divided each segment of the 16 body segments into 2-cm-thick elliptical zones. The dimensions of the elliptical zones were obtained from manually digitized photographs from frontal and sagittal views. The technique was used to estimate the segmental inertial properties of growing children (age range 4–15). The accuracy of the method was determined by comparing total body mass measured directly with the sum of all segment masses. High accuracy in the range of 2% was reported. The time-consuming nature of the photographic digitizing process motivated Sarfaty and Ladin (1993) to develop an automated version of Weinbach (1938) and Jensen's (1978) technique based on video image capturing and computer image processing.

Inverse dynamic analysis is often needed in the analysis of pathological gait. For these subjects, gross asymmetries and gross musculoskeletal deformities may be present. Under such conditions, the indirect approach based on regression equations derived from healthy subjects with assumed bilateral symmetry may not be adequate. Therefore, direct techniques such as the ones based on mathematical modelling may be essential.

The brief review provided above demonstrates that refinement and improved accuracy of inertial properties of body segments is often associated with increased time, inconvenience and health hazards. Therefore, it is necessary to address the issue of what level of accuracy is really needed when analysing the dynamics of various functional activities such as walking and running. This problem has only partially been addressed in the past (Cappozzo, 1983; Jensen, 1989) and needs to be further explored. Specifically, the presence of the inertial properties value in the various components contributing to the joint forces and joint moments can be established analytically (e.g. gravitational and inertial forces). A sensitivity error analysis should then be performed to establish the effect of assumed errors in the inertial properties. This error analysis should be conducted for various activities such as walking and running. Such studies would provide an insight into how much accuracy in estimating inertial properties is actually essential.

EQUATIONS OF MOTION AND JOINT LOADS

Deriving the resultant joint forces and moments requires combining information on the segment's kinematics, inertial properties and ground reaction forces into the equations of motion. Specifically, the data required include: the linear velocity and acceleration of the centre of mass of each segment relative to a global inertial reference frame; the angular position, velocity and acceleration of each segment relative to a global inertial reference; the location of the centre of mass, the mass and the moments of inertia of body segments relative to the segment's centre of mass; and the ground reaction forces and vertical moment acting on each foot.

Kinematic data required for the solution of the equations of motion can be derived either through differentiation of displacement data obtained through a motion analysis system or through integration of acceleration data obtained through accelerometers placed on body segments. Minimization of errors for each of these techniques is very important. The first requires proper smoothing of the displacement data prior to numerical differentiation. The second requires accurate estimation of initial conditions.

Ground reaction forces are required not only to assist in solving the equations of motion and for improving the estimates of resultant moments and forces at the joints, but also to provide information to assist with the classification of abnormal gait (Siegler and Carr, 1990). Most gait laboratories are equipped with force plates to measure the ground reaction loads simultaneously with the kinematic data. Information obtained from each force plate includes one vertical and two horizontal shear forces, the moment about a vertical axis, and the resultant point of application of the ground reaction force vector (centre of pressure).

Ground reaction loads are a reaction to the inertial forces and gravitational forces of the body transmitted through the feet. These loads need not be measured during the single limb support phase of gait, since they can be obtained directly from the equations of motion. This fact was used by Bobert, Schamhardt and Nigg (1991) to derive the ground reaction force during running, an activity without a double limb support. The major advantage of this is that estimation of ground reaction force is possible in outdoor situations relying merely on kinematic data. However, during double limb support the problem of determining the ground reaction force under each foot becomes an indeterminate problem which requires special *a priori* assumptions regarding the transfer of load from one limb to the other (McGhee et al., 1976; Pandy and Berme, 1988a). Although the ground reaction loads can be estimated from the model, the measurement of these loads is preferred, whenever possible, because they provide improved estimates of joint loads (Kuo, 1995) and are not influenced by kinematic errors, or by errors in estimating the body's inertial properties.

The equations of motion can be derived from the Newton–Euler formula applied to each body segment. The first of these equations is the translational Newton's law, which states that the sum of all forces acting on a rigid body is equal to its mass times the acceleration of its centre of mass. The second of these equations, Euler's equation, states that the moment of all forces acting on a rigid body about its centre of mass is equal to the time rate of change of the angular momentum of the rigid body about its centre of mass.

A common method for solving the equations of motion is to write them for free body diagrams of each body segment, starting with the most distal segment and proceeding to the most proximal one (Hardt and Mann, 1980; Vaughan, Davis and O'Connor, 1992).

As can be seen from Figure 4, segment i has two connections; one with a distal link at joint i and one with a proximal link at joint $i + 1$. The body-fixed, local coordinate system is located at the centre of mass and aligned with the principal axes of inertia of the segment. From Newton's law, the translational dynamic equation of equilibrium of the segment is given by

$$m_i \mathbf{a}_i = m_i \mathbf{g} + \mathbf{F}_i + \mathbf{F}_{i+1} \tag{1}$$

where m_i is the mass of the segment, $\mathbf{a}_i$ is the translational acceleration vector of the centre of the mass relative to a global inertia reference, $m_i \mathbf{g}$ is the weight vector, $\mathbf{F}_i$ is the resultant joint force on the joint i, or the ground reaction force on the foot segment, and $\mathbf{F}_{i+1}$ is the resultant force on joint $i + 1$.

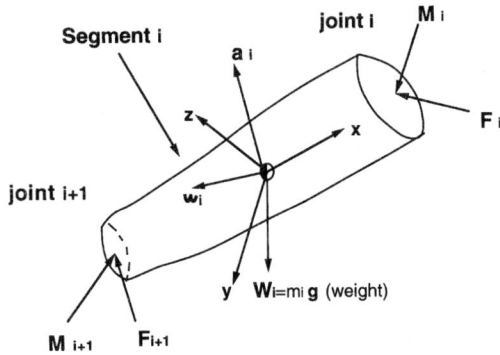

Figure 4. Free body diagram of body segment. The body-fixed, local coordinate system is located at the centre of mass and aligned with the principal axes of inertia of the segment

From Euler's equation, the rotational dynamic equation of equilibrium for the segment about the segment's mass centre will be given by

$$d/dt(l_{cm}\omega_i) = \mathbf{M}_i + \mathbf{M}_{i+1} + \mathbf{r}_i \times \mathbf{F}_i + \mathbf{r}_i \times \mathbf{F}_{i+1} \tag{2}$$

where l_{cm} is the inertia tensor about the centre of mass, ω_i is the angular velocity vector of the segment, $\mathbf{M}_i$ is the resultant moment at the joint i, or ground reaction moment at the foot, and $\mathbf{M}_{i+1}$ is the resultant moment at joint $i + 1$. $\mathbf{r}_i$ and $\mathbf{r}_{i+1}$ are the moment arms of force vectors about the centre of mass. At the foot segment, the only unknowns are the resultant force and moment at the ankle joint, which can be resolved from equations (1) and (2) above.

Alternatively, assuming that all the joints in the model are rotary joints, the system's equation can be written as (Pandy and Berme, 1988a; Apkarian, Naumann and Cairns, 1989)

$$\mathbf{T} = \mathbf{D}(\theta)\ddot{\theta} + \mathbf{H}(\theta, \dot{\theta}) + \mathbf{G}(\theta) + \mathbf{T}_{fr}$$

In the above equation, $\mathbf{T}$ is a $n \times 1$ vector of joint moments, $\mathbf{D}(\theta)$ is the $n \times n$ inertia matrix, $\mathbf{H}(\theta, \dot{\theta})$ is the $n \times 1$ vector of coriolis and centrifugal terms, $\mathbf{G}(\theta)$ is the $n \times 1$ vector of gravitational terms, and $\mathbf{T}_{fr}$ is the $n \times 1$ vector of moments due to ground reactions. An efficient algorithm that has been used to solve the above equation for the joint moments is the recursive Newton–Euler algorithm (Pandy and Berme, 1988a; Apkarian, Naumann and Cairns, 1989).

Other methods have been used to derive and solve the equations of motion, including the Lagrange method and Kane's method (Yamaguchi and Zajac, 1990; Zajac and Winters, 1990; Kuo, 1995). One interesting point to note when examining the equations of motion for the dynamics of human locomotion is that the amount of data commonly available in the gait laboratory, including segment kinematic data and ground reactions, exceeds the amount of data required to calculate the joint moments. Therefore, an opportunity exists to optimize the prediction of joint moments using the entire data set as input into an optimization algorithm. Such an optimization scheme (Kuo, 1995) can be used to minimize the effect of measurement errors as well as to minimize the effect of errors related to numerical differentiation of kinematic data, errors in inertial properties estimation, etc.

The inverse dynamic approach and the solution of the equations of motion for the joint loads are solved routinely in the gait laboratory. Despite their commonality, very few studies performed systematic analysis to identify the contribution of the various terms such as gravitational forces, inertial forces, coriolis and centrifugal forces to the resultant joint forces and joint moments developed about various joints (Cappozzo, 1983; Siegler, Seliktar and Hyman, 1982). In addition, few investigators have evaluated the effects of errors in various parameters, such as error in inertial parameter estimations, errors due to differentiation of position data, etc., on the resultant joint forces and joint moments (Cappozzo, 1983; Jensen, 1989).

The resultant joint loads obtained from inverse dynamics are 3D vectors defined in a global reference coordinate system. In this frame, it is difficult to interpret the results relative to the subject's joints, particularly if her or his direction of progression deviated from the global, laboratory-fixed coordinate frame. It is therefore a common practice to transform the joint moments and joint forces to a body-fixed, anatomical reference frame. A common and clinically accepted joint coordinate system, which has recently emerged as a proposed standard for use by the biomechanics community, was first presented by Grood and Suntay (1983) for the knee joint and later adopted by other investigators (Siegler, Chen and Schneck, 1988; Apkarian, Naumann and Cairns 1989; Vaughan, Davis and O'Connor, 1992) for other joints. Using this reference frame, the resultant moment is projected onto the three axes of the joint reference frame. One axis is fixed in one segment, the second axis is fixed in the second segment articulating at the joint, and the third axis is mutually perpendicular to the other two axes. An example of this joint reference frame for the ankle joint complex is shown in Figure 5 (Siegler et al., 1996).

The internal joint moments, calculated through the inverse dynamic approach, contribute greatly to the analysis of normal and pathological gait, since these data are the key to the determination of muscle forces, function of the passive joint structures such as ligaments, determination of joint power, and determination of bone-to-bone reaction forces.

The internal moment developed about a joint is the result of contributions from active contraction of muscles crossing the joint and from the passive resistance provided by ligaments, the joint capsule and other soft tissues. Siegler, Moskowitz and Freedman (1984) developed a technique to separate the contribution of the passive structures and the contribution of muscle contraction to the net internal moment developed about the ankle joint during level walking. In that

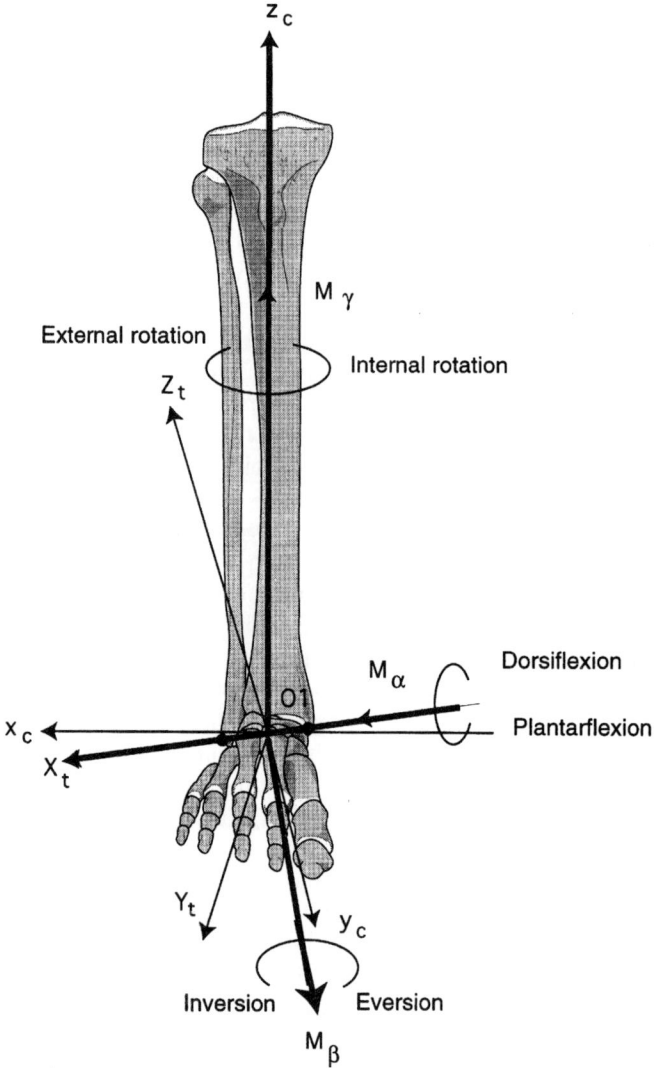

Figure 5. Definition of the joint reference frame for the ankle complex. The $X_t Y_t Z_t$ frame is a frame fixed to the tibia. The $x_c y_c z_c$ frame is a frame fixed to the calcaneus. The joint coordinate system is formed by the X_t-axis, the z_c-axis and their common perpendicular. The components of the resultant moment in the joint coordinate system are M_α, M_β, M_γ

study it was shown that the contribution of the passive structures in the sagittal plane during normal level walking to the net moment about the ankle was minimal (smaller than 6%). However, during pathological gait, contractures and other joint deformities may lead to very large contributions from the passive structures. Therefore, the net internal moment and its active and passive components can provide sensitive measures for examining pathological gait and the effects of various treatment modalities.

The internal moment developed about the joints during level walking is also the key to determining the forces developed by individual muscles. Determination of these forces provides a means to study motor control mechanisms in gait, the effects of various pathologies and the effectiveness of treatment interventions, and determination of the bone-to-bone reaction forces. It was for these and other reasons that many investigators have invested great efforts in developing techniques to estimate muscle forces from the internal moments calculated through the inverse dynamic approach. The obvious obstacle to achieving this goal was pointed out by many investigators. This is the fact that the system is redundant, since many muscles cross each joint. To overcome this problem, investigators used either a grouping technique (Morrison, 1970; Procter and Paul, 1982; Nissan, 1981; Collins, 1994, 1995) or optimization procedures (Seireg and Arvikar, 1973; Hardt, 1978; Crowninshield and Brand, 1981; Vaughan, Hay and Andrews, 1982b; Pederson et al., 1987; Yamaguchi and Zajac, 1990). However, a less obvious problem which was overlooked by many, was the error associated with attributing the entire joint moment to active muscle contraction and therefore neglecting contributions from the passive joint structures. The errors associated with such simplification may be acceptably small for 2D, sagittal plane gait analysis of healthy individuals. However, neglecting the contribution of the passive structure in a 3D gait analysis scheme, or when examining pathological gait, may lead to gross errors in predicting muscle forces.

REFERENCES

Apkarian, J., Naumann, S. and Cairns, B. (1989) A three-dimensional kinematic and dynamic model of the lower limb. *J. Biomech.*, **22**, 143–155.

Beckett, R. and Chang, K. (1968) An evaluation of the kinematics of gait by minimum energy. *J. Biomech.*, **1**, 147–159.

Bernstein, N. (1967) *The Coordination and Regulation of Movements.* Pergamon Press, Oxford, pp. 9–14.

Bobert, M. F., Schamhardt, H. C, and Nigg, B. M. (1991) Calculation of vertical ground reaction force estimates during running from positional data. *J. Biomech.*, **24**, 1095–1105.

Braune, W. and Fischer, O. (1889) Uber den Schwerpunkt des menschlichen Korpers [On the center of gravity of the human body]. *Abhandlungen der Mathematisch - Physische Klasse der Konigl. Sachsischen Gesellschaft Wissenschaft*, **15**, 561–572.

Bresler, B. and Frankel, J. B. (1950) The forces and moments in the leg during human walking. *Trans. ASME*, **72**, 27–36.

Brooks, C. B. and Jacobs, A. M. (1975) The gamma mass scanning technique for inertial anthropometric measurement. *Med. Sci. Sports*, **7**, 290–294.

Cappozzo, A. (1981) Analysis of the linear displacement of the head and trunk during walking at different speeds. *J. Biomech.*, **14**, 411–425.

Cappozzo, A. (1983) The forces and couples in the human trunk during level walking. *J. Biomech.*, **16**, 265–277.

Cappozzo, A., Leo, T. and Pedotti, A. (1975) A general computing method for the analysis of human locomotion. *J. Biomech.*, **8**, 307–320.

Cavanagh, P. R. and Gregor, R. J. (1974) The quick-release method for estimating the moment of inertia of the shank and foot. In: *Biomechanics IV* (ed. R. C. Nelson and C. A. Morehouse) University Park Press, Baltimore, pp. 524–530.

Chandler, R. F., Clauser, C. E., McConville, J. T., Reynolds, H. M. and Young, J. W. (1975) Investigation of inertial properties of the human body. *AMRL Tech. Rep.* 74–137. NTIS No. Ab-A016 485. Wright-Patterson Air Force Base, Dayton, Ohio.

Chandler, R. F., Snow, C. C. and Young, J. W. (1978) Computation of mass distribution characteristics of children. In: *Proceedings of Society of Photo-optical Instrumentation Engineers*, Paris.

Chow, C. K. and Jackobson, D. H. (1971) Studies of human locomotion via optimal programming. *Math. Biosci.*, **10**, 239–306.

Clauser, C. E., McConville, J. T. and Young, J. W. (1969) Weight, volume and center of mass of segments of the human body. *AMRL Tech. Rep.* 60–70. Wright-Patterson Air Force Base, Dayton, Ohio.

Collins, J. J. (1994) Antagonistic–synergistic muscle action at the knee during competitive weightlifting. *Med. Biol. Eng. Comput.*, **32**, 168–174.

Collins, J. J. (1995) The redundant nature of locomotor optimization laws. *J. Biomech.*, **28**, 251–267.

Crowninshield, R. D. and Brand, R. A. (1981) A physiologically based criterion of muscle force prediction in locomotion. *J. Biomech.*, **14**, 793–801.

Davy, D. T. and Audo, M. L. (1987) A dynamic optimization technique for predicting muscle forces in the swing phase of gait. *J. Biomech.*, **20**, 187–202.

de Leva, P. (1997) Adjustments to Zatsiorsky–Seluyanov's segment inertia parameters. *J. Biomech.*, in press.

Dempster, W. T. (1955) Space requirements of the seated operator: geometrical, kinematic and mechanical aspects of the body with special reference to the limbs. Wright Air Development Center, *AMRL Tech. Rep.* No. 55–159, Wright Patterson Air Force Base, WADC, Dayton, Ohio. NTIS No. AD-087892.

Grood, E. S. and Suntay, W. J. (1983) A joint coordinate system for the clinical description of three-dimensional motion: application to the knee. *J.Biomech. Eng.*, **105**, 136–144.

Hanavan, E. P. (1964) A mathematical model of the human body. *AMRL Tech. Rep.* 64–102. Behavioral Sciences Laboratory, Ohio.

Hardt, D. E. (1978) Determining muscle forces in the leg during normal human walking—an application and evaluation of optimization methods. *ASME Trans. J. Biomech. Eng.*, **100**, 72–78.

Hardt, D. E. and Mann, R. W. (1980) A five body–three dimensional dynamic analysis of walking. *J. Biomech.*, **13**, 455–457.

Hatze, H. (1975) A new method for the simultaneous measurement of the moment of inertia, the damping coefficient, and the location of the center of mass of a body segment in situ. *Eur. J. Appl. Physiol.*, **34**, 217–266.

Hatze, H. (1976) The complete optimization of a human motion. *Math. Biosci.*, **28**, 99–135.

Hatze, H. (1977) A complete set of control equations for the musculo-skeletal system. *J. Biomech.*, **10**, 799–805.

Hatze, H. (1980) A mathematical model for the computational determination of parameter values of anthropometric segments. *J. Biomech.*, **13**, 833–843.

Hill, A. V. (1938) The heat of shortening and the dynamic constants of muscle. *Proc. R. Soc.*, **126B**, 136–196.

Hinrich, R. N. (1985) Regression equations to predict segmental moments of inertia from anthropometric measurements: an extension of the data of Chandler et al. (1975). *J. Biomech.*, **18**, 621–624.

Huang, H. K. and Wu, S. C. (1976) The evaluation of mass densities of the human body in vivo from CT scans. *Comput. Biol. Med.*, **6**, 337–343.

Jensen, R. K. (1978) Estimation of biomechanical properties of three body types using photogrammetric method. *J. Biomech.*, **11**, 349–358.

Jensen, R. K. (1989) Changes in segment inertia proportions between 4 and 20 years. *J. Biomech.*, **22**, 529–536.

Kuo, A. D. (1995) A simple method for improving precision of inverse dynamics computations. In: *Nineteenth Annual Meeting of the American Society of Biomechanics*, Stanford, California.

Martin, P. E., Mungiole, M., Marzke, M. W. and Longhill, L. M. (1989) The use of magnetic resonance imaging for measuring segment inertial properties. *J. Biomech.*, **22**, 367–376.

McGhee, R. B., Koozekanni, S. H., Gupta, S. and Cheng, T. S. (1976) Automatic estimation of joint forces and moments in human locomotion from television data. In: *Proceedings of the IVIFAC Symposium on Identification and Parameter Estimation*, USSR.

Morrison, J. B. (1970) The forces transmitted by the human knee joint during activity. Unpublished PhD thesis, University of Strathclyde.

Mungiole, M. and Martin, P. E. (1990) Estimating segment inertial properties: comparison of magnetic resonance imaging with existing methods. *J. Biomech.*, **23**, 1039–1046.

Nissan, M. (1981) The use of permutation approach in the solution of joint biomechanics: the knee. *Eng. Med.*, **10**, 39–43.

Pandy, M. G. and Berme, N. (1988a) A numerical method for simulating the dynamics of human walking. *J. Biomech.*, **21** 1043–1051.

Pandy, M. G. and Berme, N. (1988b) Synthesis of human walking: a planar model for single support. *J. Biomech.*, **21**, 1053–1060.

Pandy, M. G. and Berme, N. (1989a) Quantitative assessment of gait determinants during single stance via a three-dimensional model—part 1. Normal gait. *J. Biomech.*, **22**, 717–724.

Pandy, M. G. and Berme, N. (1989b) Quantitative assessment of gait determinants during single stance via a three-dimensional model–part 2. Pathological gait. *J. Biomech.*, **22**, 725–733.

Paul, J. P. (1966) Forces transmitted by joints in the human body. *Proc. Inst. Mech. Eng.*, **181**(3F), 8.

Pederson, D. R., Brand, R. A., Cheng, C. and Arora, J. S. (1987) Direct comparison of muscle force predictions using linear and nonlinear programming. *ASME Trans. J. Biomech. Eng.*, **109**, 192–199.

Procter, P. and Paul, J. P. (1982) Ankle joint biomechanics. *J. Biomech.*, **15**, 627–634.

Rohrle, H., Scholten, R., Sigolotto, C., Sollbach, W. and Kellner, H. (1984) Joint forces in the human pelvis–leg skeleton during walking. *J. Biomech.*, **17**, 409–424.

Sarfaty, O. and Ladin, Z. (1993) A video-based system for the estimation of the inertial properties of body segments. *J. Biomech.*, **26**, 1011–1016.

Seireg, A. and Arvikar, R. J. (1973) A mathematical model for evaluation of forces in lower extremities of the musculo-skeletal system. *J. Biomech.*, **6**, 313–326.

Seireg, A. and Arvikar, R. J. (1975) The prediction of muscular load sharing and joint forces in the lower extremities during walking. *J. Biomech.*, **8**, 89–102.

Siegler, S., Lapointe, S., Nobilini, R. and Berman, A. T. (1996) A six-degrees-of-freedom instrumented linkage for measuring the flexibility characteristics of the ankle joint complex. *J. Biomech.*, **29**(7), 943–947.

Siegler, S. and Carr, E. (1990) A method for discrimination between spasticity and contracture at the ankle joint during locomotion in neurological impaired patients. In: N. a. C. Berme, A, *Biomechanics of Human Movement: Applications in Rehabilitation, Sports and Ergonomics* (ed. Berme) Bertec Corporation, Ohio.

Siegler, S., Chen, J. and Schneck, C. D. (1988) The three-dimensional kinematics and flexibility characteristics of the human ankle and subtalar joint—Part 1: Kinematics. *J. Biomech. Eng.*, **110**, 364–373.

Siegler, S., Moskowitz, G. D. and Freedman, W. (1984) Passive and active components of the internal moment developed about the ankle joint during human ambulation. *J. Biomech.*, **17**, 647–652.

Siegler, S., Seliktar, R. and Hyman, W. (1982) Simulation of human gait with the aid of a simple mechanical model. *J. Biomech.*, **15**, 415–425.

Stokes, V. P., Andersson, C. and Forssberg, H. (1989) Rotational and translational movement features of the pelvis and thorax during adult human locomotion. *J. Biomech.*, **22**, 43–50.

Townsend, M. A. (1981) Dynamics and coordination of torso motions in human locomotion. *J. Biomech.*, **14**, 727–738.

van den Bogert, A. J., Smith, G. D. and Nigg, B. M. (1994) In vivo determination of the anatomical axes of the ankle joint complex: an optimization approach. *J. Biomech.*, **27**, 1477–1488.

Vaughan, C. L., Hay, J. G. and Andrews, J. G. (1982a) Closed loop problems in biomechanics. Part I—A classification system. *J. Biomech.*, **15**(3), 197–200.

Vaughan, C. L., Hay, J. G. and Andrews, J. G. (1982b) Closed loop problems in biomechanics. Part II—An optimization approach. *J. Biomech.*, **15**(3), 201–210.

Vaughan, C. L., Davis, B. L. and O'Connor, J. C. (1992) *Dynamics of Human Gait.* Human Kinetics Publishers, Champaign, IL.

Weinbach A. P. (1938) Contour map, center of gravity, moment of inertia and surface areas of human body. *Hum. Biol.*, **10**, 356–371.

Winter, D. A. (1990) *Biomechanics of Human Movement*, 2nd edn. John Wiley & Sons, Inc., Canada.

Yamaguchi, G. T. and Zajac, F. E. (1990) Restoring unassisted natural gait to paraplegics via functional neuromuscular stimulation: a computer simulation study. *IEEE Trans. Biomed. Eng.*, **37**(9), 886–902.

Yeadon, M. R. (1989) The simulation of aerial movement—II. A mathematical inertia model of human body. *J. Biomech.*, **23**, 67–74.

Zajac, F. E. and Winters, J. M. (1990) Modeling musculoskeletal movement system: joint and body segmental dynamics, musculoskeletal actuation, and neuromuscular control. In: *Multiple Muscle Systems: Biomechanics and Movement Organization* (ed. Winters) Springer-Verlag.

Zarrugh, M. Y. (1981) Kinematic prediction of intersegment loads and power at the joints of the leg in walking. *J. Biomech.*, **14**, 713–725.

Zatsiorsky, V. M. and Seluyanov, V. N. (1983) The mass and inertia characteristics of the main segments of the human body. In: *Biomechanics VIII-B* (ed. K. K. Matsui), Human Kinetics Publishers, Champaign, IL pp. 1152–1159.

Zatsiorsky, V. M. and Seluyanov, V. N. (1985) Estimation of the mass and inertia characteristics of the human body by means of the best predictive regression equations. In: *Biomechanics IX-B* (ed. D. A. Winter), Human Kinetics Publishers, Champaign, IL, pp. 233–239.

Zatsiorsky, V. M., Seluyanov, V. N. and Chugunova, L. (1990) In vivo body segment inertial parameters determination using a gamma-scanner method. In: *Biomechanics of Human Movement: Application in Rehabilitation, Sports and Ergonomics*, Bertec Corporation, Ohio, pp. 186–202.

11

Energetic Aspects of Locomotion

ALAIN JUNQUA, FRANCK LEPLANQUAIS, JACQUES DUBOY AND
PATRICK LACOUTURE
Université de Poitiers, Poitiers, France

INTRODUCTION

Human locomotion is the result of a complex synergy between the activation muscles of the different segments in motion relative to one another. The mechanical cost in moving about has been estimated by Winter, Quanbury and Reimer (1978) in calculating the energy of the segments. The variations in the mechanical energy were assumed to reflect gait efficacy. Later, Mansour et al. (1982) associated the variation of the mechanical energy with walking speed. Furthermore, they developed a global index to normal and pathological gait efficacy.

To determine the work done by the internal forces and moments which are active in the human body, two methods are proposed in the literature. Morlock and Nigg (1991) presented several mathematical models of the foot to estimate directly the internal forces. The work done by the internal forces can then be directly calculated from the individual force or estimated from the mechanical energy. Aleshinsky (1986), in his theoretical work on mechanical energies, addressed the concept of the work done by the internal force. Recently, Duboy, Junqua and Lacouture (1994) applied the kinetic energy theorem to estimate the work of the internal forces.

This chapter reports on work done by the internal forces, $W_{F\text{int}}$, in an articulated system and its calculation by a global method. The kinetic energy theorem (Perez, 1992) is the only one which includes both the work of the system's internal forces and that of the external forces applied to the system.

The main difficulty is in the correct estimation of the work done by the external forces, $W_{F\text{ext}}$. This consists, at each instant in time, of the weight and of

Three-dimensional Analysis of Human Locomotion. Edited by P. Allard, A. Cappozzo, A. Lundberg and C. Vaughan
© 1997 John Wiley & Sons Ltd. ISBN 0 471 96949 4

the contact forces during the stance phase of gait. The work performed by the body weight is in the opposite direction to the variation of the corresponding potential energy. It depends upon the accuracy of the link segment model of the human body, the number of body segments, the errors in the kinematic data, the smoothing or filtering method used in processing the displacement data, the estimation of joint centres and the calculations of the joint and segmental angles.

The kinetic energy theorem is expressed by

$$\Delta E_{C\text{global}} = W_{F\text{int}} + W_{F\text{ext}}$$

where $\Delta E_{C\text{global}}$ is the variation in the total kinetic energy.

In order to simplify this equation, the two-dimensional (2D) calculations are presented. The reader is referred to Aissaoui et al. (1997) for its three-dimensional (3D) expression. By taking as a basis a model of 14 body segments, at each instant $E_{C\text{global}}$ can be expressed in two dimensions as

$$E_C = \sum_{i=1}^{14} \frac{1}{2} I_i \omega_i^2 + \sum_{i=1}^{14} m_i V_{i/R^*}^2 + \frac{1}{2} M V_{G/R}^2$$

where R is the Galilean reference frame of the laboratory, R^* is a translating reference frame centred in G, which is the athlete's centre of gravity, ω and $V_{i/R}$ are respectively the angular and linear velocity of the segment, m and I are the mass and the moment of inertia of the segment, and M and $V_{G/R}$ are the mass and linear velocity of the body.

ESTIMATION OF HUMAN LOCOMOTION COST

Numerous values have been obtained for the mechanical power, work and energy developed during able-bodied walking and running, and some of them are reported in Table 1. Williams and Cavanagh (1983) split them in two categories. Those which focused on the displacements of the centre of the mass conse-quently could apprehend only the external work while those which use a limb segment approach could give the internal forces work. These two different approaches can explain in part the disparity in the results shown in Table 1, which vary from 71 to 147 W for walking and from 163 to 1650 W for running.

Physiologists want to characterize the way in which the human body uses the metabolic energy in its displacements. In fact, they look for an output which describes the physical ability of the athletes and their efficiency. Here, the energy is considered as the global mechanical energy of an articulated system. First Norman et al. (1976) and then Winter (1990) have attributed to each segment the energy E_S, which can be expressed at a given time as

$$E_S = mgh + \frac{1}{2} mv^2 + \frac{1}{2} I\omega^2$$

Table 1. Mechanical powers reported in the literature (Williams and Cavanagh, 1983)

	Speed (m s^{-1})	Mechanical power (W)
Walking		
Winter, Quanbury and Reimer (1978)	1.4	147
Pierrynowski, Winter and Norman (1980)	1.5	166
Zarrugh (1981)	1.5	71
Running		
Fukunaga et al. (1978)	3.6	343
Cavagna and Kaneko (1977)	3.6	556
Norman et al. (1976)	3.6	172
Gregor and Kirkendall (1978)	3.6	163
Luhtanen and Komi (1978)	3.9	931
Luhtanen and Komi (1980)	3.9	1650

The total energy of the body is then

$$E_{\text{body}} = \sum_{i=1}^{B} E_{Si}$$

where B is the number of segments and E_{Si} the global energy of the i segment. Based on the above equations, Winter (1990) introduces the concept of flow of energy, including the mechanical energy transfer between segments. Nevertheless, like Vaughan (1984), some authors questioned the validity of this method.

THEORETICAL CONSIDERATIONS IN THE CALCULATION OF THE WORK DONE BY THE INTERNAL FORCES DURING WALKING AND RUNNING ON A TREADMILL

ON THE FLAT SURFACE

The calculation of the global work due to the external forces during a displacement such as walking or running on a flat surface is considered in the following example. Let $R = (O, x, y, z)$ be the absolute spatial reference system; it is a Galilean reference system. The contact forces between the athlete or the subject (the system S) and the ground do not perform work, since there is no displacement between the applied forces and the reaction forces during the movement. Consequently, the work performed by the external forces comes only from the weight, which is equal to $-\Delta E_{\text{p}}$. Here ΔE_{p} corresponds to the variation in the potential energy of the system between any two instants in time. The work done by the internal forces is derived from the kinetic energy theorem (Duboy, Junqua and Lacouture, 1994; Aissaoui et al., 1996) and can be expressed as

$$W(F_{\text{int},s}) = \Delta \frac{1}{2} M V^2_{G/R} + \Delta \frac{1}{2} I_i \omega_i^2 + \Delta \frac{1}{2} m_i V^2_{i/R^*} + \Delta E_p$$

between t_1 and t_2 and where R^* is a reference system linked to the global centre of mass. Usually, t_1 corresponds to the time at heel strike, while t_2 is any other instant during the walking or running cycle.

To characterize a movement, the notion of cumulative work of internal forces was proposed by Norman et al. (1976). For a given motion, the cumulative work $T_{F\text{int}}$ includes both positive work phases $W^+_{F\text{int}}$ and negative work phases $W^-_{F\text{int}}$, since these two types of work have a metabolic cost. We define the cumulative work by

$$T_{F\text{int}} = \sum_{i=1}^{n} |W_{(F\text{int},s)}|$$

where n is the number of phases successively positive and negative. The corresponding calculations are very sensitive because they are noise-dependent in the calculation of $W_{(F\text{int},s)}$. Let us now consider the work done on a treadmill and the amount of mechanical energy given by the belt.

ON A TREADMILL

The calculations of the work performed by the internal forces during walking and running on a treadmill will be considered in relation to two spatial reference frames as shown in Figure 1. Besides $R = (o, x, y, z)$, the absolute spatial

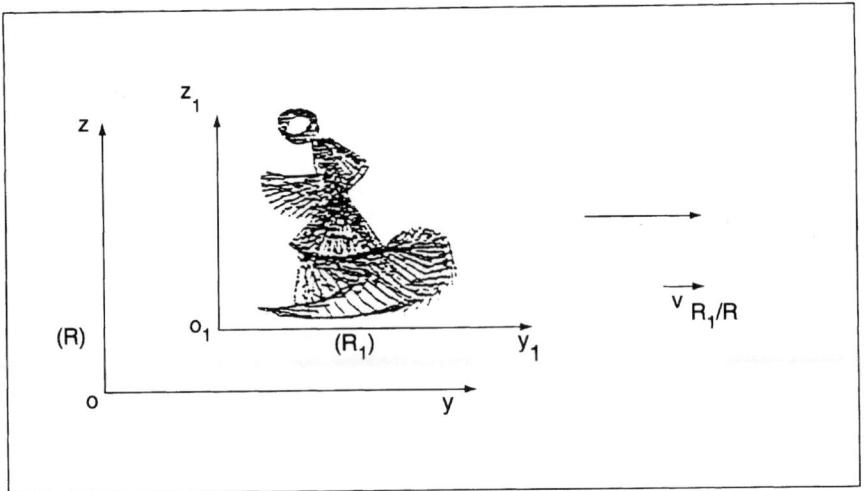

Figure 1. The two Galilean reference frames. The velocity $\mathbf{V}_{R_1/R}$ is constant

reference system associated with the laboratory, a moving reference system $R_1 = (o_1, x, y, z)$ is fixed to the belt which has a uniform and linear translation with respect to R. In this case, the corresponding velocity is $\mathbf{V}(o_1/R) = V\mathbf{y}$, with V being a positive constant and $\mathbf{y}$ a unit vector on the O_y axis. The points' velocities of the articulated system S are measured in relation to the absolute reference system R. The work done by the internal forces is calculated as if the subject was running on a flat surface by using the kinetic energy theorem. However, additional hypotheses are required. First, the runner makes contact with the ground without sliding during a stride, and second each support phase is realized without the belt's surface changing shape; in other words, the belt is considered as absolutely rigid. The following notation will be used:

- $E_{C(S/R \text{ or } R_1)}$ is the kinetic energy of the system S related to R or R_1;
- $P_{(F\text{ext},S/R \text{ or } R_1)}$ is the power of the external forces applied to S in a motion related to R or R_1;
- $P_{(F\text{int},S)}$ is the power of the internal forces of the system S whose value is independent of the reference frame.

Since the running motion is referenced in the absolute spatial reference system R, it seems natural to apply the kinetic energy theorem in relation to R, in the derivative form with respect to time:

$$\frac{\mathrm{d}}{\mathrm{d}t} E_{C(S/R)} = P_{(F\text{ext},S/R)} + P_{(F'\text{ext},S/R)} \tag{1}$$

This formulation hides one difficulty, which is that the calculation of the power corresponding to the external forces must be expressed in two terms:

$$P_{(F\text{ext},S/R)} = P_{(W,S/R)} + P_{(F\text{ext},S/R)} \tag{2}$$

The first term of the sum is the weight's power of S while the second term, giving the force's power during a moving support in R (the treadmill), is not necessarily equal to zero. Furthermore, this power cannot be evaluated directly, since the ground reaction forces are unknown. This explains in part why we used a kinematic approach, avoiding the second term in the calculation. Consequently, equation (1) is unsuitable in this form for the calculation of $P_{(F\text{int},S)}$.

With the previously formulated hypothesis, it is clear that $P_{(F\text{ext},S/R_1)} = 0$, because the centre of pressure is fixed in R_1. The kinetic energy theorem can be applied in R_1 and becomes

$$\frac{\mathrm{d}}{\mathrm{d}t} E_{C(S/R_1)} = P_{(W,S/R_1)} + P_{(F\text{int},S)} \tag{3}$$

Equation (3) is expressed in relation to R, in which the kinematic measurements are done. Then, the system's S kinetic energy is

$$E_{C(S/R_1)} = \frac{1}{2} \int_{P \varepsilon S} \mathbf{V}^2_{P/R_1} \, dm$$

where V_{P/R_1} is, as we noted, the points' P velocity of a system with respect to R_1. Using the velocity composition law:

$$\mathbf{V}_{P/R_1} = \mathbf{V}_{P/R} - \mathbf{V}_{O_1/R} = \mathbf{V}_{P/R} - \mathbf{V}_y$$

where $\mathbf{V}_y$ is the belt's velocity related to R. Thus:

$$E_{C(S/R_1)} = E_{C(S/R)} + \frac{1}{2} mV^2 - m\mathbf{V}_y \cdot \mathbf{V}_{G/R} \tag{4}$$

Since the belt's velocity is constant, equation (4) becomes, after differentiation with respect to time, and considering equation (3):

$$\frac{d}{dt} E_{C(S/R)} = P_{(W,S/R_1)} + P_{(F\mathrm{int},S)} + m\mathbf{V}_y \cdot \boldsymbol{\gamma}_{G/R} \tag{5}$$

where $\boldsymbol{\gamma}_{G/R} = \boldsymbol{\gamma}_{G/R_1}$ is the acceleration of the centre of gravity.

The power related to the pseudo-forces in relation to the treadmill's movement are developed from Newton's equation of motion for the system S given by

$$m\boldsymbol{\gamma}_{G/R} = \mathbf{R}_{(Fext \rightarrow S)}$$

where $\mathbf{R}_{(Fext \rightarrow S)}$ is the resultant of the external forces acting on the system S. But

$$\mathbf{R}_{(Fext \rightarrow S)} = \mathbf{R}_{(W \rightarrow S)} + \mathbf{R}_{(Fext \rightarrow S)}$$

By substituting $m\boldsymbol{\gamma}_{G/R}$ in the third term of equation (5), we obtain, after simplification:

$$m\mathbf{V}_y \cdot \boldsymbol{\gamma}_{G/R} = V\mathbf{R}_{(Fext \rightarrow S)} \cdot \mathbf{y}$$

This defines the term $P_{(Fext,S/R)}$ of equation (2), since the points' velocity of the system S in contact with the belt is exactly equal to $\mathbf{V}_y$.

Since the weight's power in equation (5) is

$$P_{(W,S/R_1)} = -mg\mathbf{z} \cdot \mathbf{V}_{G/R}$$

$$= \frac{d}{dt} [-mg\mathbf{OG} \cdot \mathbf{z}]$$

then the power of the internal forces in S is

$$P_{(F\mathrm{int},S)} = \frac{d}{dt} \left[E_{C(S/R)} + mg\mathbf{OG} \cdot \mathbf{z} - mV\mathbf{V}_{G/R} \cdot \mathbf{y} \right] \tag{6}$$

Finally, the global cumulative work of the internal forces in S during a period of time (t_0, t_1) can be written as

$$T_{(F\mathrm{int},S)} = \int_{t_0}^{t_1} \left[\frac{d}{dt} E_{C(S/R)} + E_P - mV\mathbf{V}_{G/R} \cdot \mathbf{y} \right] dt \tag{7}$$

since $T_{(Fint,S)} = \int_{t_0}^{t_1} |P_{(Fint,S)}| \, dt$; in this last equation the absolute magnitudes of $P_{(Fint,S)}$ appear, because the work can be positive or negative alternately.

CUMULATIVE WORK DONE BY THE INTERNAL FORCES

A 2D video analysis (50 Hz) has been carried out on the same subject walking at from 4 to 10 km/h, and running at from 10 to 20 km/h. The different values of work of the cumulative internal forces have been calculated at each trial for two consecutive steps. The internal force results of the articulated system are greatly influenced by the choice of the video frames, because of the variations in the external kinetic energy. These are slightly dependent on the anterior–posterior velocity's V_{Gy} variation, while the foot contact on the belt is directly proportional to V_{Gy}. We wish to insist on this limitation, since the steps have been defined by selecting only the frames where the subject appeared to be most in harmony with the belt's velocity. A perfect harmony, which is at each moment $V_{Gy} = 0$, is impossible to obtain, since the athlete's articulated body changes its shape at each instant. It should correspond to a periodic pattern of the V_{Gy} velocity, which is alternatively positive and negative, and with the lowest peak values.

The arbitrary choice of the sequence of frames for the identification of the two steps when the athlete seems to be in greatest harmony with the belt does not represent in any way the reality of several minutes of running, despite the athlete's ability. In fact, our eyes perceived on the video frames more or less perceptible successions of forward and backward steps of the athlete on the belt.

The velocity patterns corresponding to running at 13 km/h and 20 km/h, shown in Figures 2 and 3 respectively were the most regular of those obtained. The peak values never exceeded ± 0.25 m/s, which represented the remarkable skill of our athlete. In the internal forces' calculations according to equation (5), we were not confident of calculating $V_{Gy}(t)$ with the necessary accuracy. A high value in these peaks would have constituted a difficulty. Indeed, V_{Gy} is multiplied by a coefficient V, and the W_{Fext} associated with the works of the pseudo-forces would not have been non-negligible with respect to the other terms of equation (5).

The term W_{Fext} (Figure 4) has peak values of the same magnitude as the weight's work or the global kinetic energy's variations. It is then pre-eminent in the calculation of the internal forces' global work using the kinetic energy theorem (Figure 5). If the term W_{Fext} does not exist, which is an unrealistic assumption, the cumulative work done by the internal forces will go from 332 J (Figure 5) to 170 J as shown in Figure 6; to neglect this term is not correct.

Figure 7 illustrates the cumulative work done by the internal forces for walking on a treadmill at speeds ranging from 4 to 10 km/h. In Figure 8, the

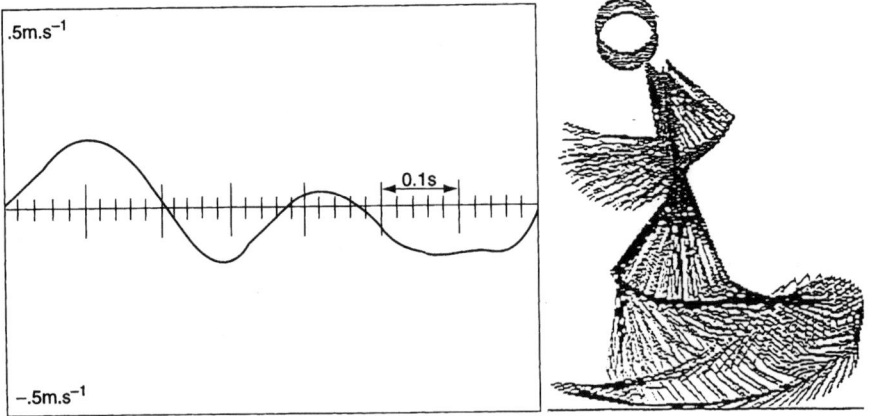

Figure 2. Time variations of V_{Gy} and the corresponding stick figure of a running step (13 km/h)

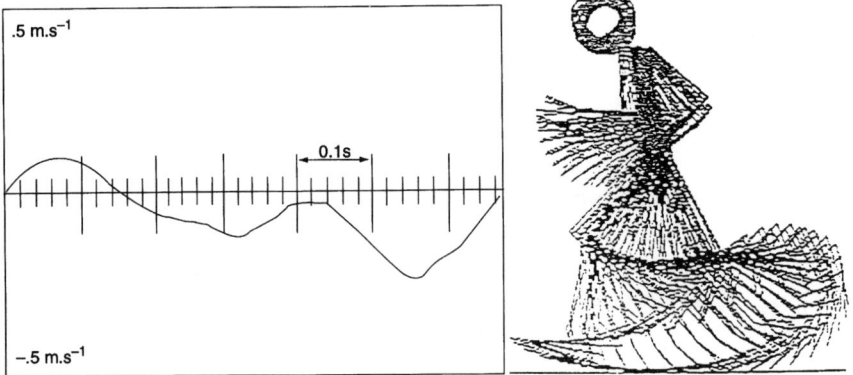

Figure 3. Time variations of V_{Gy} and the corresponding stick figure of a running step (20 km/h)

cumulative work of the internal forces is shown and appears to follow the same pattern regardless of the running speed, which ranged from 10 to 17 km/h. Results shown in Figures 7 and 8 do not show clearly that the step's duration is decreasing as the velocity of the belt is increasing. It seems more appropriate to represent the results through the evolution of the corresponding mean power as function of the treadmill's velocity (Figure 9).

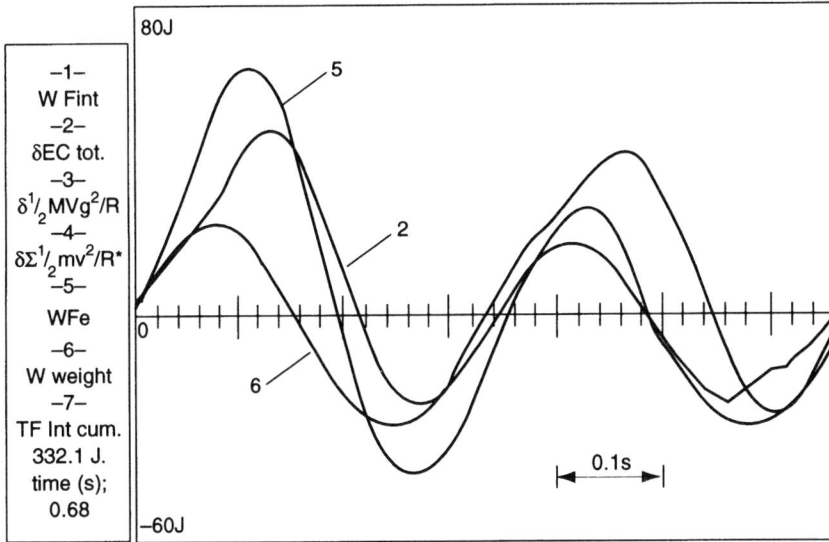

Figure 4. Evolution of the three terms in equation (5) for results obtained when the velocity of the belt is 16 km/h

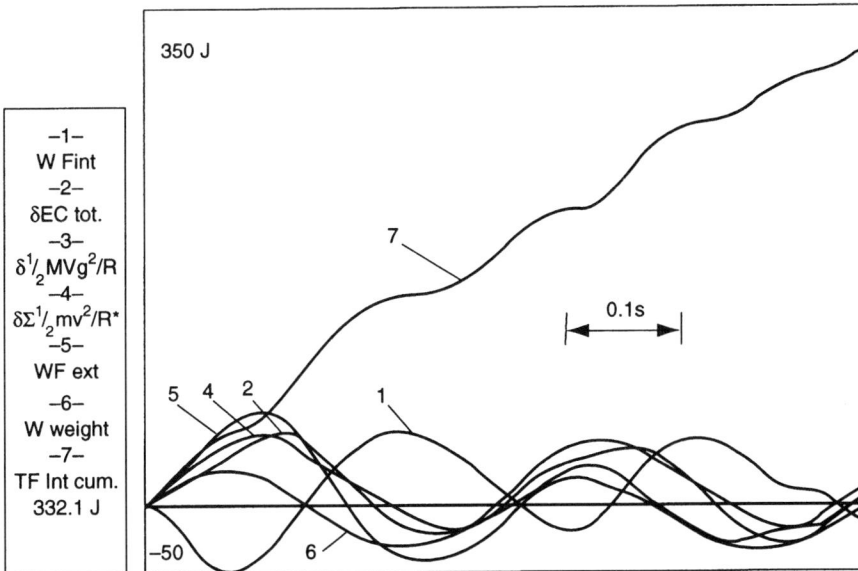

Figure 5. Cumulative work and the different constitutive energetic terms' calculations for results obtained when the velocity of the belt is 16 km/h

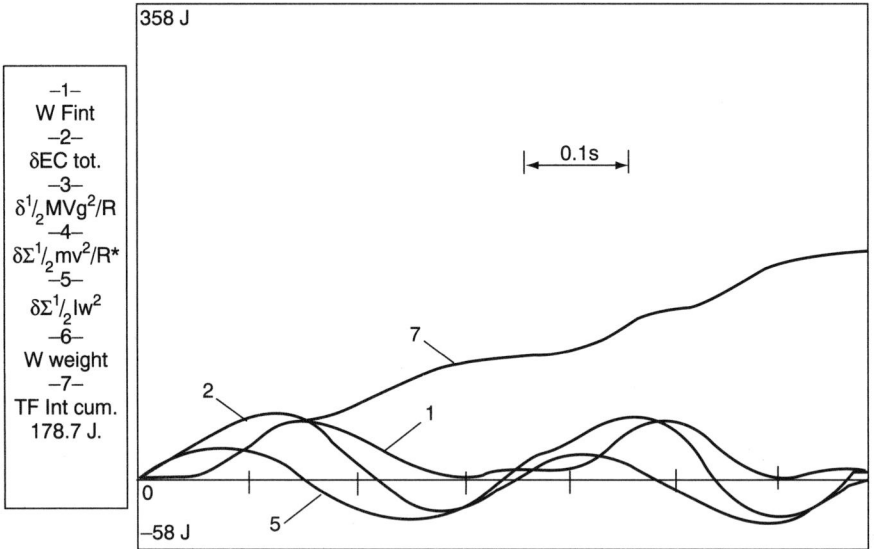

Figure 6. Evolution of the cumulative work related to the internal forces after suppressing arbitrarily the term W_{Fext} (belt's velocity: 16 km/h)

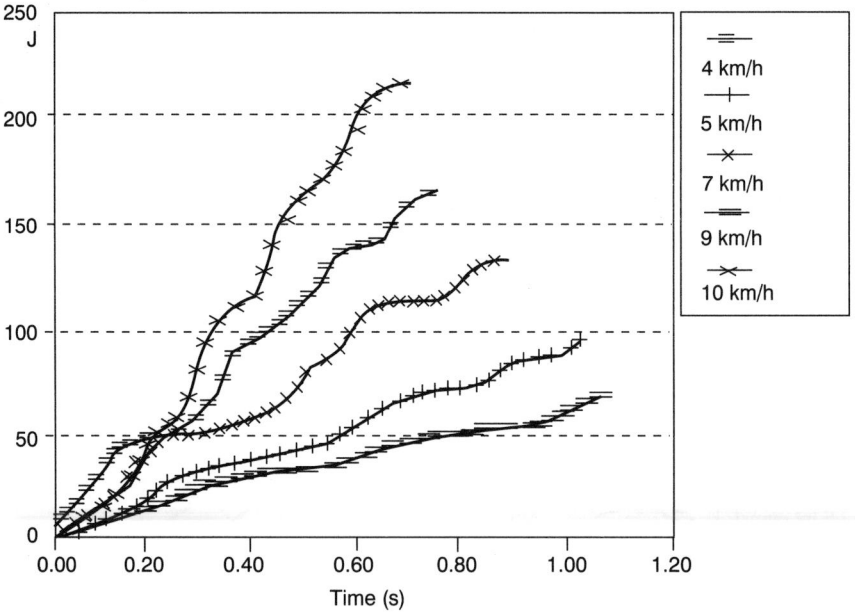

Figure 7. Evolution of the cumulative work done by the internal forces for walking on a treadmill at different running speeds

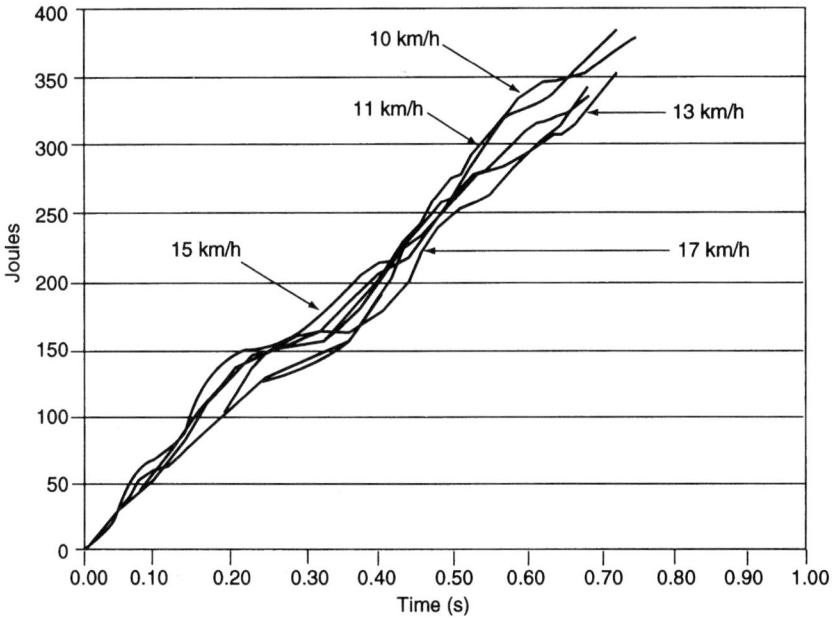

Figure 8. Evolution of the cumulative work done by the internal forces for running on a treadmill

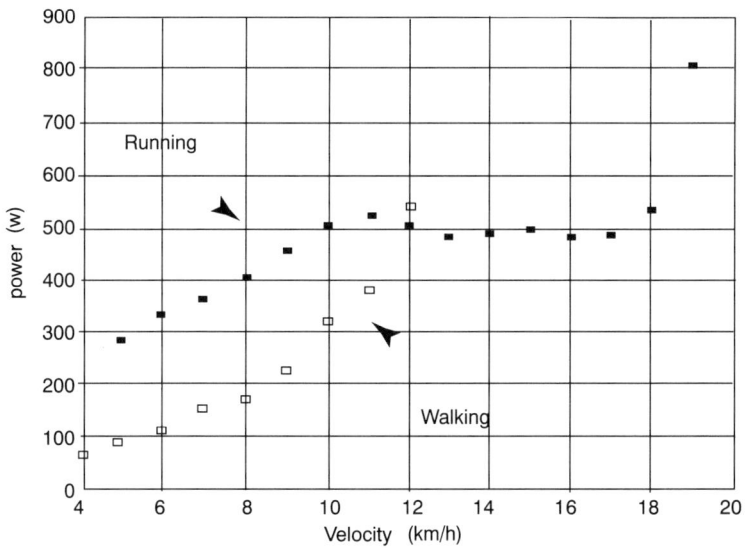

Figure 9. Mean power for a step versus the treadmill's velocity for walking and running conditions

The calculated mean power (the work done by the internal forces, related to the time unit) is approximately the same for an athlete running at 12 km/h (velocity of the mat) or walking at the same velocity on a flat surface. Note that the velocity of 12 km/h for our athlete is the highest limit for natural walking. The athlete can easily run at a speed ranging between 10 and 17 km/h; but, from 19 km/h and more, it is not a natural exercise for him, and consequently we can show an important increase of the average power from a velocity of 19 km/h upwards.

WALKING AND RUNNING ON A FLAT SURFACE: CUMULATIVE WORK DONE BY THE INTERNAL FORCES

The term $\Delta(\frac{1}{2}MV_G^2)$ behaves like $\Delta(\frac{1}{2}MV_{Gy}^2)$, and especially when V_{Gy} is high, Even if the athlete tries to maintain the velocity of his centre of mass constant, it fluctuates slightly about the anterior–posterior axis. The effect of the velocity's fluctuations will be very important in the calculation of the work done by the internal forces. This was emphasized in the discussion about the accuracy of V_{Gy}. The term corresponding to the external kinetic energy (Figure 10) predominates and its influence on the estimation of the work done by the internal forces is shown in Figure 11. However, the term $\Delta\sum_{i=1}^{14}\frac{1}{2}I_i\omega_i^2$, being very small, does not appear in these figures. This term is used to describe rotation kinetic energy for the segments.

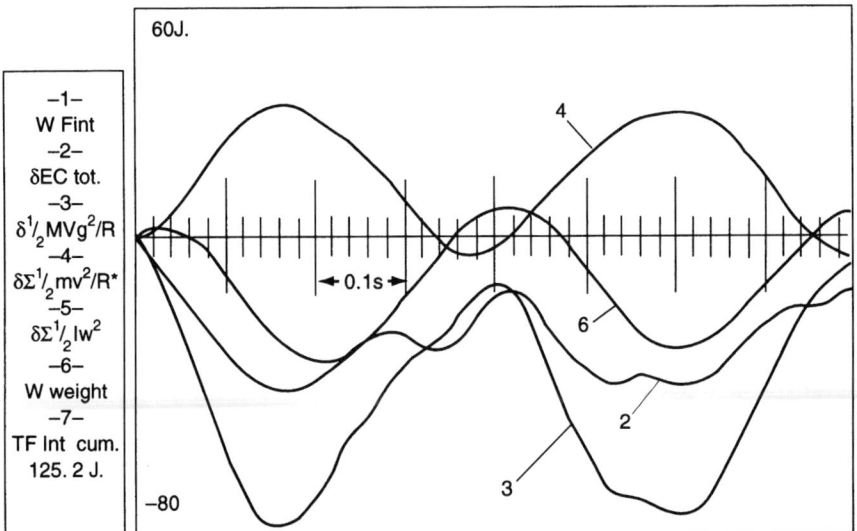

Figure 10. Variation of the kinetic and potential energies during walking at 9 km/h

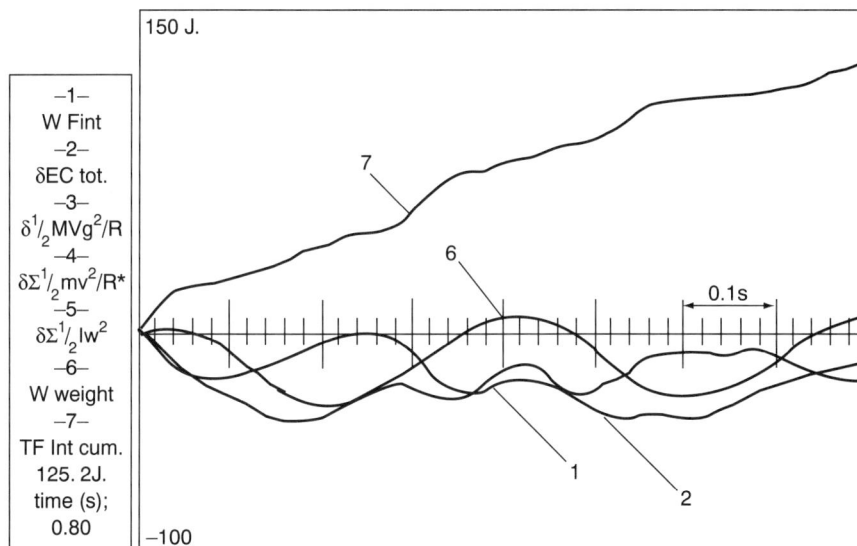

Figure 11. Cumulative work done by the internal forces and associated energy during walking at 9 km/h

Figures 12 and 13 show the variations of the mechanical energies and the work performed by the internal forces during running at 15 km/h on a flat surface. It is important to verify that the athlete is unable to keep strictly constant his anterior posterior velocity V_{Gy}, because the impulse from the reaction forces (on the ground) continually modifies this velocity; such light alterations consequently give important values for work done by internal forces. For this same reason, an abrupt style of running is very expensive.

To compare the different walking and running results, the mean power at each instant was calculated. For walking trials above a speed of 10 km/h and running speeds above 23 km/h, the results were inconsistent and were not reported. However, a running trial performed at a maximum velocity of 29 km/h is included in the results. In this trial, the mean power calculated for one step appeared spectacular: greater than 1400 W, compared to about 100 W for walking at a speed of 5 km/h (Figure 14). In Figure 14, we can see a slight increase in the power developed by this athlete up to 20 km/h but above this velocity the running is not economical for this athlete.

In Figure 15, all the results relevant to the mean power are presented for all the walking and running conditions as well as those collected on a treadmill. Only the sprinting condition (29 km/h) is not reported, in order to reduce the ordinate axis. At this scale, we notice for the same athlete that, for the same velocities, the power developed is more important for running on a treadmill than for walking. For small velocities, the values of power are similar for running and

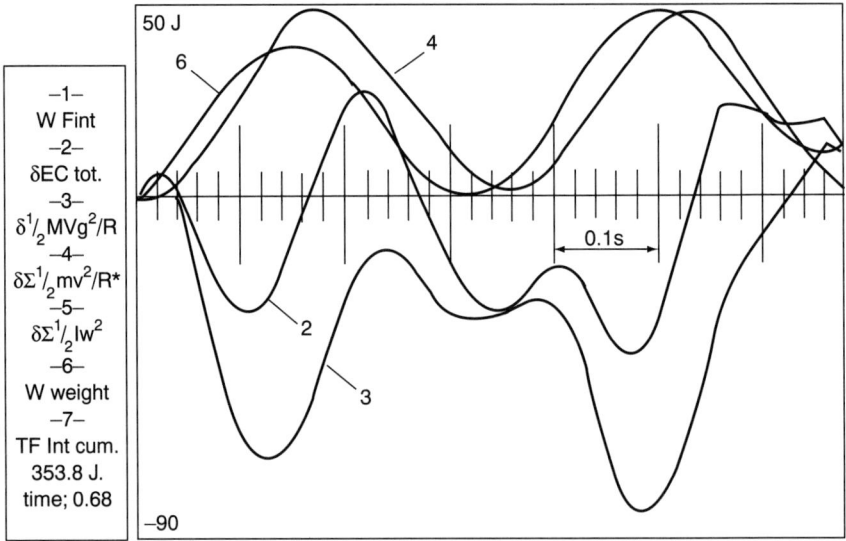

Figure 12. Variation of the kinetic and potential energies during running at 15 km/h on a flat surface

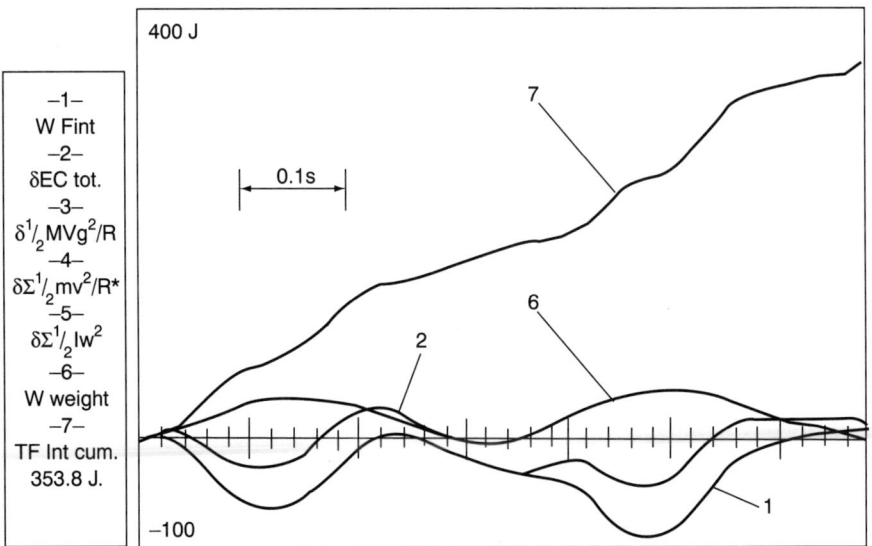

Figure 13. Cumulative work done by the internal forces and associated energy during running at 15 km/h on a flat surface

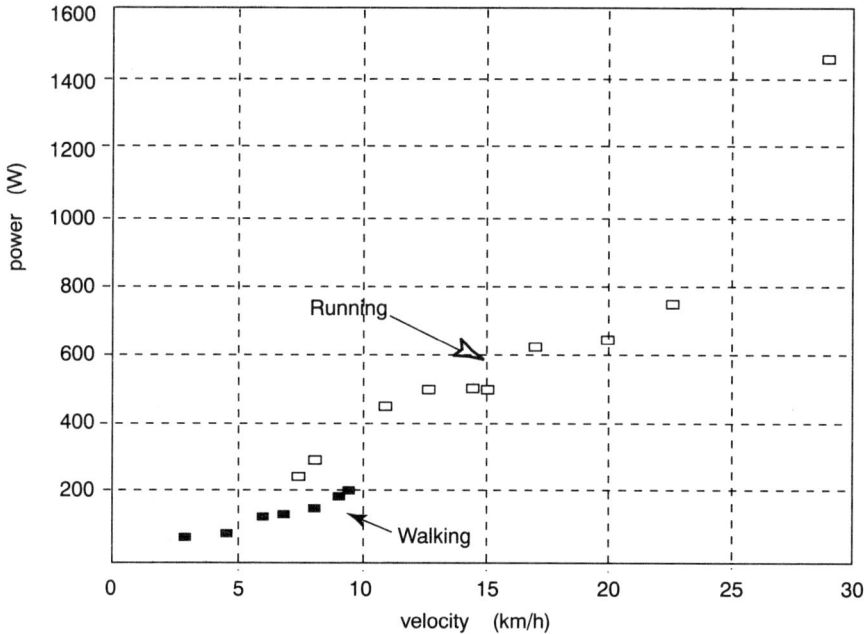

Figure 14. Powers developed for walking and running on a flat surface at different velocities

walking on a flat surface. We can observe that, for velocities between 10 and 16 km/h, the powers calculated in the case of treadmill locomotion are now comparable to those corresponding with flat surface locomotion. Finally, the powers may be more important when running at high velocities on the treadmill. Though the theoretical aspects are sound, the experimental considerations rely strongly on the present-day technological limitations of our measurement devices.

CONCLUSION

Concerning the theoretical aspects, it appears that the sum of the work done by the internal forces is indisputably dependent on the athlete's ability to either stay in harmony with the treadmill or minimize the longitudinal velocity's fluctuations of its centre of gravity during running on a flat surface. Therefore, the variation in the cumulative work with respect to the treadmill's speed or the mean running velocity on a flat surface, should be observed carefully.

Two facts must be kept in mind. First, the power corresponding to the work values calculated for a step seems to be constant between 12 km/h and 18 km/h

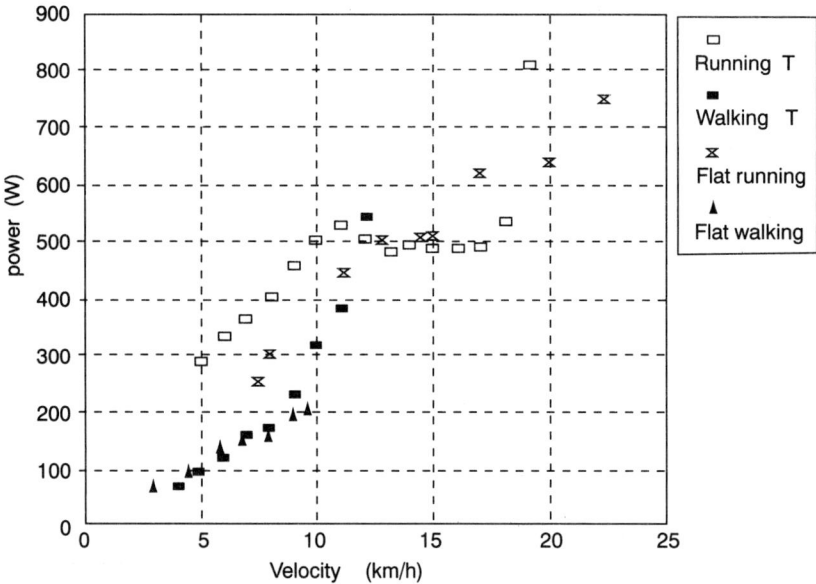

Figure 15. Walking and running on a flat surface and on a treadmill at different velocities

for our athlete. Second, the powers at small and similar speeds are less important in walking than running. Those powers are inverse at about 11 km/h on the treadmill.

ACKNOWLEDGEMENTS

We wish to express our gratitude to Mr Guy Bessonnet (Université de Poitiers, URA 861, CNRS, Robotic Group) for his contribution to the calculations related to the energetic cost on the treadmill.

REFERENCES

Aissaoui, R., Allard, P., Junqua, F., Frossard, L. and Duhaime, M. (1996) Internal work estimation in 3-D gait analysis. *Med. Biol. Eng. Comput.*, in press.

Aleshinsky, S. Y. (1986) An energy sources and fractions approach to the mechanical energy expenditure problem—1. *J. Biomech.*, **19**, 287–293.

Cavagna, G. A. and Kaneko, M. (1977) Mechanical work and efficiency in level walking and running. *J. Physiol.*, **268**, 467–481.

Duboy, J., Junqua, A. and Lacouture, P. (1994) Mécanique humaine. Éléments d'une analyse des gestes sportifs en deux dimensions. Editions revue EPS, Paris.

Fukunaga, T., Matso, A., Yuase, K., Fukimatsu, H. and Asahina, K. (1978) Mechanical power output in running. In: *Biomechanics VIB* (ed. G. Asmussen and K. Jorgensen), University Park Press, Baltimore pp. 17–22.

Gregor, R. J. and Kirkendall, D. (1978) Performance efficiency of world class female marathon runners. In: *Biomechanics VIB* (ed. G. Asmussen and K. Jorgensen), University Park Press, Baltimore pp. 40–45.

Leplanquais, F. (1995) Contribution à l'analyse mécanique des allures de la locomotion humaine. Caractérisation de tâches corporelles diverses à l'aide des travaux des efforts internes. PhD Thesis, Université de Poitiers, France.

Luhtanen, P. and Komi, P. V. (1978) Mechanical energy generated during running. *Eur. J. Appl. Physiol.*, **38**, 41–48.

Luhtanen, P. and Komi, P. V. (1980) Force, power and elasticity–velocity relationships in walking, running and jumping. *Eur. J. Appl. Physiol.*, **40**, 279–289.

Mansour, J. M., Lesh, M. D., Nowak, M. D. and Simon, S. R. (1982) A three-dimensional multi-segmental analysis of the energies of normal and pathological human gait. *J. Biomech.*, **15**, 51–59.

Morlock, M. and Nigg, B. M. (1991) Theoretical considerations and practical results on the influence of the representation of the foot for the estimation of internal forces with models. *Clin. Biomech.*, **6**, 3–13.

Norman, R. W., Sharratt, M. T., Pezzack, J. C. and Noble, E. G. (1976) Reexamination of the mechanical efficiency of horizontal treadmill running. In: *Biomechanics VB* (ed. P. V. Komi), University Park Press, Baltimore pp. 87–93.

Perez, J. Ph. (1992) *Mécanique*. Masson Editeur, Paris.

Pierrynowski, M. R., Winter, D. A. and Norman, R. W. (1980) Transfers of mechanical energy within the total body and mechanical efficiency during treadmill walking. *Ergonomics*, **23**, 147–156.

Vaughan, C. L. (1984) Biomechanics of running gait. *CRC Crit. Rev. Biomech. Eng.*, **12**(1).

Williams, K. R. and Cavanagh, P. R. (1983) A model for the calculation of mechanical power during distance running. *J. Biomech.*, **16**, 115–128.

Winter, D. A. (1990) *Biomechanics and Motor Control of Human Movement*. Wiley Interscience, New York.

Winter, D. A., Quanbury, A. O. and Reimer, G. D. (1978) Analysis of the instantaneous energy of normal gait. *J. Biomech.*, **9**, 253–257.

Zarrugh, M. Y. (1981) Power requirements and walking efficiency of treadmill walking. *J. Biomech.*, **14**, 157–165.

12

Application of Robotics Technology to the Study of Knee Kinematics

GLEN A. LIVESAY, SAVIO L-Y. WOO, THOMAS J. RUNCO
AND THEODORE W. RUDY
Musculoskeletal Research Center, University of Pittsburgh, Pittsburgh, PA, USA

INTRODUCTION

The function of the human knee is mediated by a complex interaction of the femur, tibia and patella, the ligaments and capsule, the articular cartilage and menisci, and the muscles which traverse the joint. The interdependence of these structures is such that injury or failure of any one of them can lead to deterioration of the overall function of the joint. Ligaments are particularly vulnerable, as they are subject to sprains or ruptures in virtually all injuries of the knee (Praemer, Furner and Rice, 1992). Complete ligament ruptures, especially of the anterior cruciate ligament (ACL), can cause knee instability (Johnson et al., 1992; Kaplan et al., 1991; Noyes et al., 1983, 1991; Seto et al., 1988), which may lead to subsequent tears and degeneration of the menisci and articular cartilage and increased laxity of other knee ligaments (Cerbano, Sherman and Bonamo, 1988; Finsterbush et al., 1990; Fowler and Regan, 1987; Hart, 1982; Kannus and Jarvien, 1987; McDaniel and Dameron, 1980, 1983). Because the ACL fails to heal with a quality which restores its function (Lyon et al., 1991), numerous reconstructive procedures have been developed for patients demanding high levels of knee performance. While surgical reconstruction of the ruptured ACL with a graft has vastly improved patient outcome (Buss et al., 1993; Clancy et al., 1982; Johnson, 1993; O'Brien et al., 1991; Sandberg and

Three-dimensional Analysis of Human Locomotion. Edited by P. Allard, A. Cappozzo, A. Lundberg and C. Vaughan
© 1997 John Wiley & Sons Ltd. ISBN 0 471 96949 4

Balkfors, 1988; Shino et al., 1990; Tibone et al., 1988), it fails to completely restore normal knee kinematics or to prevent the associated onset of joint degeneration (Daniel et al., 1994, 1996).

Improvements in reconstruction and rehabilitation protocols will rely on a better understanding of the role of the ACL in the intact knee. For this reason, many studies have examined knee kinematics and forces in the ACL in response to various external loads applied to the joint (Barry and Ahmed, 1986; France et al., 1983; Holden, Grood and Cummings, 1991; Holden et al., 1992, 1994; Lew et al., 1990; Lewis and Fraser, 1979; Lewis, Lew and Schmidt, 1982, 1988; Lewis et al., 1980). In our laboratory, we have developed a test system which consists of a robotic manipulator combined with a universal force–moment sensor (UFS) (Rudy et al., 1994, 1996). This system allows multiple degree of freedom (DOF) control of external loads on the joint in order to reproduce a more physiological loading condition.

The nature of the robotic manipulator makes it very attractive for use in the control of motion and loading during joint kinematics studies. This approach has certain advantages over the use of a standard materials-testing machine in the study of joint kinematics in cadaveric specimens. With multi-axial force and position control, a robotic manipulator possesses the flexibility required to apply loads and produce motions in multiple directions and to test other synovial joints with minimal changes in programming. In addition to applying external loads to the knee joint, the robotic testing system can record the resulting 6-DOF motions. The robot can then reproduce the identical path of intact knee motion in a partially dissected specimen.

In this chapter we focus on this newly developed testing system, which is designed to examine knee joint kinematics and detail the contribution from individual soft tissues (Rudy et al., 1996). We will: (1) review basic concepts in the kinematics of robotic manipulators; (2) describe the control of joint motion with the robotic–UFS testing system; (3) detail the procedure used to determine the *in situ* forces in knee ligaments with a UFS; and (4) discuss several applications of the robotic–UFS testing system to joint biomechanics.

KINEMATICS OF A ROBOTIC MANIPULATOR

The robotic manipulator is a tool which possesses the ability to control the location of its end effector relative to its base (Figure 1). This feature enables the robot to accurately control the location of an object attached to the end effector and record each location throughout a path of motion. In our laboratory, a robotic manipulator is used to apply external loads to an intact cadaveric knee and also track the resulting kinematics. The relationship between the end effector and the base of a serial manipulator can be determined using the concepts in the brief review which follows.

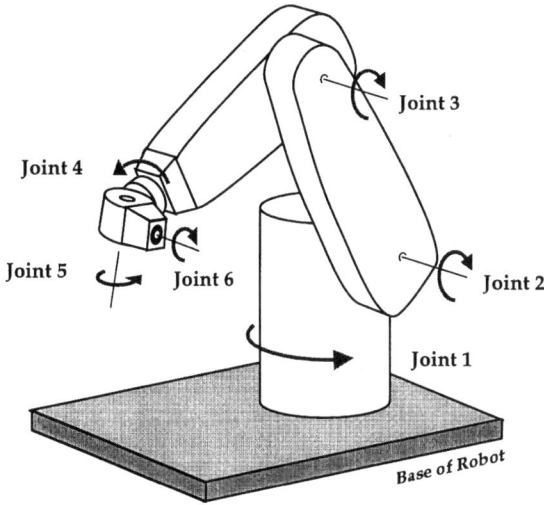

Figure 1. Typical articulated robotic manipulator showing the individual revolute joints which enable complex motions of the end effector

REFERENCES FRAMES/POSITION AND ORIENTATION

In order to describe the kinematics of a manipulator, a global reference frame must be established in which to report the location and the motion of the end effector. The global reference frame for gait studies using video analysis techniques is typically established by the positions of fixed optical markers (e.g. a frame of a known size, or hanging markers at known positions) which are used for the calibration of camera locations. Since the positions of these markers are known relative to one another, they establish a fixed reference frame in which the movement of a subject can be reported. Markers mounted on the subject's body can then be used to establish local reference frames—for the thigh, shank and foot, for example. For a robotic manipulator, the global reference frame is generally defined as being fixed at some location within its base. A local frame is then established for each of the links of the manipulator. If there is a known relationship between an object (e.g. a tibia) and the local frame of the end effector, then the object can be related to the global reference frame using the location of the end effector relative to the base.

The term 'location', as used in robot kinematics, contains information about both position and orientation. Position refers to pure translation of one frame of reference relative to another, while orientation refers to pure rotation. This is shown schematically in Figure 2 for a two-dimensional example. It can be seen that frame 1 differs from frame 0 in orientation only, while frame 2 differs in

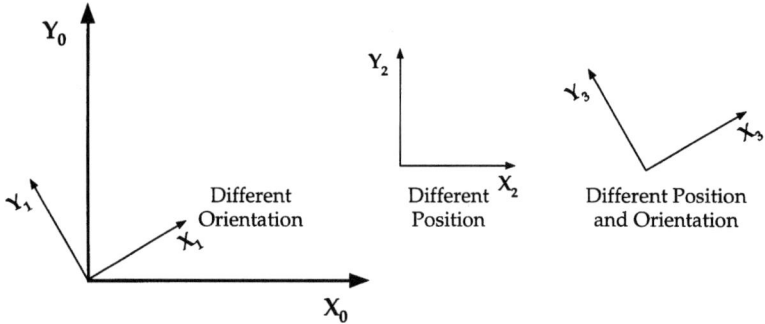

Figure 2. Schematic illustration of the distinctions between position and orientation of different reference frames

position only. The location of frame 3 differs from that of frame 0 in both position and orientation. The mathematical transformations corresponding to changes in each of these types of motion will be shown in the following section.

TRANSFORMATIONS BETWEEN FRAMES

The key to examining the kinematics of a robotic manipulator is that the overall motion of the end effector relative to the base can be resolved into simpler components. Manipulators are typically constructed of rigid links which are connected by joints that allow a particular motion between adjacent links. Pure rotation is allowed at a rotary or revolute joint, while pure translation occurs at a prismatic joint. This motion at a joint is described by the change in the position and orientation between local reference frames attached to each of the individual links. Different manipulators utilize different combinations of these joints to produce the motions of the end effector. It can be seen from Figure 1 that the 6-DOF motion of the end-effector (after joint 6) is achieved through a series of rotations which occur at the individual joints. The overall kinematics can be determined, for serial manipulators, by considering these simpler motions in a given sequence.

Making use of the 1-DOF motions between adjacent links, or local reference frames, it is a simple task to transform vectors between these frames. For the case of pure rotation, transformation of a vector from one frame to a rotated frame is achieved using a (3×3) matrix. This situation is shown in Figure 3A, and the corresponding mathematical treatment would be

$$P_0 = R_0^1 P_1 \tag{1}$$

where P_0, P_1 are (3×1) vectors which describe the location of the point P with

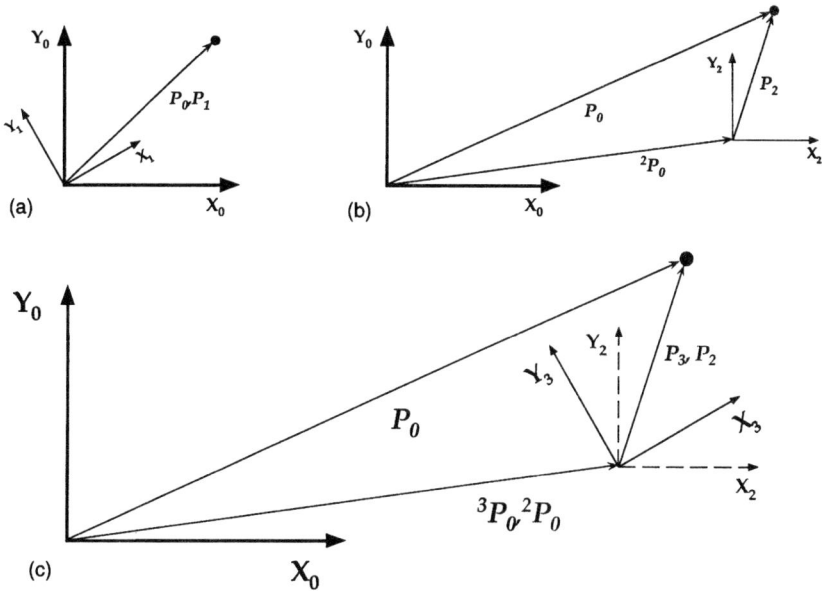

Figure 3. Sequence of diagrams demonstrating: (a) pure rotation between frames, (b) pure translation, and (c) combined translation and rotation

respect to reference frames 0 and 1, respectively. R_0^1 transforms a position described with respect to frame 1 to a position described with respect to frame 0. It is represented as a (3 × 3) matrix, with θ as the angle of rotation between the 0 and 1 frames about the z axis:

$$R_0^1 = \begin{bmatrix} c\theta & -s\theta & 0 \\ s\theta & c\theta & 0 \\ 0 & 0 & 1 \end{bmatrix} \qquad (2)$$

Transformations between reference frames which are displaced only by translations are even simpler, as shown in Figure 3B:

$$P_0 = P_2 + {}^2P_0 \qquad (3)$$

where P_0, P_2 are (3 × 1) vectors which describe the location of the point P with respect to reference frames 0 and 2, respectively. ${}^2P_0$ is also a (3 × 1) vector, and indicates the location of the origin of frame 2 relative to frame 0.

Combined translation and rotation between reference frames can be performed mathematically in one operation (Figure 3c):

$$P_0 = R_0^3 P_3 + {}^3P_0. \qquad (4)$$

This form requires that rotations and translations be considered independently. However, translations and rotations can both be represented using a (4×4) matrix notation which is commonly employed in robot kinematics, as well as for any transformation between frames (e.g. direct linear transforms in motion analysis). All (3×1) position vectors now become (4×1) vectors and carry a 1 in the last position. The transformation contained within equation (4) can now be performed through a sequence of matrix operations, translation first and then rotation

$$P_0 = T_0^3 P_3 \qquad T_0^3 = T_0^2 T_2^3 \tag{5}$$

where

$$T_0^2 = \begin{bmatrix} 1 & 0 & 0 & \Delta x \\ 0 & 1 & 0 & \Delta y \\ 0 & 0 & 1 & \Delta z \\ 0 & 0 & 0 & 1 \end{bmatrix} \qquad T_2^3 = \begin{bmatrix} c\theta & -s\theta & 0 & 0 \\ s\theta & c\theta & 0 & 0 \\ 0 & 0 & 1 & 0 \\ 0 & 0 & 0 & 1 \end{bmatrix} \tag{6}$$

The translation matrix T_0^2 contains the elements of the (3×1) vector $^2P_0$, which indicates the translation of frame 2 with respect to frame 0, while the rotation matrix T_2^3 contains the rotation information in the upper left (3×3) elements. A 1 remains in the corner position of the matrices, and for cases with no relative translation or rotation between frames, the transformation matrix becomes the (4×4) identity. The resulting transformation, T_0^3, represents this combined motion in a single (4×4) matrix:

$$P_0 = T_0^3 P_3 \qquad T_0^3 = \begin{bmatrix} c\theta & -s\theta & 0 & \Delta x \\ s\theta & c\theta & 0 & \Delta y \\ 0 & 0 & 1 & \Delta z \\ 0 & 0 & 0 & 1 \end{bmatrix} \tag{7}$$

Although this notation may appear cumbersome for this simple case, it simplifies the writing of transformations between frames and will prove useful for situations with many frames.

While the above review is not intended to be comprehensive, the notion of building complex motions from 1-DOF motions is one of the fundamental ideas for examining the forward kinematics of robotic manipulators. In addition, the above review introduces the handling of transformations between reference frames using homogeneous (4×4) matrix forms, which will be useful in the description of joint motions and forces/moments at the knee in a later section. Those with additional interest in this area should examine one of the numerous excellent introductory texts on robotics currently available (e.g. Craig, 1989; Paul, 1981).

ROBOTIC CONTROL OF JOINT MOTION

The ability to control the 6-DOF location of the end effector of the robotic manipulator is not sufficient to allow joint biomechanics studies; a method is required to enable the robot to feel the loads on the knee. As is done in most applications of force–moment control, we utilize a 6-DOF UFS mounted to the end effector to provide force–moment feedback. In this way, the measured forces and moments serve as a guide while the robot is used to move the joint. For clinical relevance, it is possible to describe and control joint motions and forces/moments with respect to an anatomical reference system. However, the axes of the anatomical reference system may not correspond to the axes of the sensor coordinate system, and transformations between these frames must be developed.

DESCRIPTION OF JOINT MOTION

Two common methods used to detail knee joint kinematics involve describing this motion: (1) in terms of general rigid-body motion; or (2) in approximately anatomical terms. The most widely used methods for describing motion at the knee using these approaches involve the helical axis representation (Beggs, 1983; Suh and Radcliff, 1978) and the joint coordinate description (Grood and Suntay, 1983). Both of these approaches can be used to completely and accurately describe the relative motion between the femur and tibia in terms of its rotation and translation components. The former represents the simplest description of general motion between two rigid bodies, while the latter attempts to provide some description of general joint motion in 'anatomical' terms. The difference between them involves the way in which overall relative motion is divided into smaller components (rotations and translations).

The helical axis system describes the motion of the tibia relative to the femur with a single translation and rotation along and about the helical axis, which is defined for each increment of joint motion (Blankevoort and Huiskes, 1991; Blankevoort et al., 1986, 1988, 1990; Hart et al., 1991; Jonsson and Karrholm, 1992, 1993; Jonsson, Karrholm and Elmqvist, 1989; de Lange, Huiskes and Kauer, 1990; de Lange et al., 1983; Lewis, Lew and Schmidt, 1988; Woltring et al., 1985). This represents the simplest description of general rigid-body motion between the femur and tibia. For the interested reader, a complete description of this method can be found in many texts on the kinematics of mechanisms (Beggs, 1983; Suh and Radcliff, 1978), and we shall not include details here.

The joint coordinate description, on the other hand, provides a convenient description of general joint motion in anatomical terms (Grood and Suntay, 1983). This description is therefore more palatable to both physicians and engineers, and it has been applied with success to the human knee (Berns, Hull and Patterson, 1990, 1992; Fujie et al., 1993). The joint coordinate description partitions general joint motion into six familiar anatomical motions of the knee: anterior–posterior (AP), medial–lateral (ML) and proximal–distal (PD) translations, and flexion–extension (FE), internal–external (IE) and varus–valgus (VV) rotations. As shown in Figure 4, the description employs an axis which is fixed in the femur, an axis which is fixed in the tibia, and a third, floating axis which is perpendicular to each of these fixed axes. The axes are established based on physical landmarks, such as the posterior aspect of the femoral condyles and the tibial long axis. The details of this description are best reviewed in the original reference by Grood and Suntay (1983).

It is important to note that both the helical axis and joint coordinate descriptions of knee motion can be used to accurately assess and report the relative motion between the tibia and femur. We utilize the joint coordinate description for its anatomical/engineering merit as well as its selection as the ISB standard for reporting kinematics data (Wu and Cavanaugh, 1995). A more practical reason will become evident in the subsequent discussion of joint forces/moments; this description of motion very naturally lends itself to use in the control of joint loading.

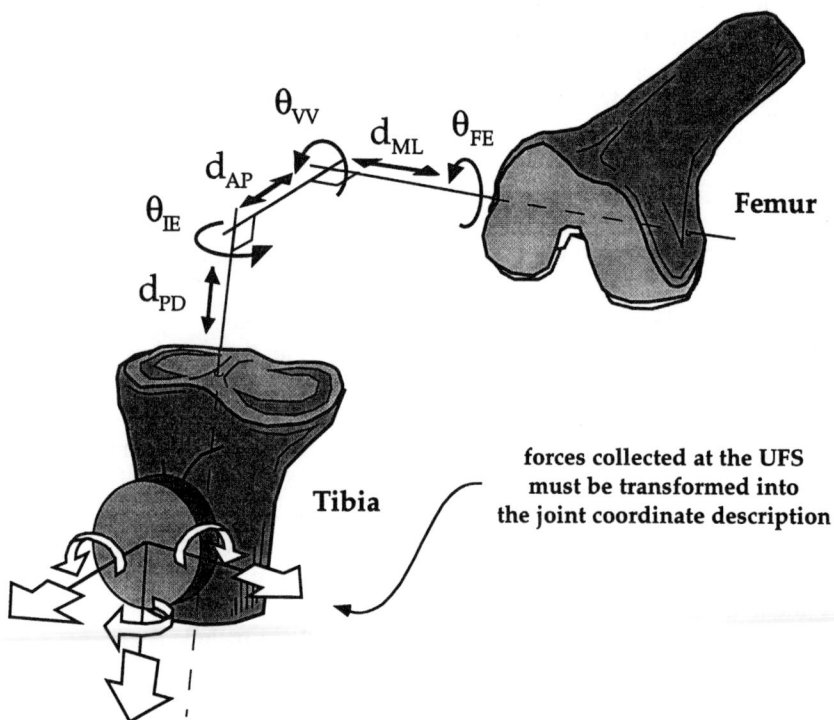

Figure 4. Diagram detailing the relationship between the sensor coordinate system and the multiple degrees of freedom comprising the joint coordinate description

DETERMINING FORCES/MOMENTS IN THE JOINT COORDINATE SYSTEM

In our testing configuration, the forces and moments generated at the joint are measured by a UFS which is rigidly mounted to the tibia (Figure 4). However, the forces and moments measured along and about the axes of the sensor coordinate system are not the same as those expressed relative to the joint coordinate system. If the latter system is utilized for reported joint motion, a method must be developed to transform the forces and moments measured by the sensor (Fujie et al., 1996).

One of the advantages of the joint coordinate system is that it may be considered as being analogous to a serial linkage between the femur and tibia. The 6-DOF end-to-end motion is represented as the result of 1-DOF motions between rigid links, each of which has its own local, orthogonal coordinate system. This allows us to describe joint motions relative to the non-orthogonal joint coordinate axes using ideas borrowed from robot kinematics. Most notably, we can consider that the total 6-DOF motion between the femur and tibia is achieved through six 1-DOF 'joints'—three revolute and three prismatic—analogous to those previously discussed for a robotic manipulator. In this system, the motions of the knee are represented (sequentially, beginning at the femur) by θ_{FE}, d_{ML}, θ_{VV}, d_{AP}, θ_{IE} and d_{PD} respectively. The individual motions allowed between links, each corresponding to an anatomical degree of freedom, are shown in Figure 4 along with their relation to the joint coordinate system. By utilizing the kinematics of these individual 1-DOF motions, we can determine the forces and moments developed in the joint coordinate system from forces and moments measured at the sensor.

We first consider the individual motions of the joint coordinate system. For example PD translation and IE rotation, indicating translation along and rotation about the tibial axis of the knee joint coordinate system, can be represented by homogeneous transformations, $Trans_{PD}$ and Rot_{IE} as:

$$Trans_{PD} = \begin{bmatrix} 1 & 0 & 0 & 0 \\ 0 & 1 & 0 & d_{PD} \\ 0 & 0 & 1 & 0 \\ 0 & 0 & 0 & 1 \end{bmatrix} \quad Rot_{IE} = \begin{bmatrix} C_{IE} & 0 & S_{IE} & 0 \\ 0 & 1 & 0 & 0 \\ -S_{IE} & 0 & C_{IE} & 0 \\ 0 & 0 & 0 & 1 \end{bmatrix} \quad (8)$$

where C_{IE} and S_{IE} represent $\cos(\theta_{IE})$ and $\sin(\theta_{IE})$, respectively.

Similarly, AP translation and VV rotation, which occur along and about the floating axis of the knee joint coordinate system, are represented by $Trans_{AP}$ and Rot_{VV} respectively, and ML translation and FE rotation, along and about the femoral axis of the knee joint coordinate system, are represented by $Trans_{ML}$ and Rot_{FE} respectively.

To obtain the homogeneous transformation describing the position and orientation of the tibial coordinate system relative to each incremental orthogonal coordinate system in the linkage, we utilize the transformation from each of the links preceding it. Therefore, the overall transformation between the femoral and tibial coordinate systems is the product of a sequence of homogeneous transformations corresponding to all three rotations and three translations:

$$X = Rot_{FE}\, Trans_{ML}\, Rot_{VV}\, Trans_{AP}\, Rot_{IE}\, Trans_{PD} \tag{9}$$

This matrix is presented below as follows:

$$X =$$

$$\begin{bmatrix} C_{FE}C_{IE} - S_{FE}S_{VV}S_{IE} & -S_{FE}C_{VV} & C_{FE}S_{IE} + S_{FE}S_{VV}C_{IE} & d_{AP}C_{FE} - d_{PD}S_{FE}C_{VV} \\ S_{FE}C_{IE} + C_{FE}S_{VV}S_{IE} & C_{FE}C_{VV} & S_{FE}S_{IE} - C_{FE}S_{VV}C_{IE} & d_{AP}S_{FE} + d_{PD}C_{FE}C_{VV} \\ -C_{VV}S_{IE} & S_{VV} & C_{VV}C_{IE} & d_{PD}S_{VV} + d_{ML} \\ 0 & 0 & 0 & 1 \end{bmatrix}$$

$$\tag{10}$$

In addition to describing knee motions in the joint coordinate system we can also use this system to express the forces and moments measured at the UFS. The UFS measures three forces and three moments relative to its Cartesian axes; these forces and moments are expressed as a single (6×1) vector, ${}^s\mathbf{F}$, in the following development. This vector is transformed into the joint coordinate description, such that the resulting forces and moments correspond to the single-DOF motions allowed between the incremental orthogonal coordinate systems. Thus we can determine the complete set of forces and moments along and about the axes of the joint coordinate description.

To determine this set of forces, a relationship is first developed in terms of differential motions. Any differential translation or rotation of the tibial coordinate system will correspond to an equivalent set of differential translations and rotations along and about the axes of the joint coordinate description. As each of the differential motions within the knee joint coordinate description is either simple translation or rotation, we can develop a Jacobian matrix which relates the differential motion with respect to the tibial coordinate description to the incremental differential motions with respect to the knee joint coordinate description. This Jacobian matrix is shown below:

$$J' = \begin{bmatrix} d_{AP}S_{VV}S_{IE} - d_{PD}C_{VV}C_{IE} & -C_{VV}S_{IE} & -d_{PD}S_{IE} & C_{IE} & 0 & 0 \\ d_{AP}C_{VV} & S_{VV} & 0 & 0 & 0 & 1 \\ -d_{AP}S_{VV}C_{IE} - d_{PD}C_{VV}S_{IE} & C_{VV}C_{IE} & d_{PD}C_{IE} & S_{IE} & 0 & 0 \\ -C_{VV}S_{IE} & 0 & C_{IE} & 0 & 0 & 0 \\ S_{VV} & 0 & 0 & 0 & 1 & 0 \\ C_{VV}C_{IE} & 0 & S_{IE} & 0 & 0 & 0 \end{bmatrix} \tag{11}$$

The computation involved in the development of the Jacobian matrix is too lengthy to show here, but a detailed explanation of the solution scheme may be found in Paul (1981). Similar ideas are commonly used to determine torque loads in serial rotary joint manipulators, and the application of this Jacobian matrix to the knee joint is discussed in Fujie et al. (1996). A second Jacobian matrix, J'', relating the tibial coordinate system to the sensor coordinate system can be similarly determined. Because the sensor coordinate system is rigidly fixed to the tibial coordinate system, this second Jacobian matrix, J'', is constant and is easily determined by measuring the relative position and orientation between the two coordinate systems. The overall Jacobian matrix, J, relating differential motion with respect to the knee joint coordinate description to motion with respect to the sensor coordinate system, can then be determined.

Through the application of the principle of virtual work, forces and moments with respect to the sensor coordinate system, $^s\mathbf{F}$, can be related to the forces and moments with respect to the knee joint coordinate description, $^J\mathbf{F}$ by

$$^J\mathbf{F} = \mathbf{J}^{T\,s}\mathbf{F} \tag{12}$$

where

$$^s\mathbf{F} = (f_x,\, f_y,\, f_z,\, m_x,\, m_y,\, m_z)^T$$

$$^J\mathbf{F} = (m_{FE},\, f_{ML},\, m_{VV},\, f_{AP},\, m_{IE},\, f_{PD})^T$$

Using this transformation, all forces and moments measured with respect to the sensor coordinate system can now be expressed as forces and moments with respect to the knee joint coordinate description.

For use in controlling the force–moment condition of a test joint, the inverse of the Jacobian matrix can also be determined. This provides a unique relationship between forces and moments along and about the axes of the knee joint coordinate description and the forces and moments measured in a Cartesian sensor coordinate description (Figure 4). The details of this mathematical approach may be found in Paul (1981) and Fujie et al. (1996). Once the desired (or target) forces and moments along and about the axes of the joint coordinate description are specified, the equivalent forces and moments required in the sensordescription—*for any configuration of the joint*—can be simply determined. This mathematical procedure can then be applied to the control of the forces and moments within the knee joint coordinate description using feedback from the UFS.

FORCE-FEEDBACK CONTROL

Once the transformation between the forces and moments measured at the sensor and those applied along and about the axes of the joint coordinate system has been established, the application of specified external loads to the intact joint is relatively straightforward. To apply a given external load, a set of desired joint forces and moments is specified in the joint coordinate description. The forces and moments to be applied to the joint are then determined in the sensor coordinate system. The desired movement of the robotic manipulator is determined by comparing the current forces and moments measured by the UFS (related to joint forces/moments) to the target, or desired, forces and moments. The robot is instructed to perform the movement in order to achieve the target forces and moments. Based on the differences between the target forces and moments and the new forces and moments measured by the UFS, a new movement is calculated and the robot instructed to move the joint accordingly. This iterative process allows the robotic testing system to move the knee through the appropriate motions, such that the desired loads are developed within the joint.

For purposes of specimen positioning, the robotic UFS test system is first used to determine a neutral path of flexion–extension for each joint. The neutral path is the motion obtained by changing the flexion angle with essentially no external force or moment applied to the knee (i.e. the target forces and moments are zero). To find this path of motion, the robot performs a 1° change in the flexion angle. If any significant forces or moments are measured by the UFS, the robot corrects the tibial position to relieve these forces (Figure 5). After finding the neutral position through a series of iterative movements, the robot moves to the next flexion angle to again find the neutral position. Determining this path is an important step, since the natural path of motion is different for every specimen. Once the path of neutral flexion–extension has been recorded, it is used to reposition the joint at each desired flexion angle for testing. This path provides the reference positions for all subsequent application of forces and moments to the knee joint. Selected applications of this procedure are described in the following sections.

IN SITU FORCES IN KNEE LIGAMENTS

Detailed understanding of a ligament's function and contribution to overall joint kinematics is dependent on an accurate determination of the *in situ* forces developed in the ligament in response to motion or external loading of the intact joint. For this reason, several approaches have previously been developed and utilized to examine the *in situ* forces in ligaments of the human body, particularly those experienced by the ACL. These methods include the use of buckle transducers (Barry and Ahmed, 1986; Jasty, Lew and Lewis, 1982; Lew et al., 1990; Lewis and Fraser, 1979; Lewis, Lew and Schmidt, 1982; Lewis et al., 1980), strain gauges affixed near the bony insertions (France et al., 1983), instrumentation of the tibial insertion (Markolf, Wascher and Finerman, 1993; Markolf et al., 1990, 1995; Wascher et al., 1993), radiographic (Vahey and Draganich, 1991) and kinematic linkage measurements (Hollis, 1988; Takai et al., 1993), and, recently, implantable transducers (Holden, Grood and Cummings, 1991; Holden et al., 1992, 1994; Korvick et al., 1992, 1993). In our laboratory,

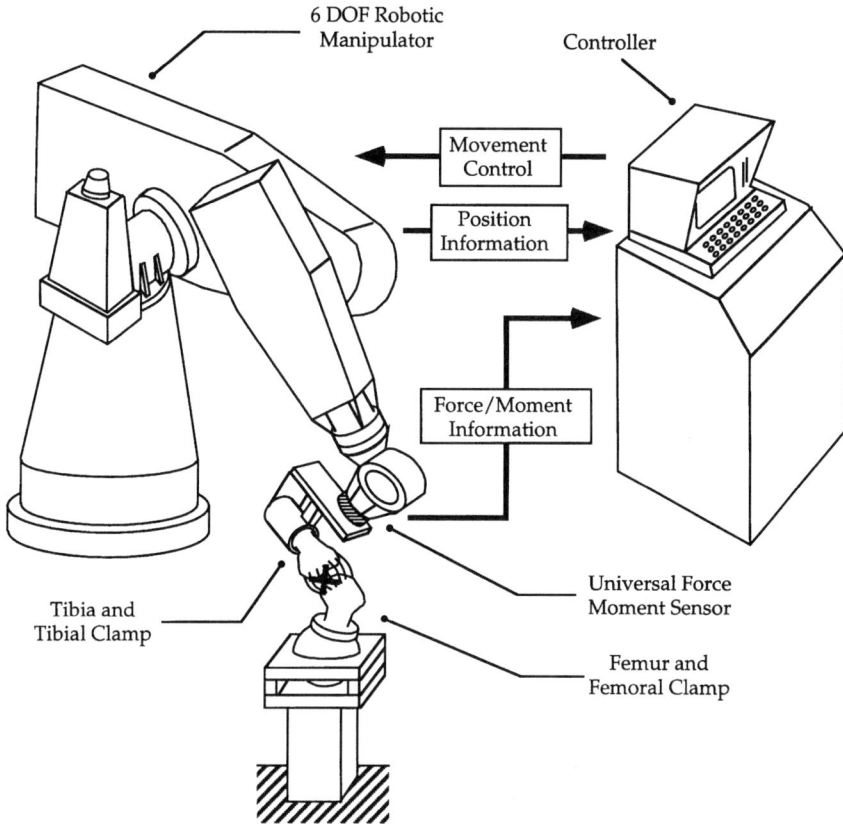

Figure 5. Schematic of the test set-up demonstrating the relationship between the robot, UFS, controller and specimen during testing

we have utilized a different approach to determine *in situ* forces in ligaments which employs a UFS combined with a force transformation scheme. This approach allows direct determination of the forces in ligamentous structures without requiring mechanical contact.

Using a UFS to determine the *in situ* forces in ligaments can be compared to using a force plate to determine the centre of pressure of foot contact on the plate surface. For the case of the force plate, a vertically directed force (a person's weight) is applied to the rigid surface of the plate, and develops forces and moments in the sensor beneath the plate (Figure 6). Based on the combination of measured forces and moments, the location of the centre of pressure on the plate surface can be calculated.

Determination of the force in a ligament is very similar, in that a force is likewise being applied to a rigid body, and all forces and moments occurring due

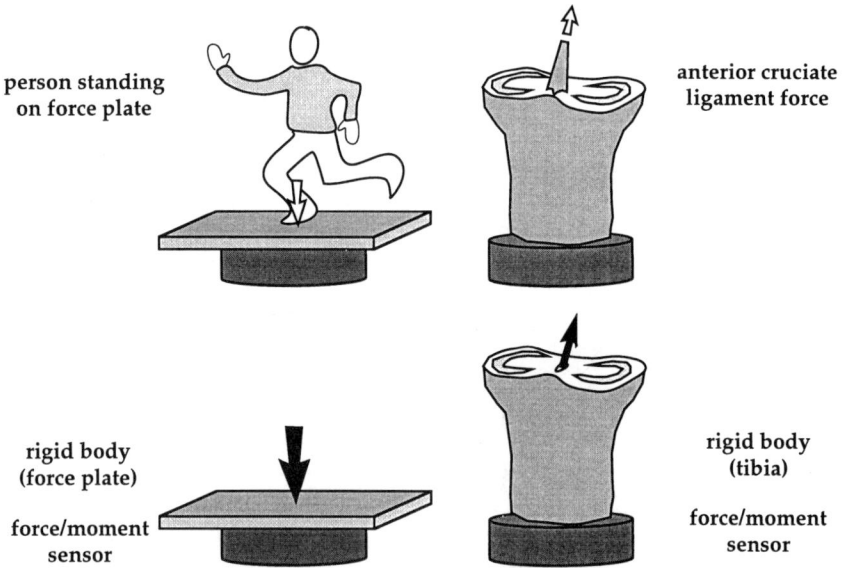

Figure 6. Diagram demonstrating the similarities between force plates and the UFS approach for determining ligament forces

to the ligament force are measured by a UFS. In this case of the force plate, the applied force is essentially vertical, and the vertical force component may be used for determination of the centre of pressure. In contrast, the direction of an applied ligament force is not known *a priori*, and a force transformation scheme is used to account for its direction. This procedure is detailed below and allows the UFS to function as a three-dimensional extension of the force plate.

Originally developed in the field of robotics for force–moment feedback at the end effector of a manipulator, a UFS is capable of measuring three forces and three moments along and about a Cartesian coordinate system fixed with respect to the sensor. A force–moment vector, $^s\mathbf{F}$, consisting of the three forces (f_x, f_y, f_z) and three moments (m_x, m_y, m_z) measured by the UFS, is defined with respect to a Cartesian coordinate system (called the sensor coordinate system):

$$^s\mathbf{F} = (f_x, f_y, f_z, m_x, m_y, m_z)^T \qquad (13)$$

If a rigid body (e.g. the tibia) is connected to the UFS (Figure 6), any external force applied to that body will be measured by the UFS as a combination of forces and moments defined with respect to the sensor coordinate system. The magnitude of the external force can be determined:

$$\text{Magnitude} = \sqrt{(f_x^2 + f_y^2 + f_z^2)} \qquad (14)$$

The direction of the external force can then be determined as follows:

$$a_x = \frac{f_x}{\sqrt{(f_x^2 + f_y^2 + f_z^2)}}, \ a_y = \frac{f_y}{\sqrt{(f_x^2 + f_y^2 + f_z^2)}}, \ a_z = \frac{f_z}{\sqrt{(f_x^2 + f_y^2 + f_z^2)}} \quad (15)$$

where a_x, a_y and a_z represent components of the direction vector, **a**, of the external force.

The point of application of the force is defined as the point where the line of action passes through the rigid body (in this case, through the tibial insertion site of the ACL). To locate the line of action, we determine the point $\mathbf{p} = (p_x, p_y, p_z)^T$ where the external force vector interests the surface of the sensor (i.e. the xy plane of the sensor coordinate system). This is similar to the calculation done for many force plate applications and involves dividing the moments by f_z:

$$p_x = -\frac{m_y}{f_z}, \ p_y = \frac{m_x}{f_z}, \ p_z = 0 \quad (16)$$

If the direction of the force is nearly parallel to the surface of the UFS (and thus f_z is very small), then the accurate determination of its point of application in this way is not possible. Even a slight measurement error will result in a large error in the location of the point **p**.

To avoid this problem, we use a force transformation scheme to rotate the sensor coordinate system mathematically such that the z' axis (of the new Cartesian coordinate system) is parallel to the direction of the external force. This transformation requires only that the direction of the z' axis be represented by the unit vector, **a**, determined from equation (13). The directions of the x' and y' axes are represented by unit orientation vectors **n** and **o**, respectively, which must be perpendicular to the z' axis. Together, these orientation vectors **n**, **o** and **a** make up the direction cosine matrix, which allows for the transformation between the primed and unprimed coordinate systems (Fujie et al., 1996).

The force–moment vector, $^s\mathbf{F}$, measured with respect to the sensor coordinate system (x, y, z), can be transformed to $^s\mathbf{F}$, described with respect to (x', y', z') using this transformation matrix. It can be seen that the line of action of the external force relative to the new coordinate system (x', y', z') can now be determined with very little error, because the external force is perpendicular to the $x'y'$ plane of this coordinate system. A point $\mathbf{p}' = (p'_x, p'_y, p'_z)^T$ along the line of action of the external force can be described with respect to the rotated coordinate system as

$$p'_x = -\frac{m'_y}{f'_z}, \ p'_y = \frac{m'_x}{f'_z}, \ p'_z = \text{arbitrary} \quad (17)$$

Once the line of action is determined with respect to the new, primed coordinate system, it can be mathematically rotated back to its original orientation using the transformation matrix. The line of action of the external force with respect to the sensor coordinate system (x, y, z) is then determined as

$$p_x = \frac{-f_z m_y + f_y m_z + f_x \sqrt{(f_x^2 + f_y^2 + f_z^2)} p}{f_x^2 + f_y^2 + f_z^2}$$

$$p_y = \frac{-f_x m_z + f_z m_x + f_y \sqrt{(f_x^2 + f_y^2 + f_z^2)} p}{f_x^2 + f_y^2 + f_z^2} \tag{18}$$

$$p_z = \frac{-f_y m_x + f_x m_y + f_z \sqrt{(f_x^2 + f_y^2 + f_z^2)} p}{f_x^2 + f_y^2 + f_z^2},$$

where p is a parameter which indicates a location along the line of action of the external force with respect to the sensor coordinate system.

When combined with information about the geometry of the surface of the rigid body (determined with respect to the sensor coordinate system), the point of application of the external force can be determined as the intersection between the line of action of the external force (p) and the surface of the rigid body.

Because the parametric equations that describe the force vector are symmetric, there is no dependence upon the direction of the applied force. The error remains quite small even when the measured force is applied nearly parallel to the surface of the UFS. This technology enables accurate determination of the line of the external force on the tibia such that we can locate the precise point of application of the ligament force, even as it moves during loading.

SELECTED APPLICATIONS IN JOINT BIOMECHANICS

The various factors involved in the study of joint biomechanics using a robotic manipulator are detailed in the previous sections of this chapter. An understanding of the kinematics of the manipulator allows position-based control of the motion of the end effector. The application of known external loading conditions, however, requires the ability to measure the forces and moments applied to the joint. The loads applied to the knee can be further specified and controlled in terms of the joint coordinate system with the appropriate transformation of forces and moments. Evaluation of the *in situ* forces carried by various structures within the intact joint (e.g. the ACL) can then be performed to determine their individual contributions to joint kinematics during loading of the knee. This approach is detailed in several references on the application of robotics to the control of joint kinematics (Fujie et al., 1989, 1990, 1993) and in recent studies involving the determination of ligament forces in addition to the control of joint kinematics (Rudy et al., 1994, 1996).

To illustrate the utility of the robotic–UFS testing system, four selected applications of this system will be presented. These recently completed studies demonstrate different strengths of this approach, including unconstrained motion during testing, determination of the force distribution within ligaments, and the evaluation of reconstructive procedures. The final application presents another advantage of the present testing system; the ability to reproduce intact knee motions in a dissected specimen allows the measurement of *in situ* surface strains on ligaments using optical techniques.

EFFECT OF JOINT CONSTRAINTS ON THE *IN SITU* FORCES OF THE ACL

The complex function of the ACL motivated a study of the effects of different joint constraints placed on knee motion during testing. Specifically, we sought to determine whether excess constraints significantly increased the non-uniformity of the *in situ* force distribution within the ACL (Livesay et al., 1996, 1997). Eight porcine knee joints were dissected free of musculature, leaving the joint capsule and knee ligaments intact. The tibia and femur were cut 20 cm from the joint line and secured within thick-walled aluminium cylinders, and the femoral cylinder was rigidly fixed relative to the base of the robot. The UFS was mounted between the tibial cylinder and the end effector of the robot to serve a role in force–moment feedback (Rudy et al., 1994, 1996) (Figure 5).

After a passive flexion–extension path was determined, the knee was placed in 30° flexion and the robot was used to apply AP loading to ±100 N at a rate of 20 mm/min to the intact joint while allowing 5-DOF (unconstrained) motion through force–moment control (e.g. ML translations were allowed). All joint positions and orientations were recorded by the robot and UFS forces and moments from the with a personal computer. Testing of the intact knee was then performed at 60° and 90° flexion. Next, all tissues except the ACL were removed, and the *identical* path of intact knee motion under this unconstrained tibial loading at 30° flexion was reproduced. Using the recorded UFS outputs, the *in situ* forces in the ACL at 30° flexion were then determined by superposition (Fujie et al., 1995; Livesay et al., 1995). Similar testing was performed at 60° and 90° flexion to determine forces in the ACL at these angles. In addition, the effect of more constrained testing on the *in situ* forces and force distribution within the ACL was evaluated through identical testing with the joint limited to 1-DOF motions (AP translation only). The motions allowed during the unconstrained (5-DOF) and constrained (1-DOF) testing are detailed in Figure 7.

The total tibial translation under ±100 N AP loading was approximately 1.5 times greater for the unconstrained case than for the constrained case. However, the magnitudes of *in situ* forces developed in the ACL were not significantly different; under anterior loading of 100 N at 30° flexion, the *in situ* forces were 108.2 ± 7.9 N and 116.5 ± 9.2 N, respectively. This was also true at 60° and 90°

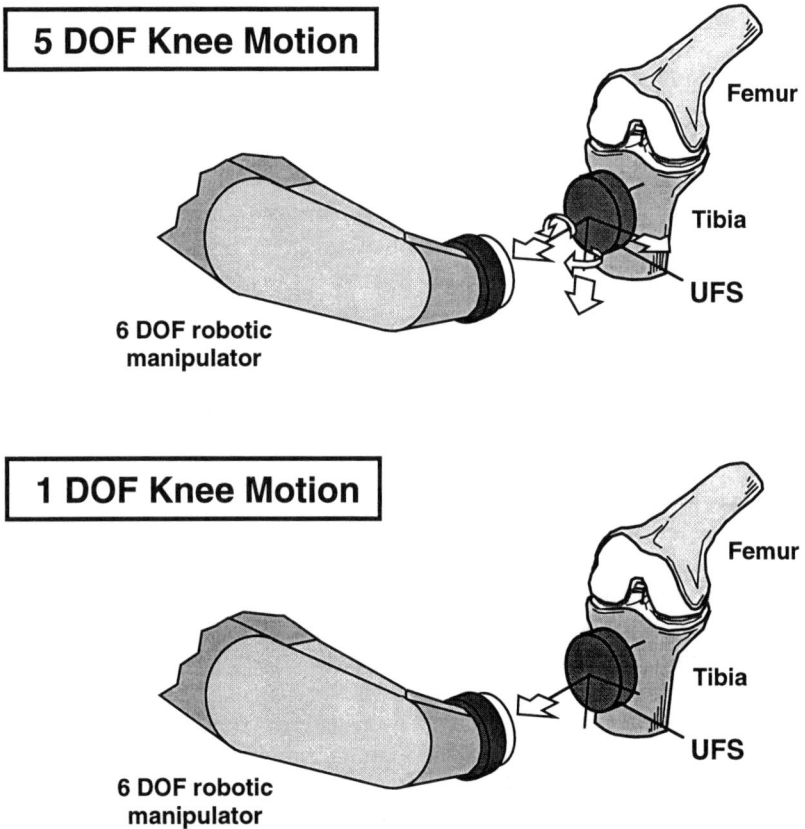

Figure 7. Diagrams detailing the joint motions allowed under 5-DOF and 1-DOF anterior tibial loading. From Livesay et al. (1997) with permission

of flexion. In contrast, the direction of the *in situ* force in the ACL was shown to be significantly different between the unconstrained and constrained cases ($p < 0.05$). The force vector was closer to the tibial plateau, and more in line with the applied load for the unconstrained case (Figure 8). The point of application of the force in the whole ACL was noted to be centrally located within the ACL insertion for the unconstrained case for all flexion angles, but was found to shift anteriorly at higher flexion angles for the constrained case (Figure 9).

This study demonstrates that the constraints placed on the knee lead to differences in the apparent role of the ACL. Through multiple tests on each specimen (one of the primary advantages of this system), the direction of the *in situ* force in the ACL was shown to be significantly affected by more constrained testing, although the magnitude of *in situ* force in the ACL was not (Livesay et al., 1996, 1997).

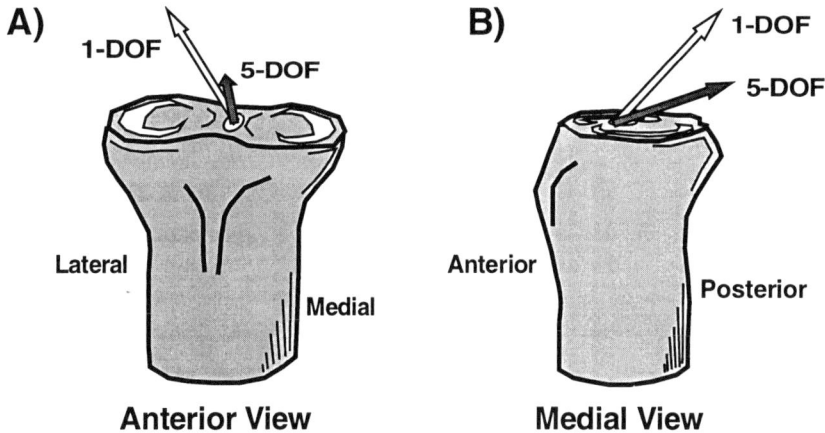

Figure 8. Schematic diagram of the directions of the *in situ* forces in the ACL in a right knee for 100 N anterior tibial loading at 30° flexion under 5-DOF and 1-DOF constraints. From Livesay et al. (1997) with permission

IN SITU FORCE DISTRIBUTION WITHIN THE HUMAN ACL

While the human ACL is too complex to be considered a single uniform rope attached to the tibia and femur, little is known about the distribution of forces within this ligament. The force distribution between the AM and PL bundles of the ACL would also be expected to change with knee flexion angle and the applied external loads during multi-DOF knee motion. To determine force distribution within the human ACL under various external loading conditions, nine human cadaveric knees were prepared and mounted to the robotic–UFS test system and tested with 5-DOF knee motion as described for the previous example (Sakane et al., 1997).

The robot was used to apply AP tibial loading to ±110 N to the intact knee at 0°, 15°, 30°, 60° and 90° of flexion. Resulting knee kinematics were recorded for later reproduction. The AM bundle was identified as the portion of the ACL which was under tension during passive flexion of the knee to 90°, and was transected through a medial parapatellar incision. The robot then repeated the previously recorded 5-DOF motions of the intact knee at the selected knee flexion angles in position control. Resultant forces after the AM bundle was transected were recorded by the UFS, and the vector difference between these forces and the forces of the intact knee were used to determine the *in situ* force of the AM bundle. To determine the *in situ* force in the PL bundle, the remainder of the ACL was then transected and the identical testing procedure repeated. By subtracting the forces after the AM bundle was transected from the forces after the PL bundle was transected, the *in situ* force within the PL bundle was determined as the vector difference. The *in situ* force in the whole ACL required

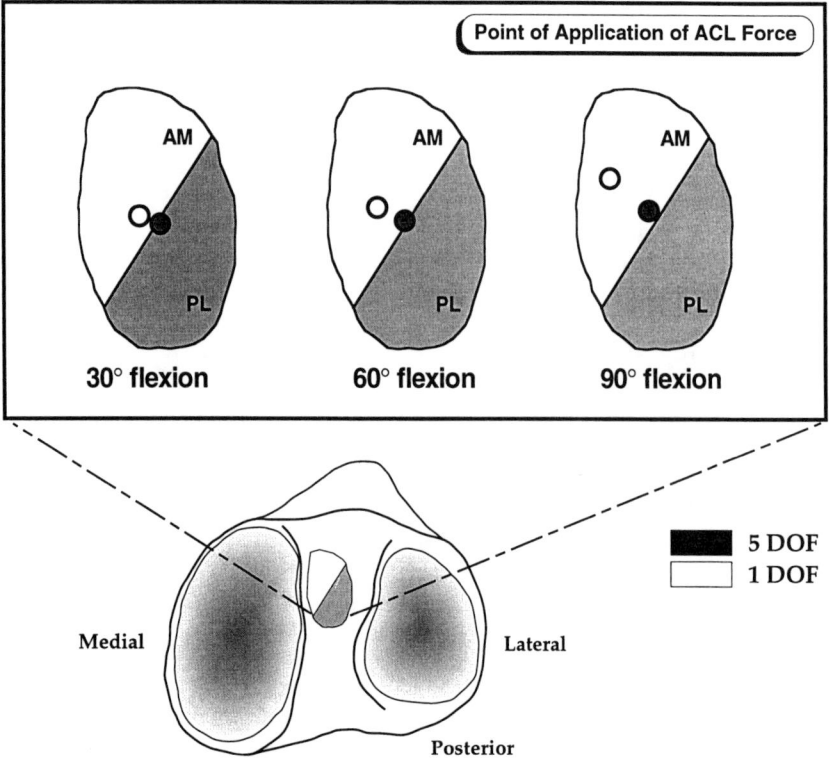

Figure 9. Superior view of the tibial plateau of a right knee, with exploded views showing the point of application of the *in situ* force within the ACL for a representative specimen under 100 N applied anterior tibial loading. From Livesay et al. (1997) with permission

comparison of the force data from the UFS for the intact knee with that from after transection of the PL bundle.

The magnitude of the *in situ* force in the whole ACL under 110 N of anterior tibial loading varied from a high of 110.6 ± 14.8 N at 15° of knee flexion to a low of 71.1 ± 29.5 N at 90° of knee flexion (Sakane et al., 1997). Magnitudes of the *in situ* force in the AM and PL bundles are shown with the magnitudes of the *in situ* force in the whole ACL in Figure 10. The changing magnitude of the *in situ* force in the PL bundle with flexion angle parallels the changing magnitude of the *in situ* force in the whole ACL. The AM bundle, on the other hand, remains relatively constant throughout the range of motion for this loading.

These observations of *in situ* forces in human specimens indicate that both the AM and PL bundles are both important to ACL function. They also confirm the non-uniformity of the ACL, and help to improve our understanding of the relationship between its anatomical complexity and functional contribution.

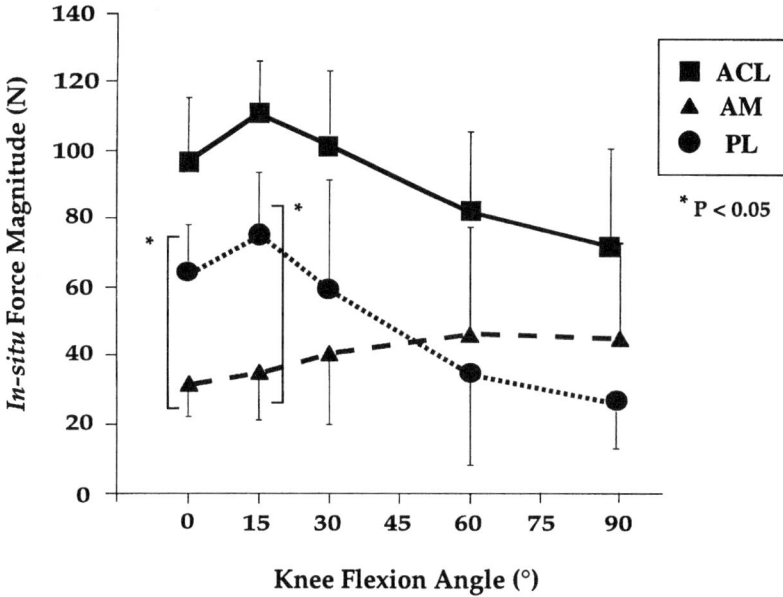

Figure 10. Magnitude of the *in situ* force in the human intact ACL, AM bundle and PL bundle under unconstrained, 110 N anterior tibial loading. From Sakane et al. (1997) with permission.

EFFECT OF TIBIAL FIXATION LEVEL ON THE *IN SITU* FORCES OF AN ACL GRAFT

Despite its current popularity and relative success, endoscopic reconstruction of the ACL using a bone–patellar tendon–bone (BPTB) graft has not yet been perfected. Using a recently developed robotic–UFS testing system (Rudy et al., 1996), we assessed the overall stability of porcine knees following ACL reconstruction with different levels of tibial graft fixation (Ishibashi et al., 1997). Because the different fixation levels could all be evaluated within each specimen, interspecies variability was minimized.

Porcine cadaver knees ($n = 20$) were dissected down to joint capsule and mounted in the robotic system as described previously. Robotic testing of the intact knee to ± 110 N AP tibial loading at $30°$, $60°$ and $90°$ knee flexion was performed first to determine the normal AP displacements. The ACL was then arthroscopically transected, and the path of intact joint motion was reproduced by the robot while forces/moments were measured by the UFS. The *in situ* forces of the ACL were calculated as the difference in forces and moments measured before and after transection of the ACL. The knee was then reconstructed with a BPTB autograft in a simulation of the endoscopic technique, and the distal end of the graft was fixed sequentially at three different levels in each specimen—proximal, middle and distal (Figure 11). Identical AP testing as done for the intact knee was then performed using the robot for each fixation level, and the resulting kinematics and the *in situ* forces of the grafts were compared to the intact case.

The level of tibial fixation was demonstrated to have a significant effect on the resulting anterior displacement and internal rotation of the tibia as well as the *in situ* forces of the graft (Ishibashi et al., 1997). Proximal fixation produced the most stable knee (AP displacements reduced to 120% of intact 30° and 170% at 90°), becoming significantly less stable with more distal fixation. Under 110 N anterior loading at 30° flexion, the forces in the replacement graft relative to the normal ACL forces were $102 \pm 22\%$, $93 \pm 24\%$ and $86 \pm 22\%$ (mean $\pm$ SD) for the proximal, middle and distal fixations, respectively. Proximal fixation also most closely reproduced the *in situ* forces of the normal ACL at 60° and 90° flexion, although these values were reduced. These results suggest that proximal graft fixation may provide the most acute stability of the reconstructed knee.

EVALUATION OF *IN SITU* STRAIN OF THE ACL

Quantitative data on the strain and strain distribution in the ACL during knee motion can also help to characterize the complex role of this ligament. Measuring ligament surface strains in a non-contact manner during normal kinematics requires two things: (1) the ability to locate points on the ligament surface in 3D; and (2) the ability to reproduce intact joint motion in a specimen which is partially dissected to allow visualization of the ligament. For the current work, joint loading and motion reproduction were controlled using a robotic–UFS testing system recently developed in our laboratory (Rudy et al., 1996). Tracking

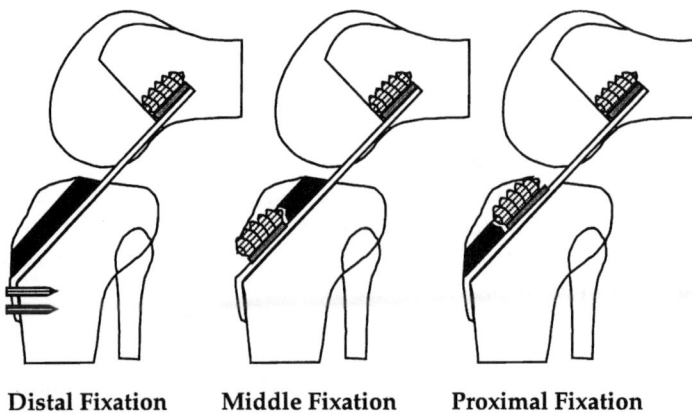

Distal Fixation   Middle Fixation   Proximal Fixation

Figure 11. Schematic diagram of the fixation level of the ACL replacement graft at the tibia. From Ishibashi et al. (1997) with permission

of the 3D positions of markers on the surface of the ACL was performed using a three-camera video digitizing system (Motion Analysis Corp., Santa Rosa, CA) (Harner et al., 1995; Lee and Danto, 1992). For our overall testing configuration, we can track the 3D positions of the surface markers to within 0.1 mm, which corresponds to a strain measurement resolution of at least 0.3% (Runco, 1996; Runco, Sakane and Woo, 1996).

Porcine knees ($n = 8$) were dissected free of musculature and mounted in the robotic system (Rudy et al., 1996). Under force–moment control, the passive flexion–extension path of the intact joint was then learned by the robot–UFS system. AP loading of the intact knee to ±200 N was then applied by the robot at 30°, 60° and 90° flexion. Joint positions were recorded such that they could be reproduced in a later, partially dissected specimen. All soft tissues except the ACL were then removed, and the entire medial femoral condyle (and part of the medial tibial plateau) were removed to allow complete visualization of the ligament surface. A row of seven elastin stain markers (approximately 1.5 mm in diameter, 7 mm apart) was placed longitudinally along the surface of the AM bundle of the ACL, with marker 1 near the tibial insertion and marker 7 near the femoral insertion (Figure 12). The robot then reproduced the previous path of intact knee motion while the video digitizing system was used to track the three-dimensional (3D) positions of the surface markers.

The entire ACL, including insertion sites and small bone blocks, was then dissected out of the joint. The ligament was placed on a smooth, flat surface, hydrated with 0.9% saline solution, and allowed to seek its own rest position.

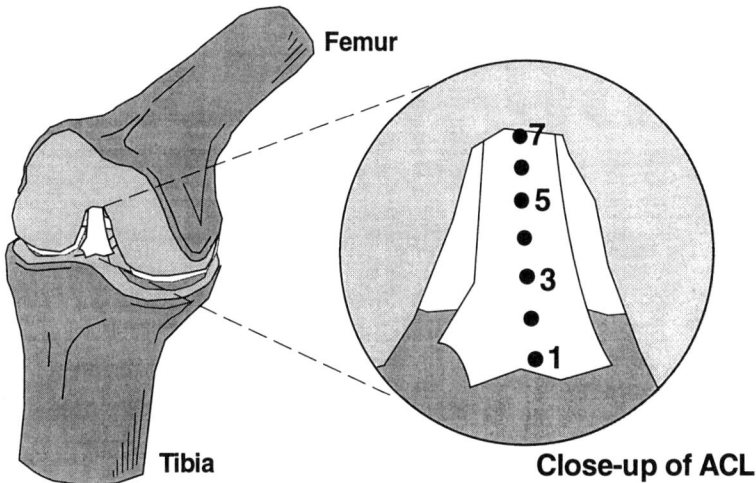

Figure 12. Schematic diagram of the placement of the elastin stain markers along the AM surface of the ACL for non-contact strain measurement. From Runco, Sakane and Woo (1996) with permission

The video digitizing system was then used to record the 3D positions of the markers to determine reference lengths for three subregions along the ACL (markers 1–3, 3–5 and 5–7). Ligament surface strains were then calculated and compared as a function of surface region and flexion angle using a two-way ANOVA. The increase in strain under the 200 N anterior loading was compared using a Fischer PLSD test.

Significant variations in strain between the three regions of the AM bundle were observed for all motions (Runco, 1996; Runco, Sakane and Woo, 1996). Strains were significantly higher under anterior tibial loading than in the passive condition ($p < 0.01$), and variations in strain between the different regions was also more pronounced. The proximal region experienced significantly higher strains at all flexion angles ($p < 0.05$) with a maximum strain of 6.5% under 200 N anterior loading at 60° flexion. The *increase* in strain in this region associated with anterior loading was significantly greater at 90° than at 30° ($p < 0.05$). The middle and distal regions showed successively lesser strains than the proximal region at all flexion angles. The distal region experienced a significant increase in strain due to the anterior tibial loading at both 30° and 60° flexion, but essentially no change at 90°. Significant differences were observed between the strains of all three regions under anterior tibial loading at 90° flexion ($p < 0.05$).

These results clearly demonstrate that the surface strain along the AM bundle of the ACL varies with longitudinal position as well as knee flexion angle. The ability to learn and repeat the intact joint motion using a robotic manipulator allowed the measurement of *in situ* strain of soft tissues in a dissected specimen. When the system is used in conjunction with a 3D video digitizing system, the regional variation of strain for the normal ACL can be measured without any mechanical contact to the ligament.

DISCUSSION AND FUTURE DIRECTIONS

The development of the robotic–UFS testing system has enabled us to examine the kinematics of the knee joint in response to external loads and to assess the function of individual structures within the knee. This has further allowed evaluation of various ligament reconstructions and determination of the surface strains in a non-contact manner. While the applications discussed here have focused on the knee, with particular emphasis on ACL function, the techniques employed can be extended to examine other synovial joints. This flexibility allows the robotic–UFS system to be adapted to apply known external loads and measure the resulting kinematics for a particular joint with minimal changes in programming and clamping. Additionally, the contributions of individual structures within a joint can be evaluated.

One of the most interesting aspects of our biomechanics research as it relates to the field of 3D analysis of human locomotion is that similar problems are studied using different approaches. For example, a gait study might measure the kinematics of the knee during a particular activity (e.g. walking) and then calculate joint loading using information obtained from a floor-mounted force plate. In contrast, we utilize robotic manipulation to apply known external loadings to the knee and obtain the resulting kinematics of the joint. In the future, we anticipate the application of kinematics recorded during gait analysis to a cadaveric specimen in the laboratory. Since the robotic–UFS system can reproduce the joint motion, this approach would enable the direct measurement of joint loads and also allow the function of individual structures to be examined.

Currently, the robotic–UFS system is being used to examine more complex loadings of the knee, such as are produced in sports and during rehabilitation. This allows us to examine the response of the ACL and PCL during the resulting motions, and will improve our understanding of the mechanisms of injury and help in developing protocols for restoring ligament function. We are also adapting this testing procedure to examine other synovial joints, including the human shoulder and spine, where motions are much more complex.

ACKNOWLEDGEMENTS

The authors gratefully acknowledge the financial support of NIH grant #39683 and the University of Pittsburgh Medical Center. The authors are also grateful to Hiromichi Fujie PhD for his significant contributions to the development of this system, and Ross Fox MD, Masataka Sakane MD, Yasuyuki Ishibashi MD and Guoan Li PhD for their work on the sample applications described.

REFERENCES

Barry, D. and Ahmed, A. M. (1986) Design and performance of a modified buckle transducer for the measurement of ligament tension. *J. Biomech. Eng.*, **108**, 149–152.

Beggs, J. S. (1983) *Kinematics*. Hemisphere Publishing Corporation, Washington, DC.

Berns, G. S., Hull, M. L. and Patterson, H. A. (1990) Implementation of a five degree of freedom automated system to determine knee flexibility in vitro. *J. Biomech. Eng.*, **112**, 392–400.

Berns, G. S., Hull, M. L. and Patterson, H. A. (1992) Strain in the anteromedial bundle of the anterior cruciate ligament under combination loading. *J. Orthop. Res.*, **10**, 167–176.

Blankevoort, L. and Huiskes, R. (1991) Ligament–bone interaction in a three-dimensional model of the knee. *J. Biomech. Eng.*, **113**, 263–269.

Blankevoort, L., Huiskes, R. and de Lange, A. (1986) Helical axes along the envelope of passive knee joint motion. *Trans. ORS*, **11**, 410.

Blankevoort, L., Huiskes, R. and de Lange, A. (1990) Helical axes of passive knee joint motions. *J. Biomech.*, **23**, 1219–1229.

Blankevoort, L., Huiskes, R. and de Lange, A. (1988) The envelope of passive knee joint motion. *J. Biomech.*, **21**, 705–720.

Buss, D. D., Warren, R. F., Wickiewicz, T. L., Galinat, B. J. and Panariello, R. (1993) Arthroscopically assisted reconstruction of the anterior cruciate ligament with use of autogenous patellar–ligament graft. *J. Bone Joint Surg.*, **75A**, 1346–1355.

Cerbono, F., Sherman, M. F. and Bonamo, J. R. (1988) Patterns of meniscal injury with acute anterior cruciate ligament tears. *Am. J. Sports Med.*, **16**, 603–609.

Clancy, W. G., Nelson, D. A., Reider, B. and Narechania, R. G. (1982) Anterior cruciate ligament reconstruction using one-third of the patellar ligament, augmented by extra-articular tendon transfers. *J. Bone Joint Surg.*, **64A**, 352–359.

Craig, J. J. (1989) *Introduction to Robotics*, 2nd edn. Addison-Wesley Publishing Company, New York.

Daniel, D. M., Stone, M. K., Dobson, B. E., Fithian, D. C. Rossman, D. J. and Kaufman, K. R. (1994) Fate of the ACL-injured patient. A prospective outcome study. *Am. J. Sports Med.* **22**, 632–644.

Daniel, D. M., Fithian, D. C., Stone, M. K., Dobson, B. E., Luetzow, W. F. and Kaufman, K. R. (1996) A ten-year prospective outcome study of the ACL-injured patient. In: *Proceedings of the Annual Meeting of the American Academy of Orthopedic Surgeons (AAOS)*, Atlanta, GA, p. 77.

Finsterbush, A., Frankl, U., Matan, Y. and Mann, G. (1990) Secondary damage to the knee after isolated injury of the anterior cruciate ligament. *Am. J. Sports Med.* **18**, 475–479.

Fowler, P. J. and Regan, W. D. (1987) The patient with symptomatic chronic anterior cruciate ligament insufficiency: results of minimal arthroscopic surgery and rehabilitation. *Am. J. Sports Med.* **15**, 321–325.

France, P. E., Daniels, A. U., Goble, M. E. and Dunn, H. K. (1983) Simultaneous quantitation of knee ligament force. *J. Biomech.*, **16**, 553–564.

Fujie, H., Mabuchi, K., Tsukamoto, Y., Yamamoto, M., M. Sasada, T. and Arai, S. (1989) Application of robotics to palpation of injury of ligaments: development of a new method of knee instability test. *Tissue Eng.*, **BED-14**, 119–122.

Fujie, H., Mabuchi, K., Tsukamoto, Y., Yamamoto, M. and Sasada, T. (1990) A new method of human knee diagnosis using robotics. In: *Proceedings of the 1st World Congress of Biomechanics* (ed. S. L-Y. Woo), La Jolla, CA, Volume 1, p. 58.

Fujie, H., Mabuchi, K., Woo, S. L-Y., Livesay, G. A., Arai, S. and Tsukamoto, Y. (1993) The use of robotics technology to study human joint kinematics: a new methodology. *J. Biomech. Eng.*, **115**, 211–217.

Fujie, H., Livesay, G. A., Woo, S. L-Y., Kashiwaguchi, S. and Blomstrom, G. (1995) The use of a universal force–moment sensor to determine in-situ forces in ligaments: a new methodology. *J. Biomech. Eng.*, **117**, 1–7.

Fujie, H., Livesay, G. A., Fujita, M. and Woo, S. L-Y. (1996) Forces and moments in six-DOF at the human knee joint: mathematical descriptions for control. *J. Biomech.*, **29**, 1577–1585.

Grood, E. S. and Suntay, W. J. (1983) A joint coordinate system for the clinical description of three-dimensional motions: applications to the knee. *J. Biomech. Eng.*, **105**, 136–144.

Harner, C. D., Xerogeanes, J. W., Livesay, G. A. et al. (1995) The human posterior cruciate ligament: an interdisciplinary study. *Am. J. Sports Med.* **23**, 736–745.

Hart, J. A. L. (1982) Meniscal injury associated with acute and chronic instability of the knee joint. *J. Bone Joint Surg.*, **64B**, 119.

Hart, R. A., Mote, C. D. J. and Skinner, H. B. (1991) A finite helical axis as a landmark for kinematic reference of the knee. *J. Biomech. Eng.*, **113**, 215–222.

Holden, J. P., Grood, E. S. and Cummings, J. F. (1991) The effects of flexion angle and tibial rotation on measurement of anteromedial band force in the goat ACL. *Trans. ORS*, **16**, 588.

Holden, J. P., Korvick, D. L., Grood, E. S., Cummings, J. F. and Bylski-Austrow, D. L. (1992) In-vivo forces in the anterior cruciate ligament during walking and trotting in a quadruped. In: *The Second North American Congress on Biomechanics* (ed. L. Draganich, R. Wells and J. Bechtold), Chicago, IL, pp. 133–134.

Holden, J. P., Grood, E. S., Korvick, D. L., Cummings, J. F., Butler, D. L. and Bylski-Austrow, D. I. (1994) In vivo forces in the anterior cruciate ligament: direct measurements during walking and trotting in a quadruped. *J. Biomech.*, **27**, 517–526.

Hollis, J. M. (1988) *Development and Application of a Method for Determining the In-situ Forces in Anterior Cruciate Ligament Bundles*. PhD thesis, University of California, San Diego (UCSD).

Ishibashi, Y., Rudy, T. W., Livesay, G. A., Stone, J. D., Fu, F. H. and Woo, S. L-Y. (1997) The effect of anterior cruciate ligament graft fixation site at the tibia on knee stability: evaluation using a robotic testing system. *Arthroscopy*, **13**(2), 177–182.

Jasty, M., Lew, W. D. and Lewis, J. L. (1982) In-vitro ligament forces in the normal knee using buckle transducers. *Trans. ORS*, **7**, 241.

Johnson, L. L. (1993) The outcome of a free autogenous semitendinosus tendon graft in human anterior cruciate reconstructive surgery: a histological study. *Arthroscopy*, **9**(2), 131–142.

Johnson, R. J., Beynnon, B. D., Nicholas, C. E. and Renstrom, P. A. (1992) Current concept review: the treatment of injuries to the anterior ligament. *J. Bone Joint Surg.*, **74A**, 140–151.

Jonsson, H. and Karrholm, J. (1992) Kinematics of the weight-bearing knee with and without ACL injury. *Trans. ORS*, **17**, 664.

Jonsson, H. and Karrholm, J. (1993) Helical axis positions during motion of the ACL injured and normal knees. *Trans. ORS*, **18**, 349.

Jonsson, H., Karrholm, J. and Elmqvist, L.-G. (1989) Kinematics of active knee extension after tear of the anterior cruciate ligament. *Am. J. Sports Med.* **17**, 796–802.

Kannus, P. and Jarvien, M. (1987) Conservatively treated tears of the anterior cruciate ligament: long term results. *J. Bone Joint Surg.*, **69A**, 1007–1012.

Kaplan, M. J., Howe, J. G., Fleming, B., Johnson, R. J. and Jarvinen, M. (1991) Anterior cruciate ligament reconstruction using patellar tendon graft. Part II. A specific sport review. *Am. J. Sports Med.* **19**(5), 458–462.

Korvick, D. L., Rupert, M. P., Holden, J. P., Grood, E. S. and Cummings, J. F. (1992) Peak in vivo forces in the anterior cruciate ligament and patellar tendon during various activities: preliminary studies in a goat. *Adv. Bioeng.*, **22**, 95–97.

Korvick, D. L., Holden, J. P., Grood, E. S. and Cummings, J. F. (1993) Relationships between patellar tendon, anterior cruciate ligament and ground reaction forces; an in-vivo study in goats. *Trans. ORS*, **18**, 336.

de Lange, A., Huiskes, R. and Kauer, J. M. G. (1990) Measurement errors in roentgen-stereophotogrammetric joint-motion analysis. *J. Biomech.*, **23**, 259–269.

de Lange, A., van Dijk, R., Huiskes, R. and van Rens, T. J. G. (1983) Three-dimensional experimental assessment of knee ligament length patterns in-vitro. *Trans. ORS*, **8**, 10.

Lee, T. Q. and Danto, M. I. (1992) Application of a continuous video digitizing system for tensile testing of bone–soft tissue–bone complex. *ASME Adv. Bioeng.*, **BED-22**, 87–90.

Lew, W. D., Engebretsen, L., Lewis, J. L., Hunter, R. E. and Kowalczyk, C. (1990) Method for setting total graft force and load sharing in augmented ACL grafts. *J. Orthop. Res.*, **8**(5), 702–711.

Lewis, J. L. and Fraser, G. A. (1979) On the use of buckle transducers to measure knee ligament forces. In: *Biomechanics Symposium*, Proceedings of the Joint ASME-CSME Applied Mechanics, Fluids Engineering, and Bioengineering Conference (ed. W. C. Van Buskirk), Niagara Falls, NY, AMD-32, pp. 71–730.

Lewis, J. L., Lew, W. D. and Schmidt, J. (1982) A note on the application and evaluation of the buckle transducer for knee ligament force measurement. *J. Biomech. Eng.*, **104**, 125–128.

Lewis, J. L., Lew, W. D. and Schmidt, J. (1988) Description and error evaluation of an in-vitro knee joint testing system. *J. Biomech. Eng.*, **110**, 238–2489.

Lewis, J. L., Jasty, M., Schafer, M. and Wixson, R. (1980) Functional load directions for the two bands of the anterior cruciate ligament. *Trans. ORS*, **5**, 307.

Livesay, G. A., Fujie, H., Kashiwaguchi, S., Morrow, D. A., Fu, F. H. and Woo, S. L-Y. (1995) Determination of the in-situ forces and force distribution within the human anterior cruciate ligament. *Ann. Biomed. Eng.*, **23**, 467–474.

Livesay, G. A., Morrow, D. W., Sakane, M., Rudy, T. W., Fu, F. H. and Woo, S. L-Y. (1996) Evaluation of the effect of joint constraints on the force distribution within the ACL. *Trans. ORS*, **21**, 1996.

Livesay, G. A., Rudy, T. W., Woo, S. L-Y., Sakane, M., Li, G. and Fu, F. H. (1997) Evaluation of the effect of joint constraints on the force distribution in the anterior cruciate ligament. *J. Orthop. Res.*, **15**, 278–284.

Lyon, R. M., Akeson, W. H., Amiel, D., Kitabayashi, L. R. and Woo, S. L-Y. (1991) Ultrastructural differences between the cells of the medial collateral ligament and the anterior cruciate ligaments. *Clin. Orthop. Relat. Res.*, **272**, 279–286.

Markolf, K. L., Wascher, D. C. and Finerman, G. A. M. (1993) Direct in vitro measurement of forces in cruciate ligaments. Part II: The effect of section of the posterolateral structures. *J. Bone Joint Surg.*, **75A**, 387–394.

Markolf, K. L., Gorek, J. F., Kabo, J. M. and Shapiro, M. S. (1990) Direct measurement of resultant forces in the anterior cruciate ligament. *J. Bone Joint Surg.*, **72A**, 557–567.

Markolf, K. L., Burchfield, D. M., Shapiro, M. M., Shepard, M. F., Finerman, G. A. M. and Slauterbeck, J. L. (1995) Combined knee loading states that generate high anterior cruciate ligament forces. *J. Orthop. Res.*, **13**, 930–935.

McDaniel, W. J. J. and Dameron, T. B. J. (1980) Untreated ruptures of the anterior cruciate ligament: a follow-up study. *J. Bone Joint Surg.*, **62A**, 696–705.

McDaniel, W. J., Jr and Dameron, T. B., Jr. (1987) The untreated anterior cruciate ligament rupture. *Clin. Orthop.*, **172**, 158–163.

Noyes, F., Matthews, D., Mooar, P. and Grood, E. (1983) The symptomatic anterior cruciate deficient knee: Part 1. The long-term functional disability in athletically active indivi-duals. *J. Bone Joint Surg.*, **65A**, 154–162.

Noyes, F. R., Cummings, J. F., Grood, E. S., Walz-Hasselfeld, K. A. and Wroble, R. R. (1991) The diagnosis of knee motion limits, subluxation, and ligament injury. *Am. J. Sports Med.* **19**(2), 163–171.

O'Brien, S. J., Warren, R. F., Pavlov, H., Panariello, R. and Wickiewicz, T. L. (1991) Reconstruction of the chronically insufficient anterior cruciate ligament with the central third of the patellar ligament. *J. Bone Joint Surg.*, **73A**, 278–286.

Paul, R. P. (1981) *Robot Manipulators: Mathematics, Programming, and Control*. The MIT Press, Cambridge, MA, pp. 217–220.

Praemer, A., Furner, S. and Rice, D. (1992) *Musculoskeletal Conditions in the United States*. American Academy of Orthopaedic Surgeons, Rosemont, IL.

Rudy, T., Livesay, G. A., Xerogeanes, J. W., Takeda, Y., Fu, F. H. and Woo, S. L-Y. (1994) A combined robotics/UFS approach to measure knee kinematics and determine in-situ ACL forces. ASME *Adv. Bioeng.*, **BED-28**, 287–288.

Rudy, T., Livesay, G. A., Woo, S. L-Y. and Fu, F. H. (1996) A combined robotics/universal force sensor approach to determine in-situ forces of knee ligaments. *J. Biomech.*, **29**, 1257–1360.

Runco, T. J. (1996) *A Study of the Non-uniform In-situ Strains on the Surface of the Anterior Cruciate Ligament.* MS thesis, University of Pittsburgh.

Runco, T. J., Sakane, M. and Woo, S. L-Y. (1996) Non-contact evaluation of the in-situ surface strain distribution of the anterior cruciate ligament. *ASME Adv. Bioeng.*, **BED-33**, 173–174.

Sakane, M., Fox, R. J., Woo, S. L-Y., Livesay, G. A., Li, G. and Fu, F. H. (1997) In situ forces in the anterior cruciate ligament and its bundles response to anterior tibial loads. *J. Orthop. Res.*, **15**, 285–293.

Sandberg, R. and Balkfors, B. (1988) The durability of anterior cruciate ligament reconstruction with patellar tendon. *Am. J. Sports Med.* **16**, 341–343.

Seto, J. L., Orofino, A. S., Morrissey, M. C., Medeiros, J. M. and Mason, W. J. (1988) Assessment of quadriceps/hamstring strength, knee ligament stability, functional and sports activity levels five years after anterior cruciate ligament reconstruction. *Am. J. Sports Med.* **16**, 170–180.

Shino, K., Inoue, M., Horibe, S., Hamada, M. and Ono, K. (1990) Reconstruction of the anterior cruciate ligament using allogenic tendon. *Am. J. Sports Med.* **18**(5), 457–465.

Suh, C. H. and Radcliff, C. W. (1978) *Kinematics and Mechanisms Design.* John Wiley, New York.

Takai, S., Livesay, G. A., Woo, S. L-Y., Adams, D. J. and Fu, F. H. (1993) Determination of the in-situ loads on the human anterior cruciate ligament. *J. Orthop. Res.*, **11**, 686–695.

Tibone, J. E., Antich, T. J., Perry, J. and Moynes, D. (1988) Functional analysis of untreated and reconstructed posterior cruciate ligament injuries. *Am. J. Sports Med.* **16**, 217–233.

Vahey, J. W. and Draganich, L. F. (1991) Tensions in the anterior and posterior cruciate ligaments of the knee during passive loading: predicting the ligament loads from in-situ measurements. *J. Orthop. Res.*, **9**, 529–538.

Wascher, D. C., Markolf, K. L., Shapiro, M. S. and Finerman, G. A. M. (1993) Direct in vitro measurement of the forces in cruciate ligaments. Part I: The effect of multiplane loading in the intact knee. *J. Bone Joint Surg.*, **75A**, 377–386.

Woltring, H. J., Huiskes, R., de Lange, A. and Veldpaus, F. E. (1985) Finite centroid and helical axis estimation from noisy landmark measurements in the study of human joint kinematics. *J. Biomech.*, **18**, 379–389.

Wu, G. and Cavanaugh, P. R. (1995) ISB recommendations for standardization in the reporting of kinematic data. *J. Biomech.*, **28**, 1257–1261.

13

Neural Network Models of the Locomotor Apparatus

CHRISTOPHER L. VAUGHAN

Department of Biomedical Engineering, University of Cape Town, Cape Town, South Africa

INTRODUCTION

Human gait is characterized by smooth, regular and repeating movements. Such coordinated motion requires a complex underlying control system. A central question facing those who focus on the locomotor apparatus is: what is the nature of this control system? Clearly, the interaction between the central nervous system (CNS), the peripheral nervous system (PNS) and the musculoskeletal effector system is crucial for coordinated movement to occur. The cause-and-effect sequence of events, illustrated in Figure 1, may be summarized as follows (Vaughan, Davis and O'Connor, 1992):

1. registration and activation of the gait command in the CNS;
2. transmission of the gait signals to the PNS;
3. contraction of muscles that develop tension;
4. generation of forces at, and torques across, synovial joints;
5. regulation of joint forces and torques by the rigid skeletal segments based on their anthropometry;
6. displacement (i.e. movement) of the segments in a manner that is recognized as functional gait;
7. generation of ground reaction forces.

While much of the research on human gait has tended to concentrate on events (4) to (7), and there has been some effort to understand the relationship between

Three-dimensional Analysis of Human Locomotion. Edited by P. Allard, A. Cappozzo, A. Lundberg and C. Vaughan
© 1997 John Wiley & Sons Ltd. ISBN 0 471 96949 4

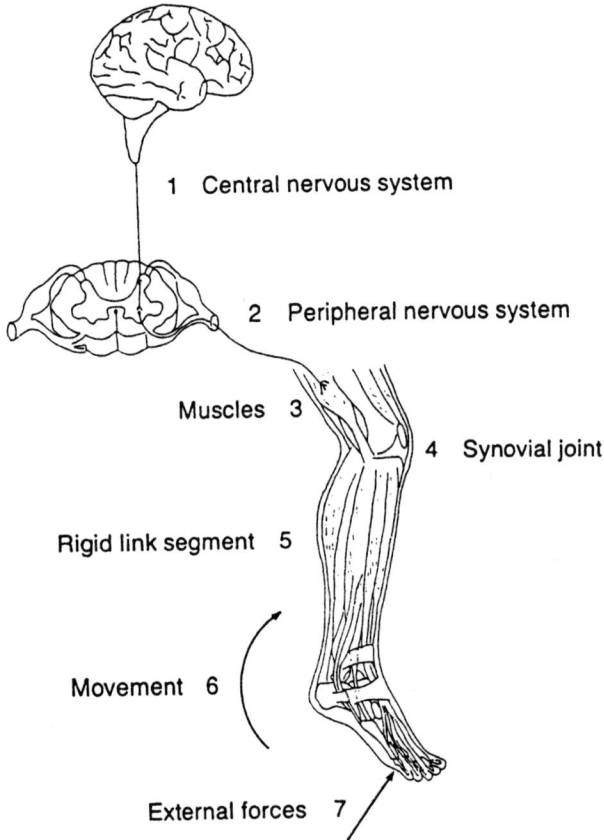

Figure 1. The cause-and-effect sequence of events that occurs in the neural, muscular and skeletal systems when a person walks. Note that this illustration focuses only on the feedforward mechanisms; feedback has not been included. Reproduced from Vaughan, Davis and O'Connor (1992) with permission

(3) and (4), there have been very few attempts to study events (1) to (3) (Vaughan and Sussman, 1993). The cascade of events (1) to (7) emphasizes the top-down control of human gait, but the role played by sensory feedback (e.g. baroreceptors, joint receptors, muscle spindles) should not be overlooked (Loeb, Levine and He, 1990). It was the realization that a true understanding of human locomotion depended on a model of the nervous system that led us to the field of artificial neural networks (Sepulveda, Wells and Vaughan, 1993).

Despite the fact that artificial neural networks (ANNs) have been in existence for over 50 years (McCulloch and Pitts, 1943), it is only in the last decade that they have emerged as a serious and accepted area of research endeavour. An article in *Time* by Gorman, Brown and McCarroll (1988) provided legitimacy to the field in the eyes of the lay public, although there were a few instances of hyperbole. One enthusiastic supporter of the technology proclaimed that 'these machines will be the steam engines of the 21st century' while a detractor argued that 'the only thing they [ANNs] have in common with the human brain is the word neural' (Gorman, Brown and McCarroll, 1988). From their inception, the design of ANNs has been motivated by the recognition that the brain computes in an entirely different manner from the conventional digital computer (Haykin, 1994).

The fundamental unit of the nervous system (including the brain) is the neuron. Typical neurons found in the CNS have a nucleus, or *soma*, with simple processing abilities; *dendrites* (backward extensions) that incorporate *synapses* (or connection sites) for incoming units; and a forward extension known as an *axon*, through which information is carried to other neurons by means of an action potential (Figure 2). The output from each neuron is determined by the nuclear processing, utilizing the transfer or activation function, and the net summation of the incoming excitatory and inhibitory stimuli. The excitatory level of a connection site is also known as the synaptic weight and is thought to be the variable that determines the actual behaviour of a group of neurons.

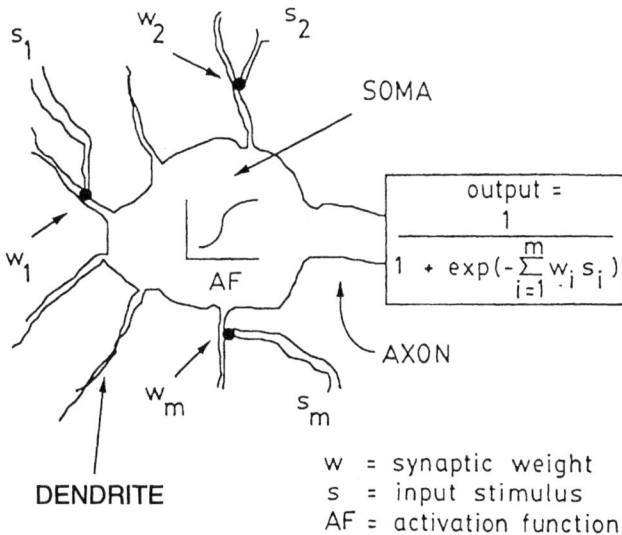

$$\text{output} = \frac{1}{1 + \exp\left(-\sum_{i=1}^{m} w_i \, s_i\right)}$$

w = synaptic weight
s = input stimulus
AF = activation function

Figure 2. The fundamental unit of the nervous system is the neuron. The output from the neuron may be modelled by an exponential function (Sepulveda, Wells and Vaughan, 1993)

An artificial neural network, as the name implies, is a group of many neurons, or processing elements, that are interconnected and distributed in layers. The following definition has been adapted from Haykin (1994):

An ANN is a massively parallel distributed processor that has a natural propensity for storing knowledge based on experience and making this available for later use. The ANN resembles the brain and the CNS in two respects:

1. *Knowledge is acquired by the network through a learning process.*
2. *Interneuron connection strengths, known as synaptic weights, are used to store the knowledge.*

Haykin (1994) suggested that ANNs offer the following useful properties:

1. *Nonlinearity.* This is important, particularly if the underlying physical mechanism responsible for an input signal is nonlinear.
2. *Input–output mapping.* The ANN learns from examples by constructing an input–output map.
3. *Adaptivity.* The ANNs have a built-in capability to adapt their synaptic weights to changes in the surrounding environment and can thus be easily retrained.
4. *Evidential response.* In pattern recognition, an ANN not only makes a selection but it can also provide information about the reliability of the decision made.
5. *Contextual information.* Every neuron in the ANN is affected by the global activity of all other neurons in the network, thereby dealing naturally with contextual information.
6. *Fault tolerance.* When ANNs implemented in hardware are damaged, performance is merely degraded rather than catastrophic failure being precipitated.
7. *Hardware implementation.* The massively parallel nature of ANNs enables them to be implemented using VLSI (very large scale integration) technology.
8. *Uniformity of analysis and design.* The neurons are common to all ANNs, allowing for the sharing of theories and learning algorithms and integration of modules.
9. *Neurobiological analogy.* ANNs can guide the research of biologists and engineers can look to neurobiology for new ideas to solve complex problems.

All of these properties are believed to have relevance for those involved in the modelling of the locomotor apparatus, particularly as this relates to the control of bipedal gait.

DIFFERENT ARTIFICIAL NEURAL NETWORK (ANN) ARCHITECTURES

As described in the previous section, a neuron is the information-processing unit that is fundamental to the operation of an ANN. Figure 3 illustrates a nonlinear model of a neuron. This model has three basic elements (Haykin, 1994):

1. A set of synapses, each of which is characterized by a weight. Specifically, a signal x_j at the input of synapse j which is connected to neuron k is multiplied by the synaptic weight w_{kj}. The weight w_{kj} is positive if the synapse is excitatory and negative if the synapse is inhibitory.
2. An adder for summing the input signals.
3. An activation function for limiting the amplitude of the neuron's output. The model also includes an externally appplied threshold θ_k that alters the input to the activation function. A bias term, the negative of the threshold, is sometimes employed. Typically, the output is in the range [0, 1].

Mathematically, the model may be described by three equations:

$$u_k = \sum_{j=1}^{p} w_{kj} x_j \tag{1}$$

$$y_k = \varphi(u_k - \theta_k) \tag{2}$$

$$v_k = u_k - \theta_k \tag{3}$$

where $x_1, x_2, \ldots, x_p$ are the input signals, $w_{k1}, w_{k2}, \ldots, w_{kp}$ are the synaptic weights of neuron k, u_k is the linear combiner output, θ_k is the threshold, $\varphi(.)$ is the activation function, y_k is the output signal of the neuron, and v_k is the

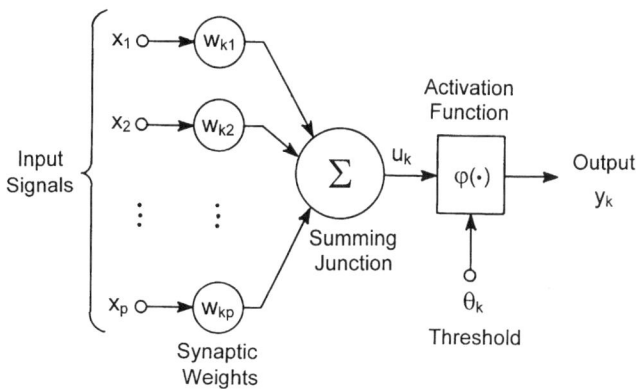

Figure 3. The basic building block of a simple neural network consists of a nonlinear model of a neuron (Haykin, 1994). It includes multiple input signals and synaptic weights which feed into a summing junction, an activation function, threshold and output

activation potential of neuron k. While the activation function is sometimes represented by a simple threshold function, or a piecewise-linear function, the most common form is the sigmoidal function. This can be represented either by a logistic function or a hyperbolic tangent function:

$$\varphi(v) = 1/[1 + \exp(-v)] \tag{4}$$

or

$$\varphi(v) = \tanh(v/2) \tag{5}$$

In general, there are three main classes of network architecture: single-layer feedforward networks; multilayer feedforward networks; and recurrent networks (Haykin, 1994).

In the simplest form of layered network, an input layer of source nodes projects onto an output layer of neurons in a feedforward arrangement (Figure 4a). Since no computation is performed by the input nodes, this network is referred to as a *single-layer feedforward network*. A common application for this network is a linear associative memory: an output pattern (or vector) is associated with an input pattern (vector).

A second class of feedforward ANNs is distinguished by the addition of one or more hidden layers, and the computation nodes are referred to as hidden units or hidden neurons (Figure 4b). The addition of these layers enables the *multilayer feedforward network* to extract higher-order statistics. The ANN is said to be fully connected (as illustrated in Figure 4b) when every node in each layer of the network is connected to every other node in the adjacent forward layer. There are instances, however, when some of the synaptic connections are missing, and the network is said to be partially connected. Such networks are designed and constructed to reflect prior information about the characteristics of the activation pattern being classified (Haykin, 1994).

A *recurrent network* is distinguished from a feedforward ANN in that it has at least one feedback loop (Figure 4c). The presence of these feedback loops has a profound influence on both the learning capability of the ANN and its performance. The unit-delay operators (cf. Figure 4c) produce nonlinear dynamical behaviour (Haykin, 1994) which is an important characteristic of biological systems.

This section would not be complete without mention of the learning process. An ANN has the ability to learn from its environment and to improve its performance through learning. This process implies the folowing sequence of events (Haykin, 1994):

Figure 4. Neurons may be combined in a variety of architectures: (a) a feedforward network with a single layer of neurons; (b) a fully connected feedforward network with one hidden layer and output; and (c) a recurrent network with hidden neurons (Haykin, 1994)

Input Layer
of Source
Nodes

Layer of
Output
Neurons

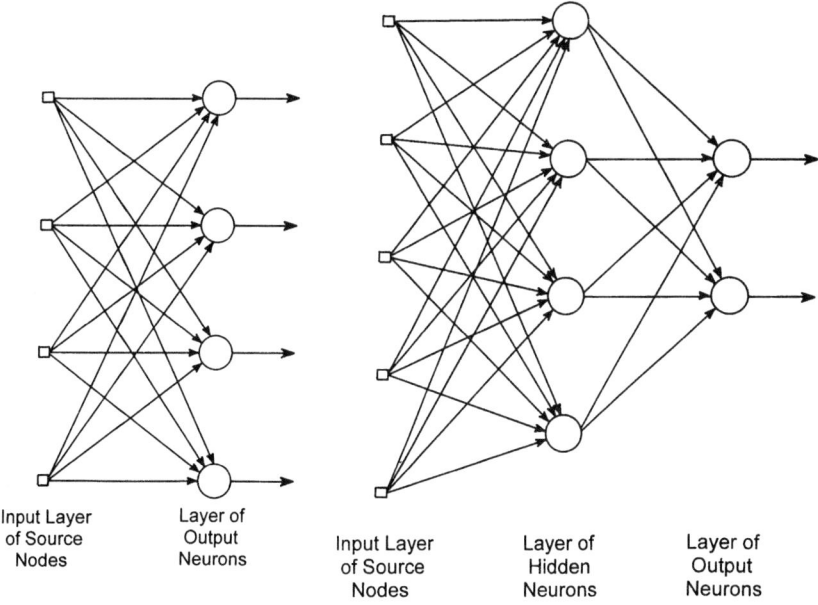

(a) Feedforward network with a
single layer of neurons

Input Layer
of Source
Nodes

Layer of
Hidden
Neurons

Layer of
Output
Neurons

(b) Fully connected feedforward network with
one hidden layer and output

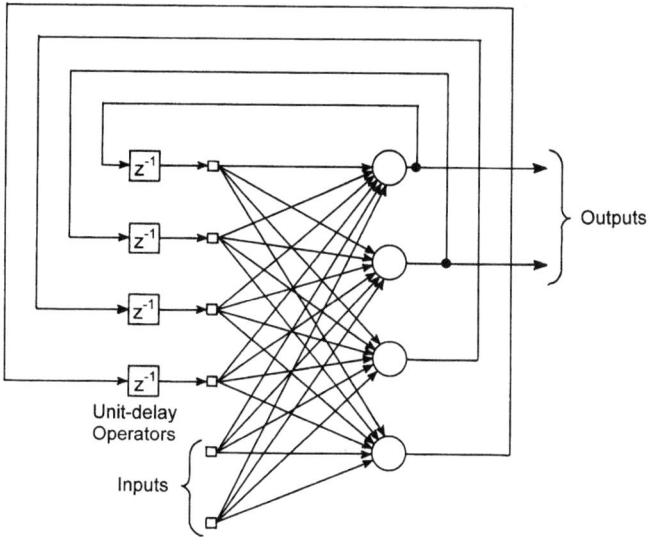

Outputs

Unit-delay
Operators

Inputs

(c) Recurrent network with hidden neurons

1. The ANN is stimulated by an environment.
2. The ANN undergoes changes as a result of this stimulation.
3. The ANN responds in a new way to the environment as a direct result of the changes that have occurred in its internal structure.

Simply put, the purpose of learning is to find the optimum set of synaptic weights w_{kj} which will enable the ANN to generate the output vector y_k when presented with the input vector x_j. There are four basic learning algorithms (or rules): error correction, Hebbian, competitive and Boltzmann. *Error-correction learning* is based on traditional optimal design principles. In contrast, *Hebbian* and *competitive learning* are inspired by the lessons of neurobiology, while *Boltzmann learning* is based on information theory (Haykin, 1994). The learning paradigm refers to the manner in which an ANN relates to its environment, and there are three basic classes of paradigm: supervised learning (under the watchful eye of an external 'teacher'); reinforcement learning (involving the use of a 'critic' through a trial and error process); and self-organized learning (where there is no supervision).

Having introduced and defined ANNs, described some of the basic mathematics, and discussed different ANN architectures and learning processes, it is now appropriate to address the application of ANNs to the locomotor apparatus. Specifically, the areas to be covered include: modelling human locomotion; functional electrical stimulation and rehabilitation; classification of gait patterns; and future opportunities.

ANNs AND MODELLING HUMAN LOCOMOTION

The field of robotics utilized ANNs to replace the traditional inverse dynamics algorithms almost a decade ago (Kawato, Furukawa and Suzuki, 1987). Shortly thereafter, the application of ANNs to study the movement of real biological systems appeared (Massone and Bizzi, 1989). We extended our work on reaching in the upper extremity (Wells and Vaughan, 1989) and applied it to human locomotion (Sepulveda, Wells and Vaughan, 1993). The model was based on a multilayer feedforward network architecture (cf. Figure 4b). We hypothesized that the input signals (electromyography or EMG) could be used to predict the output signals (joint moments). The ANN was trained with the back-propagation algorithm (supervised learning), which was composed of two stages: a feedforward step, where neuronal outputs were specified; and a feedback stage, where the synaptic weights and bias terms were updated. The two steps were repeated for several patterns (sets of input and output values) until the difference between the predicted output and the expected values was below a specified tolerance value. This learning phase is illustrated in the left-hand flow chart of Figure 5. The input signals were the EMG values for 16 muscles (gluteus medius and

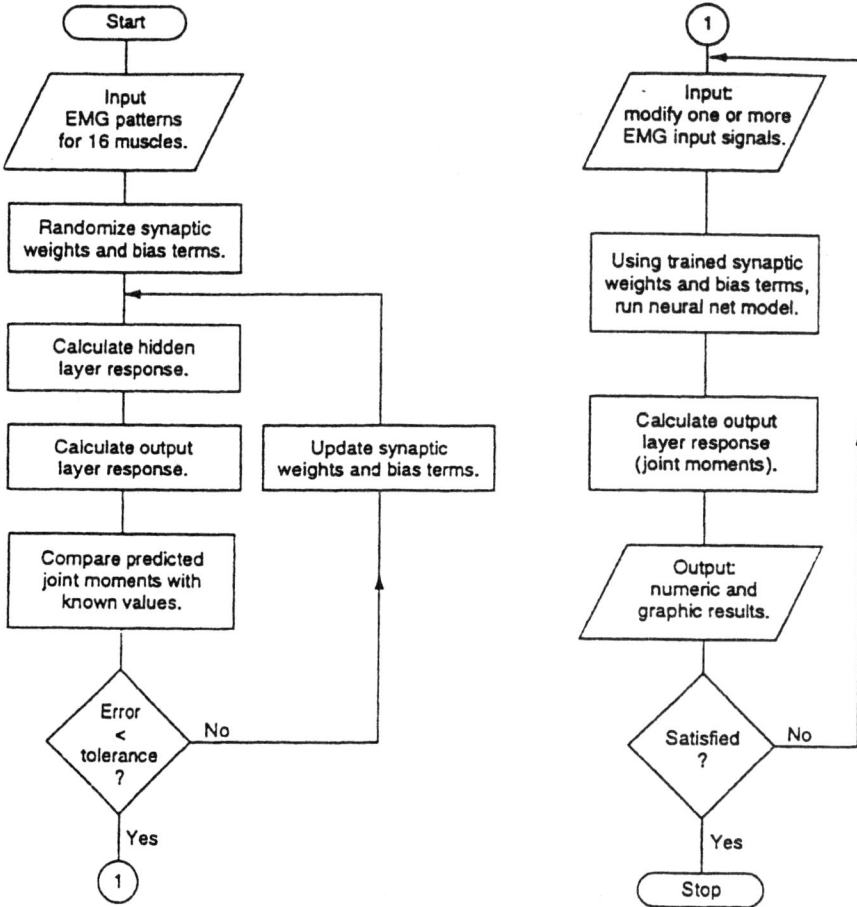

Figure 5. The back-propagation algorithm used to train an ANN may be illustrated in flow chart form (left-hand side), while the actual running of the network is shown on the right (Vaughan, Brooking and Olree, 1996)

maximus, semitendinosus, biceps femoris, erector spinae, sartorius, rectus femoris, vastus lateralis, adductor longus and magnus, tibialis anterior, extensor digitorum longus, medial and lateral gastrocnemius, soleus and peroneus) and the output signals were the three sagittal plane moments at the hip, knee and ankle joints (Winter, 1987). Successful training of the ANN (left-hand side of Figure 5) took approximately 100 000 iterations, while simulating any abnormality required no more than a fraction of a second for a single feedforward loop (right-hand side of Figure 5).

Two abnormalities were simulated: a 30% reduction in soleus activity, and a complete elimination of the rectus femoris (Sepulveda, Wells and Vaughan, 1993). Figure 6 depicts the network prediction for the reduction in soleus activity during the entire gait cycle. It shows a clear decrease in plantarflexor (i.e. extensor) ankle moment during most of the stance phase and especially just prior to toe-off. This prediction is consistent with what we might expect, since the soleus, which is an important plantarflexor, reaches maximum activity between 40% and 60% of the gait cycle. Despite the success of our ANN, we nevertheless recognized a number of shortcomings: there were no feedback loops; the physical arrangement of muscles was ignored; and time did not appear as an explicit variable (Vaughan and Sussman, 1993).

Heller et al. (1993) took a different approach from ours (Sepulveda, Wells and Vaughan, 1993). They used joint angles as the input to their ANN and EMG as the output. They gathered unilateral data (hip and knee angles and the EMG for

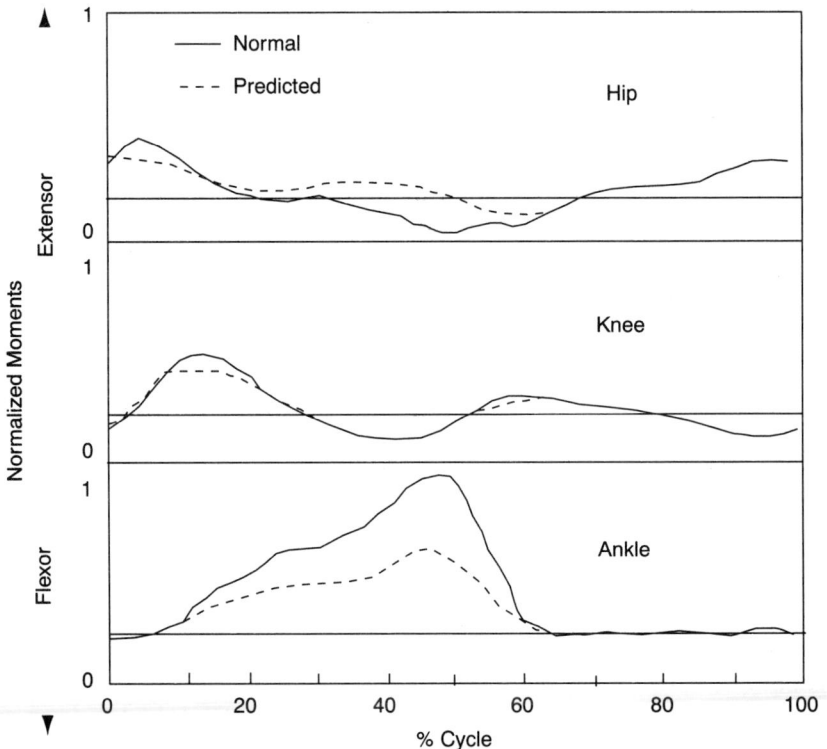

Figure 6. An abnormality—30% reduction in soleus activity—may be simulated using an ANN (Sepulveda, Wells and Vaughan, 1993). The solid lines (normal) show the moments at the hip, knee and ankle joints before simulation. The dashed lines (predicted) are after the simulated reduction in soleus electromyography

the ipsilateral semitendinosus and vastus medialis muscles) for a normal male at two walking speeds. Their ANNs were based on two multilayer feedforward networks (Figure 4b)—one with a single hidden layer and another with two hidden layers—and training was accomplished with the back-propagation algorithm. They compared their ANN predictions with those from symbolic learning based on fuzzy set theory. Both of these inductive approaches were successful in reproducing the correct timing in the EMG patterns, although there were some problems with predicting the correct magnitudes. While Heller et al. (1993) felt that their ANN showed some promise in modelling cyclic movements such as locomotion, it did not provide them with biomechanical insight.

One of the most intriguing models of the locomotor apparatus is that developed by Taga, Yamaguchi and Shimizu (1991) and Taga (1994). They linked computational neuroscience (i.e. ANN theory) with biomechanics from the perspective of nonlinear dynamical theory. Their bipedal model consisted of eight segments, 10 degrees of freedom, and 19 muscle actuators. There was a pair of neural rhythm generators for the trunk, and the left and right hip, knee and ankle joints. Each of these generators was modelled by an ANN which integrated sensory and motor information. Once the model had been trained, it not only produced level gait under normal conditions, but it was also able to adapt to various environmental perturbations such as uneven terrain or increased carrying load. Taga (1994) has demonstrated that the speed of walking could be controlled by a single parameter which drove the neural oscillators, and the step cycle could be entrained by a rhythmic input to the neural oscillators.

Although it would obviously be advantageous to study ANNs directly in humans (Eke-Okoro, 1994), this approach has serious methodological drawbacks because any direct measurements would, of necessity, be invasive, and the sheer complexity of such a study would be enormous. Savelberg and Herzog (1995) circumvented this problem by using ANNs to predict muscle forces in cats. Both EMG and muscle tension were measured *in vivo* for the soleus and gastrocnemius muscles in three animals walking and trotting at four different speeds on a motor-driven treadmill. They used their ANNs (multilayer feedforward networks based on back-propagation training) to test the generalizability of the models to session, speed and subject. They showed that their model provided extremely good force predictions for the different conditions (correlation coefficients 0.88–0.96) and concluded that the ANN approach had considerable promise.

FUNCTIONAL ELECTRICAL STIMULATION AND REHABILITATION WITH ANNs

For patients with a lesion to the spinal cord, the loss of voluntary control of their muscles provides significant challenges for locomotion. Functional electrical stimulation (FES), in which the nerves or muscles are artificially stimulated, has provided some hope to paraplegics that they might one day walk again. The challenge for the engineer designing such an FES system is to decide how best to manage the *control* of the muscle activation patterns. One of the first groups to recognize the potential of ANNs for this task was Popovic et al. (1993) at the University of Alberta in Canada. Their basic experimental work was conducted on chronic, spinalized cats: the inputs to their adaptive logic ANN were the neural recordings from the superficial peroneal and tibial nerves, while the output was the EMG of the medial gastrocnemius. They were satisfied with the success of their ANN model but were cautious about its generalizability for implementation in human patients (Popovic et al., 1993).

Another FES group, this one based in Chicago, has applied an ANN controller to produce ambulation by paralysed patients with spinal cord injuries (Graupe and Kordylewski, 1995). Based on their prior work with FES, and apparently oblivious of the basic research done by the Alberta group (Popovic et al., 1993), Graupe and Kordylewski (1995) based their ANN on adaptive resonance theory. The inputs to their controller were the EMG patterns for trunk muscles above the level of the spinal cord lesion, while the four outputs were the FES patterns to the quadriceps and hamstrings on both legs (Figure 7). Use of a 'vigilance parameter' enabled the authors to test the operation of their ANN controller on a paralysed patient and they concluded that their system allowed safe ambulation in the face of system errors.

Building on our work together at Clemson University (Sepulveda, Wells and Vaughan, 1993), Sepulveda and Cliquet (1995a,b) have also applied ANN theory to the restoration of gait in spinal cord-injured patients. This Brazilian group used knee angles as the input to their ANN, and the stimulation signals to the femoral and peroneal nerves served as output. During the 'example-gathering' phase (Figure 8), the patient was allowed to walk with an untrained ANN. An observer made the desired changes to the FES amplitude via computer keyboard based on gait cycle-to-cycle observations. This information was used to train the ANN. The 'ready-to-use' ANN was later employed in automatic control, with a unique patient-specific network. While they had some success with their approach, Sepulveda and Cliquet (1995b) recommended separate ANNs for the stance and swing phases as well as a safety circuit to override the automatic controller. They implemented this feature by means of voice recognition technology.

In a rehabilitation context, Guiraud (1994) has designed an ANN for the control of an active external orthosis of the lower limbs (Figure 9). Employing a classic three-layer feedforward ANN architecture (Figure 4b), and a supervised learning algorithm, the system used six joint angles as the input and six voltage control signals to the motors as output (cf. Figure 9). Although the ANN was successfully implemented on the 'walking machine', it would appear that the system has not yet been applied to an actual patient. Despite the preliminary

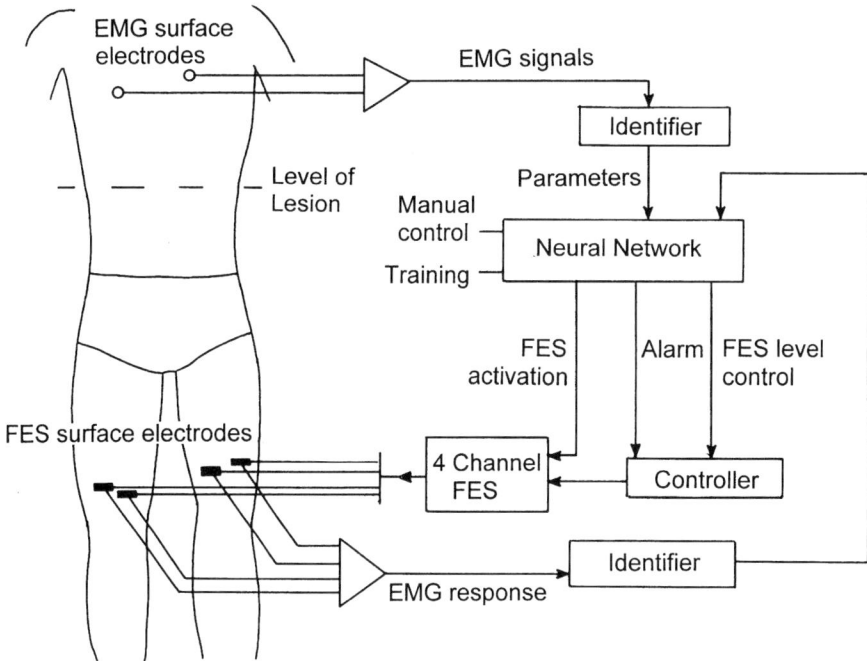

Figure 7. The electromyographic patterns from muscles above the lesion in a paralysed person serve as input to the ANN that controls the functional electrical simulation (output) of the quadriceps and hamstrings (Graupe and Kordylewski, 1995)

nature of the work, Guiraud (1994) felt that the flexibility and interesting properties of ANNs augured well for the future.

USING ANNs TO CLASSIFY LOCOMOTION PATTERNS

In contrast to the two previous sections, where ANNs were used for a purely biological purpose, there have been pattern recognition applications of ANN theory for classifying human gait. Biafore et al. (1991) were interested in identifying the age at which the development of mature gait stabilizes. Since the temporal patterns of biomechanical variables (e.g. sagittal plane knee angle) change in children between 1 and 7 years of age, and identification of the age at stabilization, is a non-trivial task, this group from San Diego developed an ANN to address the problem. Their data consisted of 12 different kinematic variables, normalized to 50 time increments per gait cycle, and based on 415 studies of

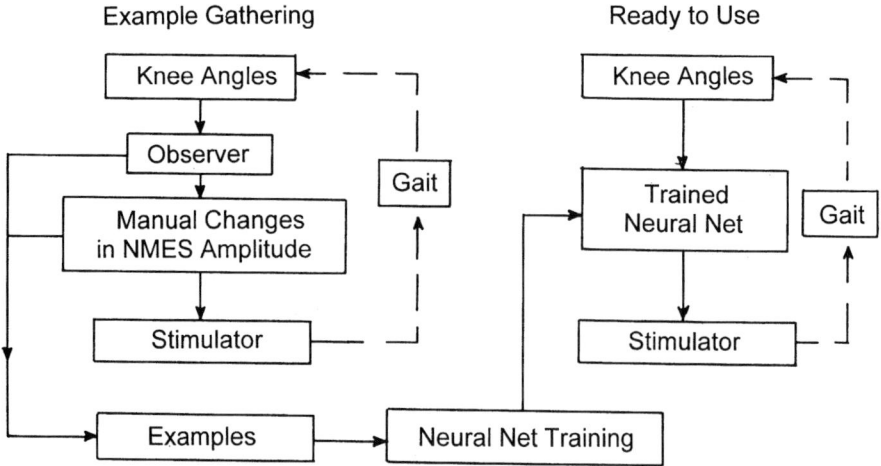

Figure 8. Knee joint angles served as the input to this ANN which provided stimulation signals to the femoral and peroneal nerves to assist spinal cord-injured patients to improve their walking (Sepulveda and Cliquet, 1995a)

normal children. They used a standard three-layer fully connected feedforward network (cf. Figure 4b) and the back-propagation learning algorithm. The ANN was trained to predict the 12 parameters at time $t + 1$ based on the input of the 12 parameters at time t. When a new gait pattern was presented to the trained ANN, the resulting error measured the difference between the new gait pattern and the learned gait pattern. When data for children aged 1–6 years of age were presented to a network trained to recognize normal gait in 7-year-old children, the resulting error graph appeared to suggest that gait stabilizes between 3.5 and 4.0 years (cf. Figure 10).

Holzreiter and Kohle (1993) argued that ANNs offered an opportunity to find the balance between gait analysis which was based on mechanics or statistics and traditional observation by humans. They chose the three-layer feedforward network (Figure 4b) and a back-propagation training algorithm. To train this ANN, they used the vertical ground reactions, based on force plate measurements, as the input. The data set comprised multiple footstrikes of 94 normal subjects and 131 patients (71 with a calcaneus fracture, 12 with artificial limbs and the balance with various diseases). When 80% of the total number of subjects (i.e. 180 people including both normals and patients) were used for training, the network achieved a 95% success rate in classifying the other subjects. While this study was admittedly a first start, it is debatable whether this application of ANN is really warranted. After all, the clinician, even one with limited experience, would have little difficulty in achieving a success rate of

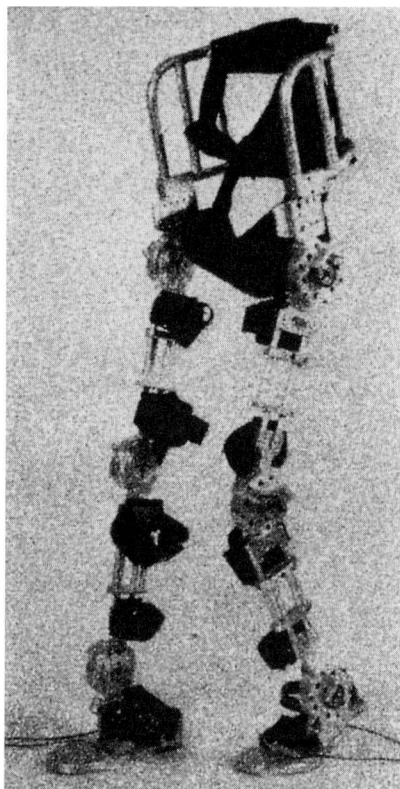

Figure 9. An ANN has been used to control an active external orthosis of the lower limbs. Reproduced from Guiraud (1994) with permission

close to 100%. Perhaps a more useful approach might be to use the ANN to stratify the pathological patterns or, as recently accomplished using fuzzy clustering of cerebral palsy children by O'Malley, Abel and Damiano (1995), assign different memberships for each patient.

In a series of three papers, all covering much the same ground, Aminian et al. (1993, 1995a,b) have used an ANN model to identify level, downhill and uphill walking. Four accelerometers were attached to the trunk and right foot of six subjects. The accelerations were recorded on a portable datalogger while the subjects walked. Training data were gathered on a treadmill which allowed for variations in the incline (-15% to $+5\%$) and the walking speed (3.5 to 7.0 m/s). After the four accelerations had been parameterized, they were fed into a three-layer feedforward ANN which was trained via back-propagation. The outputs from the ANN were incline and speed. Once the ANN had been trained using the

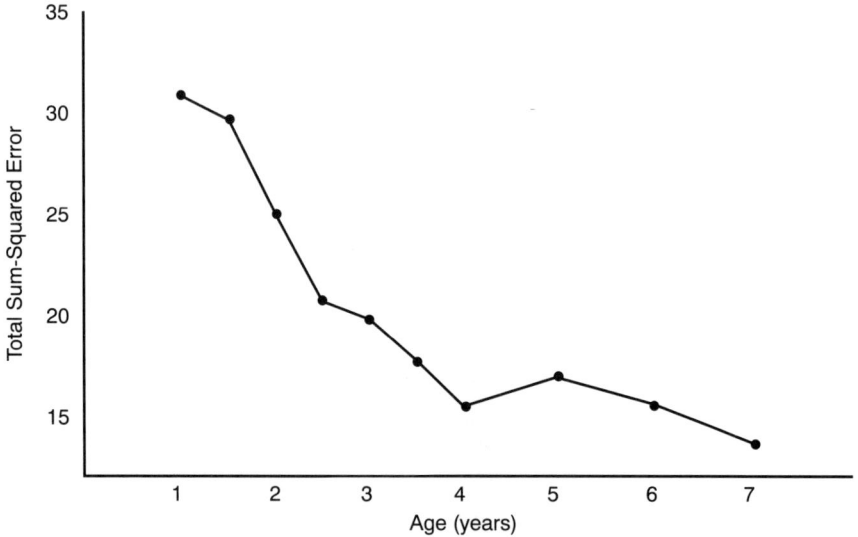

Figure 10. Gait data for different ages was presented to an ANN that had been trained to recognize the patterns of normal 7-year-old children (Biafore et al. 1991). The reduction in the error suggests that mature gait stabilizes between 3.5 and 4.0 years

treadmill data, the subject was sent to walk on an outdoor test circuit. The elevations of this circuit (i.e. the inclines) were known, as were the distances along the track (Figure 11). Since the times to traverse the separate distances were measured by the datalogger, the speeds were also known. When the ANN predictions were compared with the known values, there was remarkable agreement (Figure 12). Because the parameterization was subject specific, the ANN provided a simple and elegant method to record the energy expenditure of a walker (or runner) on an outdoor course.

FUTURE OPPORTUNITIES FOR ANNs IN HUMAN GAIT

While some have argued that ANNs cannot represent causal knowledge and therefore cannot provide insight into decisions made (Lapham and Bartlett, 1995), it is our firm belief that the applications for ANNs in human gait are many and the surface of opportunities has hardly been scraped. As highlighted in the preceding sections, the vast majority of ANN applications in human locomotion to date have utilized the feedforward network architecture. Figure 13 illustrates the current knowledge of mammalian spinal cord circuitry: both the propriocep-

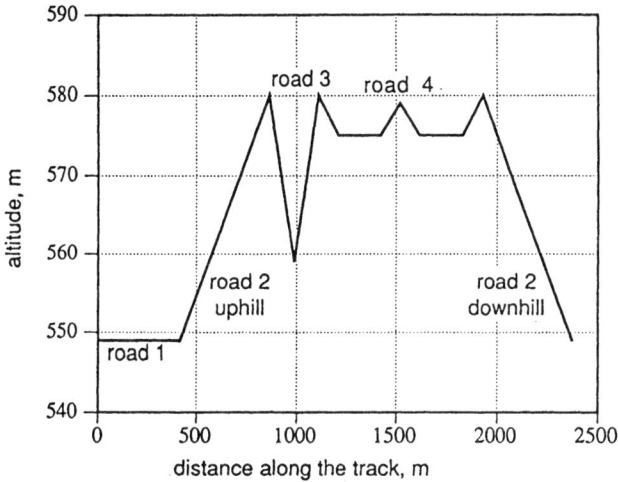

Figure 11. A cross-country running track has been plotted as altitude above sea level versus distance along the track (Aminian et al., 1995b). Note that the vertical and horizontal axes, although both in metres, are drawn to different scales

tive feedback and the descending commands converge on a matrix of interneurons (Loeb, Levine and He, 1990). Among these interneurons are at least some distinct subtypes with predictable input patterns (e.g. Renshaw cells, propriospinal cells, Ib inhibitory interneurons, reciprocal inhibitory interneurons) that allow a complex distribution of input signals. We believe that this matrix of connections between the spinal interneurons, which link the sensors and actuators, can provide the necessary framework that will enable the building of a new ANN model of bipedal gait (Vaughan, Brooking and Olree, 1996).

One area of future research focus that we have identified is the artificial neuron itself. We have proposed a new type of artificial neuron that is based on real biological neurons: the input and output signals are frequency modulated (rather than amplitude modulated), fatigue and excitability at the synaptic junction have been added, and time variations are allowed (Brooking and Vaughan, 1993). This new neuron has been incorporated in an ANN that utilizes feedback structures (cf. Figure 4c) and the system has been trained to reproduce an oscillating output that mimics a central pattern generator (CPG) (Vaughan, Brooking and Olree, 1996). For a detailed description of CPGs see Chapter 3.

Loeb, Levine and He (1990) have suggested that researchers should consider the possible matrix of interneurons in Figure 13 as a blank slate, to be explored without preconceived notions. We have adapted their proposed hierarchical model of the relationship between an open-loop controller and a closed-loop regulator for motor control in human locomotion (Figure 14). We have modelled

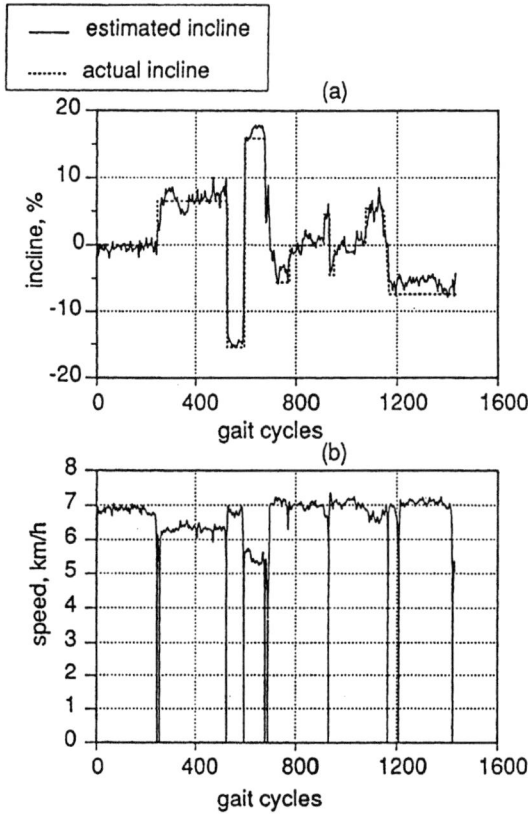

Figure 12. An ANN, trained on data from runners performing on an inclined treadmill, provides very good prediction for the inclines of the cross-country track illustrated in Figure 11. Reproduced from Aminian et al. (1995b) with permission

the matrix of interneurons in the spinal cord by a cascade of three ANNs: a spinal oscillating network (or CPG); a trajectory-generating network; and a trajectory-regulating network. Although these three networks are conceptually distinct processes from a control engineering perspective, and should be separately implemented in a model, they might well reflect different emergent properties of the same group of interneurons. The ouput from the trajectory-regulating network would be the muscle activation signals which, in turn, would serve as input to a Newtonian model of the musculoskeletal effector system (Vaughan, Davis and O'Connor, 1992). The sensory feedback from this model includes the joint angles and joint angular velocities, while each muscle group would have a sensor for combined length and velocity (Ia muscle spindle

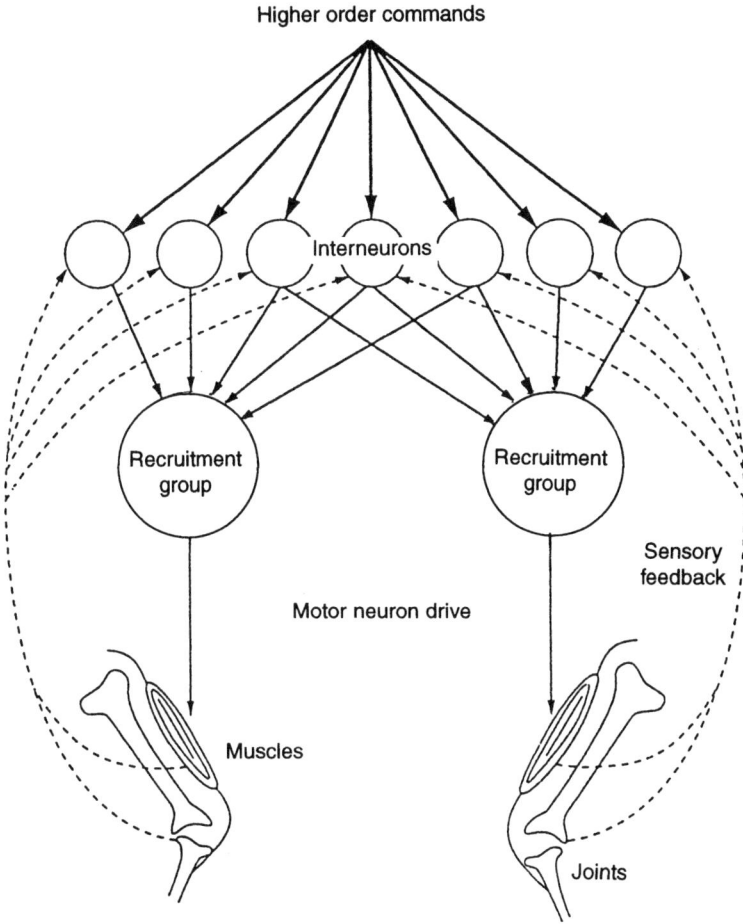

Figure 13. The current knowledge of mammalian spinal cord circuitry: both the proprioceptive sensory feedback and the descending higher-order commands converge on a matrix of interneurons

efferent), force (Ib Golgi tendon organ) and activation (efference copy via the Renshaw cells).

Very few theories of human gait have been proposed, and those theories have not been scrutinized very closely (Vaughan and Sussman, 1993). Such theories should not concentrate on the musculoskeletal system in isolation, but serious efforts should also be made to account for the role of the neural control system. It seems clear that ANNs, particularly those that involve feedback loops, should play a pivotal role in advancing our knowledge in this important field.

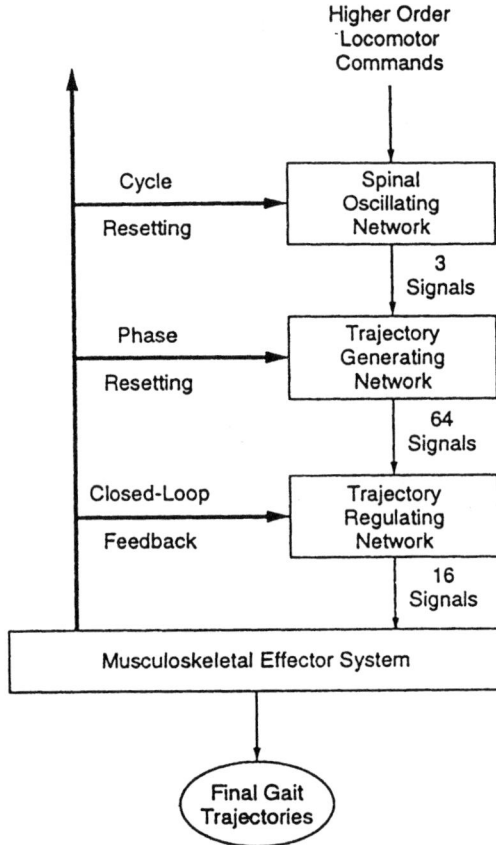

Figure 14. The matrix of interneurons in the spinal cord may be modelled by a cascade of three ANNs: a spinal oscillating network (also known as a central pattern generator); a trajectory-generating network; and a trajectory-regulating network (Vaughan, Brooking and Olree, 1996)

ACKNOWLEDGEMENTS

This work was supported in part by a grant from the National Institutes of Health (R01HD30134). In addition, I gratefully acknowledge the assistance and input of my colleagues at Clemson University (Derek Wells and Francisco Sepulveda) and the University of Virginia (Gary Brooking, Ken Olree, Chip Levy and Stephanie Goar).

REFERENCES

Aminian, K., Robert, P., Jequier, E. and Schutz, Y. (1993) Level, downhill and uphill walking using neural networks. *Electron. Lett.*, **29**, 1563–1565.

Aminian, K., Robert, P., Jequier, E. and Schutz, Y. (1995a) Estimation of speed and incline of walking using neural-network. *IEEE Trans. Instrum. Measurem.*, **44**(3), 743–746.

Aminian, K., Robert, P., Jequier, E. and Schutz, Y. (1995b) Incline, speed, and distance assessment during unconstrained walking. *Med. Sci. Sports Exer.*, **27**(2), 226–234.

Biafore, S., Quatro, G., Focht, L., Kaufman, K., Wyatt, M. and Sutherland, D. (1991) Neural network analysis of gait dynamics. *Trans. Orthop. Res. Soc.*, **15**, 255.

Brooking, G. D. and Vaughan, C. L. (1993) Control of human movement with artificial neural nets: a biologically based neuron. In: *Proceedings of IVth International Symposium on Computer Simulation in Biomechanics* (ed. B. Lanjerit). École Nationale Supérieure d'Arts et Metiers, Paris, France, pp. CSB 10–13.

Eke-Okoro, S. T. (1994) Evidence of interaction between human lumbosacral and cervical neural networks during gait. *Electromyogr. Clin. Neurophysiol.*, **34**(6), 345–349.

Gorman, C., Brown, S. and McCarroll, T. (1988) Putting brainpower in a box. *Time*, 8 August, 59.

Graupe, D. and Kordylewski, H. (1995) Artificial neural network control of FES in paraplegics for patient responsive ambulation. *IEEE Trans. Biomed. Eng.*, **42**(7), 699–707.

Guiraud, D. (1994) Application of an artificial neural network to the control of an active external orthosis of the lower leg. *Med. Biol. Eng. Comput.*, **32**(6), 610–614.

Haykin, S. (1994) *Neural Networks. A Comprehensive Foundation*. Macmillan, Englewood Cliffs, NJ, pp. 1–89.

Heller, B. W., Veltink, P. H., Rijkhoff, N. J. M., Rutten, W. L. C. and Andrews, B. J. (1993) Reconstructing muscle activation during normal walking: a comparison of symbolic and connectionist machine learning techniques. *Biol. Cybernet.*, **69**, 327–335.

Holzreiter, S. H. and Kohle, M. E. (1993) Assessment of gait patterns using neural networks. *J. Biomech.*, **26**(6), 645–651.

Kawato, M., Furukawa, K. and Suzuki, R. (1987) A hierarchical neural-network model for control and learning of voluntary movement. *Biol. Cybernet.*, **57**, 169–185.

Lapham, A. C. and Bartlett, R. M. (1995) The use of artificial intelligence in the analysis of sports performance: a review of applications in human gait analysis and future directions for sports biomechanics. *J. Sports Sci.*, **13**, 229–237.

Loeb, G. E., Levine, W. S. and He, J. (1990) Understanding sensorimotor feedback through optimal control. *Cold Spring Harbour Symp. Quant. Biol.*, **LV**, 791–803.

Massone, L. and Bizzi, E. (1989) Generation of limb trajectories with a sequential network. In: *International Joint Conference on Neural Networks*, Vol. 2, IEEE Press, New York, pp. 345–349.

McCulloch, W. S. and Pitts, W. (1943) A logical calculus of the ideas immanent in nervous activity. *Bull. Math. Biophys.*, **5**, 115–133.

O'Malley, M. J., Abel, M. and Damiano, D. (1995) Fuzzy clustering of temporal distance and kinematic parameters for cerebral palsy children. *Gait Posture*, **3**(2), 92.

Popovic, D. B., Stein, R. B., Jovanovic, K. L., Dai, R., Kostov, A. and Armstrong, W. W. (1993) Sensory nerve recording for closed-loop control to restore motor functions. *IEEE Trans. Biomed. Eng.*, **40**(10), 1024–1031.

Savelberg, H. H. C. M. and Herzog, W. (1995) Artificial neural networks used for the prediction of muscle forces from EMG-patterns. In: *Proceedings of XVth Congress of the International Society of Biomechanics* (ed. K. Hakkinen, K. L. Keskinen, P. V. Komi and A. Mero), University of Jyvaskyla, Finland, pp. 810–811.

Sepulveda, F. and Cliquet, A. (1995a) An artificial neural system for closed loop control of locomotion produced via neuromuscular electrical stimulation. *Artif. Organs*, **19**(3), 231–237.

Sepulveda, F. and Cliquet, A. (1995b) Gait generation via neuromuscular electrical stimulation controlled by an artificial neural prototype. In: *Proceedings of XVth Congress of the International Society of Biomechanics* (ed. K. Hakkinen, K. L. Keskinen, P. V. Komi and A. Mero). University of Jyvaskyla, Finland, pp. 830–831.

Sepulveda, F., Wells, D. M. and Vaughan, C. L. (1993) A neural network representation of electromyography and joint dynamics in human gait. *J. Biomech.*, **26**(2), 101–109.

Taga, G. (1994) Emergence of bipedal locomotion through entrainment among the neuro-musculo-skeletal system and the environment. *Physica D*, **75**, 190–208.

Taga, G., Yamaguchi, Y. and Shimizu, H. (1991) Self-organized control of bipedal locomotion by neural oscillators in unpredictable environment. *Biol. Cybernet.*, **651**, 147–159.

Vaughan, C. L., Brooking, G. D. and Olree, K. S. (1996) Exploring new strategies for controlling multiple muscles in human locomotion. In: *Human Motion Analysis: Current Applications and Future Directions* (ed. G. Harris and P. Smith), IEEE Press, New York, pp. 93–113.

Vaughan, C. L., Davis, B. L. and O'Connor, J. C. (1992) *Dynamics of Human Gait*. Human Kinetics Publishers, Champaign, IL.

Vaughan, C. L. and Sussman, M. D. (1993) Human gait: from clinical interpretation to computer simulation. In: *Current Issues in Biomechanics* (ed. M. Grabiner), Human Kinetics Publishers, Champaign, IL, pp. 53–68.

Wells, D. M. and Vaughan, C. L. (1989) A 3D transformation of a rigid link system using back propagation. In: *International Joint Conference on Neural Networks*, Vol. 2, IEEE Press, New York, p. 630.

Winter, D. A. (1987) *The Biomechanics and Motor Control of Human Gait*. University of Waterloo Press, Ontario.

14

Simulation of Human Movement: Goals, Model Formulation, Solution Techniques and Considerations

SCOTT BARNES, ELENA OGGERO, GUIDO PAGNACCO
AND NECIP BERME
Ohio State University, Columbus, Ohio, USA

INTRODUCTION

The term simulation can have different meanings. Relevant definitions (American Heritage Electronic Dictionary, 1992) are: (1) imitation or representation, as of a potential situation or in experimental testing; and (2) representation of the operation or features of one process or system through the use of another. These definitions illustrate the importance of intent with regard to simulation. The significance of purpose is especially pertinent when considering the simulation of human movement.

There are many reasons to simulate human movement. These diversified objectives have included, to name just a few, evaluation of hypotheses regarding the mechanics of locomotion (Beckett and Chang, 1968; Chen, Hines and Hemami, 1986; Gubina, Hemami and McGhee, 1974; Mochon and McMahon, 1980; Onyshko and Winter, 1980; Winter, 1979), understanding and insight regarding specific neuromuscular disorders (Hemami, 1985; Saunders, Imman and Eberhart, 1953), interventional treatment decisions and rehabilitation (Delp

Three-dimensional Analysis of Human Locomotion. Edited by P. Allard, A. Cappozzo, A. Lundberg and C. Vaughan
© 1997 John Wiley & Sons Ltd. ISBN 0 471 96949 4

et al., 1990; Hoy, Zajac and Gordon, 1990; Winter, 1984, 1990; Yamaguchi and Zajac, 1990), understanding and subsequent desire to enhance athletic performance (Hatze, 1981; Pandy and Berme, 1988a; Pandy and Zajac, 1991; Pandy et al., 1990), understanding balance or posture (Dinneen and Hemami, 1993; Kuo and Zajac, 1993a, b; Hemani and Farnsworth, 1977; Mackinnon and Winter, 1993), and evaluation of muscle function and neuro-control hypotheses (Hatze, 1977a, b, 1978, 1981b; Hatze and Venter, 1981; Hemami and Dinneen, 1993; Hemami, Robinson and Ceranowicz, 1980; Hogan, 1984; van Dijk, 1978; Zajac, 1993). In this chapter the concepts of creating a biomechanical simulation are outlined in conjunction with the presentation of an illustrative example.

In the context of biomechanics, simulation has been viewed as the formulation and solution of the forward dynamics problem (Nigg and Herzog, 1995; Pandy and Berme, 1988). While we agree with this historical perspective of biomechanical simulation, this restricted thinking could limit the extension of simulation technology and the creation of more empowering biomechanical simulations. The reality of the human biomechanical systems is not restricted to combined groups of rigid bodies, simplified actuators and classic controls (such as the case with robotics, where forward dynamics alone has had significant impact on advancing that discipline). Conversely, the living biomechanical system is an integrated system combining motion generation components (musculoskeletal), adaptive control, reflexes, self-analysis (inverse dynamics) and learning. The point is not to criticize the development of forward dynamics as related to biomechanical simulation, but to highlight the limitations of treating simulation of human movement as only the posing and solution of the forward dynamics problem without consideration of these other key components.

As an example, consider how significant modelling and simulation could be in the case of juvenile cerebral palsy (Pandy and Berme, 1989a). Assume that the goal is to create a model complete enough to predict the gait of afflicted children and examine the effects of possible interventions. Assume also that the model is verified by comparing simulation output with actual gait data. Now suppose that tendon transfers and tendon lengthening are simulated with the model to help the surgeon make interventive decisions. The surgery is performed based on simulation results and, if the modelling is accurate enough, the gait analysis of the patient (after therapy) will be as the simulation predicted. Even from this example, one can surmise the broad spectrum of analytical, mathematical, computational and artificial intelligence tools that are required to simulate complex phenomena, and it is no wonder that such simulations have not yet been created. Forward dynamics is one significant part of the toolbox that must be integrated with other tools in order to realize such solutions.

A few definitions of terms used throughout this chapter may be appropriate before proceeding.

Inverse Dynamics

This can be described as solving for the unknown kinetic variables from a known set of kinematic data. A mathematical representation for a n degree-of-freedom system of body segments would be:

$$\{\tau\} = [I]\{\ddot{\vartheta}\} + \{L\}$$

$[I(\vartheta)]$	$n \times n$	Inertial matrix as a funtion of generalized displacements, ϑ
$\{\ddot{\vartheta}\}$	$n \times 1$	Vector of generalized accelerations
$\{L(\vartheta, \dot{\vartheta})\}$	$n \times 1$	Vector of all load terms as a function of displacements and velocities
$\{\tau\}$	$n \times 1$	Vector of applied forces

Related to human motion, inverse dynamics is the treatment of the kinematic data obtained in quantitative human movement studies to analyse the kinetics of the observed motion. Here the kinematics of a system of links (the body segments) is recorded in time. The measured positions are the input parameters, and the forces/joint torques are the output quantities of such an analysis.

Forward Dynamics

This can be stated as solving for the unknown accelerations and then integrating for velocities and positions from a known set of kinetic input parameters. A mathematical representation would be:

$$\{\ddot{\vartheta}\} = [I]^{-1}(\{\tau\} - \{L\})$$

This terminology, also called direct dynamics, refers to the generation or synthesis of motion for a given system of linked bodies. Here a model of the system is formulated and its motion is driven by the input of the forces/joint torques. The output is the motion of the system.

Simulation

This is a representation or model of a physical system or particular situation for the purpose of example. In the biomechanical context, simulation is the abstraction of a system and its function, generally a simplification, created in an effort to study and understand components of the actual system.

Model

The simplified and generalized representation of the actual system is often called a model. The model is the representation of the original system. Modelling is the act of formulating the model.

CREATING A SIMULATION

In light of these remarks and perspectives, the process of creating a model and using it for simulation can be examined. One could say that all models are wrong, while some are useful. The reason for creating a simulation should remain paramount when considering the utility of a model, or specific simulation. A typical progression of events related to the creation of a simulation is: (1) identify the goal or purpose; (2) create a model to represent the system of interest in such a way that the original goals can be attained, and state explicitly any simplifying assumptions; (3) use the model to simulate the system's behaviour; (4) use empirical data to verify the performance of the model.

This chapter presents a systematic methodology for formulating and implementing biomechanical system simulation. This methodology is broken down into manageable components, as follows.

1. Conceptual model formulation of all subsystems to be considered:
   (a) mechanical subsystem;
   (b) actuator subsystem;
   (c) control subsystem.
2. Mathematical formulation and integrated solutions for corresponding subsystems.

An illustrative example implementation is also embedded into this discussion in order to demonstrate the power of a modular yet integrated approach to goal-oriented biomechanical modelling and to illustrate applications of the methods as they are described.

One point that must be stressed here is the importance of designing and implementing the model in a modular way so that refinements in specific areas can be made without affecting the overall formulation. This is especially important for simulations developed for research purposes, where different hypotheses are to be tested. Also, a significant advantage of the modular approach is that parts of models can be re-used in different simulations. This is becoming increasingly important as simulations grow in complexity.

An Application Example: Squat to Stand

The development of a simulation for a specific task is considered: rising from a stationary squatted position to a stable upright position under the influence of gravity. The goals of this simulation are not only to reproduce the movement, but to shed some light on how balance is maintained in the presence of external perturbations. In addition, the evolution from a very crude to a more biofidelic model is presented.

MODEL FORMULATION

The first step in building a model for a simulation is making an analysis of the relevant parameters of the event to be reproduced. It is important to consider possible assumptions that would make the problem treatable. Several hypotheses may initially be made to simplify the problem at hand as much as possible. Some of them can be negated later and substituted by others to create a simulation closer to reality. As previously stated, model development must be guided by the end goals of the simulation. For this reason, it is important to make sure that the assumptions used to simplify the model and its implementation will not prevent the fulfilment of the goals.

For example, the act of ambulating is so common that one might think it is easily simulated. However, upon closer inspection, the complexity of the human neuromusculoskeletal system is revealed and the goal of a *general* simulation of human ambulation becomes very ambitious. In this sense, the term general refers to a simulation that would by applicable to many types of ambulation. Optimally, the model used would be easily modified to fit individual anthropometric parameters. With these capabilities, the simulation would be applicable to a wide range of uses, the nature of which could be clinical, rehabilitative, research, sports-related or even artistic. It becomes obvious that incorporating generalities in a simulation complicates the model formulation. On the other hand, very specific models and simulations have limited applicability.

When modelling complex human movements, decisions must be made as to how to represent the mechanical, actuation and control elements for the simulation. Where the boundaries are drawn and to what extent these subsystems are modelled relates directly to the goal or purpose of the simulation at hand. This perspective could lead to a modular model, with the advantage that subsystems can be exchanged and evaluated in a comparative way. Thoughtful model formulation allows inclusion of coupling without sacrificing modularity.

MECHANICAL SUBSYSTEM

Specific inspection of the mechanical subsystem leads to choices that will affect the whole model. How to represent these components must be decided. Rigid-link representations have been shown to be quite useful for most locomotion and ergonomic simulations. With a few assumptions, the governing equations of motion can be derived via direct or indirect methods. Another abstraction is the concept of deformable bodies: a model that includes body deformation is chosen in place of rigid links. For example, the long bones can be represented as deformable beams. Other choices include ellipsoids, finite-element models, and other derivations from continuum mechanics. Variations of these formulations include using hybrid forms of deformable bodies and rigid links. Simulations of specific structures, such as the foot, benefit from the use of deformable body representations, but more complete simulations can become overburdened. The trade-off between improved accuracy and additional computational expense must be explored when making this decision.

In conjunction with the body segment formulation, it must be considered how to represent the intersegmental joints. It could be argued that most major human joints have six degrees of freedom (DOF), some with limited range of motion. Often these joints function with some combination of rolling and sliding contact of articular surfaces. This means that to model just the relative motions for each joint, six independent variables (or coordinates) would be required (generally representing three translations and three rotations). Assumptions regarding individual joint function and their specific relevance to the goals can lead to simplifications. Such assumptions often include two-dimensional (2D) versus three-dimensional (3D) modelling, bilateral symmetry, simple hinge behaviour, and no friction. Several kinematic representations for anatomical joints are often employed in modelling. They are the simple pin joint (1 DOF), sliders (1 DOF), combined slider and revolute (2 DOF), and ball and socket (3 DOF). The issues of friction and range of motion (e.g. joint locking) have to be considered. More complicated models have been formulated that include the contact of rolling and sliding surfaces of irregular geometry. These are usually 3D, contain the ligaments as well as other anatomical structures, represent the geometry of the contacting surfaces, and are often formulated for dynamic (versus static or quasi-static) simulations. Ligaments are often modelled as infinite- or high-stiffness cables, and can be used to limit the range of motion of joints. A more accurate model for ligaments would include their nonlinear stiffness and rate-dependent deformation characteristics.

Mass and inertial properties are less intuitive to model, but very significant in the overall formulation. When the rigid-link approach is used to approximate the body segments, the mass may be considered to be concentrated at one point. More sophisticated mechanical formulations may employ a distributed mass model or actual anthropometic measurements for subject-specific modelling. In the human system, the mass distribution is dynamic: as muscles contract and relax, their individual lengths and centres of mass move in such a way that they can greatly affect the inertial properties of a specific body segment, e.g. the quadriceps during leg extension. The inclusion of such effects is an important consideration during model formulation.

Squat to Stand Example: Mechanical Subsystem

This activity can be assumed to be completely symmetric, and represented with a planar model. Theoretically, this does not decrease the generality of the model, but it does introduce great simplifications into the model formulation. Another simplifying hypothesis is that the upper body is completely passive. This presumes that there are no active movements of the arms or the trunk to provide postural control. This hypothesis, while significantly simplifying the model by permitting the upper body to be treated as a lumped mass, introduces a strong limitation on the generality of the model. In fact, as one would experience in performing this movement, the upper body is commonly employed to increase the stability and facilitate the action. This hypothesis would have been unacceptable if a goal of the simulation were to study the effects of upper body motion on the movement being simulated. These two hypotheses facilitate the selection of the mechanical part of the model. Since the activity does not involve deformation of body segments, the idea of representing the system as a linkage appears to be promising. Assuming that during the simulation the feet stay completely in contact with the ground, there is no need to include them in the linkage. Under these conditions the mechanical section of the model can be a two-segment planar linkage, which is usually described as an inverted double pendulum.

To obtain the equations describing the mechanical linkage, any passive elements and joint definitions in the model need to be specified. These are the springs and dampers that can be introduced to represent specific behaviour of elements of the real system such as ligaments, tendons, and joint friction. For the moment, let us leave these unspecified. Since no hypotheses have been formulated on this subject, it would not yet be possible to derive the equations of motion.

Initially it can be assumed that no passive force-generating elements, such as springs and dampers, are present. This is a common assumption in this kind of model, as these elements are usually modelled with the muscles. That still leaves the joints undefined.

The joints present in the model are the ankle and the knee. These joints have complicated geometry and motions. As a first approximation, both can be considered as pin joints. Naturally, certain limitations in their range of motion need to be implemented. In this specific simulation, only the knee joint is likely to reach the end of its motion range. This will happen when the model represents the upright position. Considering that in reality the knee is not only prevented from hyperextending, but it also locks in the extended position under an axial load to provide stability, the task of modelling this part of the human body can become arduous. In a first approximation, it is possible to assume that no locking occurs and that the knee simply comes to a stop when the extended position is reached. Under this assumption the stop can be simulated by forcing the rotational velocity of the joint to zero. This imposes a theoretical limitation on the validity of the simulation only if the knee reaches its maximum extension with a significant angular speed. Furthermore, if the locking of the knee is not considered, the model does not correctly represent a very important element used for maintaining a stable, upright posture. We will see the impact of these assumptions when we apply the corresponding mathematical relationships after we derive the equations of motion for the finalized model.

ACTUATOR SUBSYSTEM

The formulation of the actuator subsystem can include models of diverse intricacy. Some have met their goal with only passive elements and no active actuators present. By representing muscles or a whole limb system as a mass–spring–damper complex, good approximations of ground reaction forces can be obtained for simulating some activities (Pandy and Berme, 1988a). If the goals become more sophisticated, active elements are required. Depending on the goal of the simulation, there are a number of simplifying considerations that can be made. It is sometimes advantageous to use resultant torque values as a first approximation. In doing so, the lines of action of a muscle are not required, so there is no need to define insertion points. When considering a single joint, the model would generate a specific torque value about the axis of concern for each time step in the simulation. Eventually, the agonistic and antagonistic muscle groups for a joint can be considered as independent torque generators. In addition, models that include the anatomical structure of the muscle can be incorporated. These muscle models can have active and passive elements. They may include the insertion points and determine the lines of action of the muscle forces. These components obviously become highly integrated with the geometry selected for the physical representation of the body segments. It is important to realize that some muscles and muscle groups act over more than one joint. These multiarticular muscles are difficult to model, since their pulley-like action over the intermediate joints has to be included. Another complicating factor results from the redundancies presented by multiple muscles acting on single joints (Sepulveda, Derek and Vaughan, 1993; Yamaguchi, Moran and Si, 1995).

Even more complicated models of individual muscles can be formulated. Many models have been postulated to understand muscle function, such as the significant work by Gasser and Hill (1924). Even individual muscle spindles can be modelled to achieve certain simulation goals (Gielen and Houck, 1987). At this level, the behaviour of individual motor units can be studied. Issues such as their recruitment and neuromuscular phenomena are now possible subjects for simulation. It is, of course, debatable whether a simulation of walking needs such a detailed series of micro-muscle models. However, understanding exactly how the muscles function will benefit all forms of motion simulation (Yamaguchi and Zajac, 1989).

Squat to Stand Example: Actuator Subsystem

Several different actuator assumptions will be tested in the example. Migrating from a very simplified actuator, or muscle model, to one that is complex and contains many of the characteristics of actual human systems illustrates the power of modular model formulation. Choosing no actuator is also an option: early simulations of walking implemented a passive or ballistic model of the leg during swing using a simple inverted pendulum (Mochon and McMahon, 1980). This is not an effective option for this example, given the desire to rise from a squat to standing position; this consideration of the goals rules out the possibility of a ballistic simulation.

If it is assumed that all muscular actions can be represented by torques applied at the joints (ankle and knee), then the actuation source can be considered a pure torque generator, one at the ankle and one at the knee. This is a significant simplifying assumption. By having all actuators acting on the mechanical linkage only by means of the resultant torques, the solver for the mechanical model can be completely separated from the actuators and control subsystems. Introducing active elements into the model also means to introduce a control. In fact, even if it might not appear so obvious, every decision on how an active element behaves is indeed the introduction of a control model and its strategy. This coupling is present for all active actuators, but becomes more of an implementation concern with more elaborate actuator models.

A more elaborate actuation model can be implemented to compare the pure torque generators with a more physiologically based muscle and control model. This model includes agonistic and antagonistic muscle groups and individual muscle spindle physiology. Here the additional complexity of coupling between the actuator and control subsystems becomes unavoidable. These details are presented later.

CONTROL SUBSYSTEM

Even if it is decided that no control strategy will be formulated, a control has been imposed. The simplest strategy has no actuators to control. In this purely ballistic approach, only the initial conditions are given and the simulation predicts the outcome. As a control model, the easiest one to implement is, quite obviously, an open-loop controller with a torque that is exclusively a function of time (Pandy and Berme, 1988b,c, 1989a,b). This also removes all the complexities associated with computing the kinematics of the insertion points of the muscles (Delp et al., 1990). How the joint torque changes with respect to time depends mainly on the goal of the simulation. One way is to solve the inverse dynamic problem using data collected during an experiment in which a subject performs the same actions that the simulation is trying to reproduce (Pandy and Berme, 1989a,b; Nigg and Herzog, 1995). If this is the case, the geometric and inertial parameters used in the mechanical model also have to be consistent with such a subject. Another way to obtain the torque is to impose an optimization criterion (Yamaguchi, Moran and Si, 1995; Sepulveda, Derek and Vaughan, 1993). For instance, this could involve finding the torque functions that minimize the time of the movement, the torque peak value, the necessary mechanical energy, or a combination of these.

The use of such a simplified actuator and control model, however, has a substantial limitation: there is no way to compensate for external perturbations, and this approach provides no means for actively maintaining balance when equilibrium conditions are desired. This is not a serious limitation if the interest of the simulation is concentrated on certain movements where balance or postural stability is not a concern or goal.

When force and torque application is a function of position, velocity or other generalized variables in the model, the control becomes closed loop. Closed-loop control is a very extensive subject in its own right. Reflexes, for example, are a form of closed-loop control. Relevant to simulation are muscle- and spinal-level reflexes. Simulation of higher-level controls, such as the cerebellum and brain, becomes more challenging. More adaptive approaches, including neural networks and artificial intelligence techniques, can be applied to model these interactions and relate other biomechanically relevant parameters to control (Aminian, Jequier and Schutz, 1994; Salation and Zheng, 1992; Sepulveda, Derek and Vaughan, 1993; Wan and Yip Pak, 1991).

Another idea that is emerging can be called 'feedforward control'. An over-simplified description of the feedforward concept could be that the controller (brain) predicts where the controlled system (body) is going to be at some time in the future. At the same time, the system (probably some portion of the brain) is analysing the current body position and comparing it to the previous forward prediction. When discrepancies are noted, appropriate corrective signals are issued. A key point of this concept is that it is a dynamic process, and this same idea can be adapted to explain some processes of motor learning and perform-ance enhancement. Being dynamic means that as a certain motor task is repeated (training), the feedforward point keeps getting further and further away in the future, as is probably the case with highly trained athletes, dancers, and other very skilled movers. Also, as increasingly possible perturbations have been experienced during the repetitions or training process, patterned responses are learned. The feedforward concept implies that a very accurate and rapid motion analysis is built into the system (the moving body). In general, most simulations of human movement do not incorporate movement analysis capability into the formulation of the problem. Hopefully, further development of the feedforward concept will lead to more useful simulations if the fundamental idea is verified or disproved.

Squat to Stand Example: Control Subsystem

Initially, an open-loop control consisting of torque profiles as functions of time was chosen as the method to control the resultant ankle and knee joint torques for the squat to stand example. This was attractive from a simplicity and implementation point of view and, when optimized, achieved one of the initial goals of being able to rise from the squat. However, this methodology will fail to maintain a stable standing position, let alone allow recovery from any sort of perturbation.

A possible step to improve the mode behaviour is to introduce a feedback control system. A direct and easy way to accomplish this is to define the torques as functions of the joint angles instead of time. Delays in the control loop can be introduced to provide further improvement. Keeping in mind all the assumptions made, a simula-tion using a model with such improvements can provide, especially if compared with the previous open-loop control, some insights into the strategy used for this kind of movement. The open-loop control supposes a preplanning of the actuators' activity and cannot compensate for external disturbances. On the other hand, the closed-loop control allows for rapid adjustments to external perturbations, providing that these are not too large in magnitude and the overall action is not too fast compared to the delays in the control circuit.

MATHEMATICAL MODEL FORMULATION

To implement any of the mechanical, actuation and control structures, the formulation needs to be expressed mathematically. The resulting expression must then be solved to produce the simulation. Typically, the simulations of interest here are solved numerically.

Two general classes of methods for generating the dynamic equations for the mechanical model can be identified. One uses a direct formulation, while the other is based on indirect analytical methods. Direct formulation utilizes the d'Alembert principle of dynamic equilibrium in conjunction with free body diagram analysis. The indirect analytical methods of interest are the Lagrangian and Newton–Euler formulations. Direct derivation can be applied to linkages, but it becomes impractical for systems with more than two links. The indirect approach can obtain the equations for large systems, but can mask the physics of the system. The advantages and limitations of each method are highlighted here.

The direct application of d'Alembert's principle and free body diagram analysis for the formulation of the system equations is summarized by Roth (1989). Despite the importance of this type of formulation, few textbooks present this methodology in a concise, methodological, detailed, systematic, logical and explicit way. This is especially true for the case of rigid bodies in motion (dynamics). The advantages of this direct approach include the intuitive nature of the process. In formulating a problem in this way, one acquires good insight into the physics of the system. Moreover, if the system is simple enough for this method to be used, there will probably be a closed-form solution. In this case, a computer implementation would be very fast. One limitation of this method is the difficulty encountered when many (practically more than two) bodies are considered. Unfortunately, most biomechanical systems of interest fall into this category. For example, the fairly straightforward derivation for the inverted double pendulum shown in Figure 1a takes more than 10 pages of algebraic manipulation to arrive at the final equations.

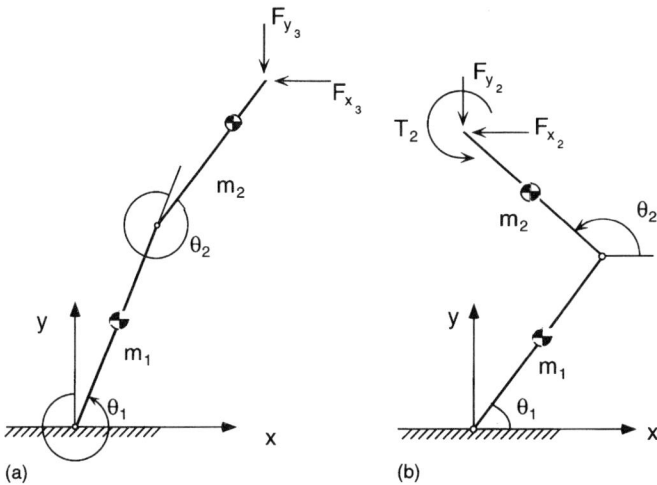

Figure 1. (a) Inverted double pendulum: nomenclature. (b) Rising from a squatted position: mechanical linkage used

The developments of the analytical methods of Lagrange and Newton–Euler were motivated, in part, by the limitations above. Generally speaking, the Lagrangian method is very useful for obtaining the equations in analytical form, while the Newton–Euler method lends itself to numerical solutions. It is important to note that the Lagrangian and recursive Newton–Euler formulations yield equivalent simulation results, although their mathematical forms and implementations are different.

Lagrange's equations of motion represent a specialized expression of Hamilton's principle and are stated as follows:

$$\frac{d}{dt}\left(\frac{\partial T}{\partial \dot{q}_j}\right) - \frac{\partial T}{\partial q_j} + \frac{\partial V}{\partial q_j} = Q_j \quad \text{for} \quad j = 1, 2, \ldots m$$

yielding m equations for the m generalized coordinates q_j. If the system is holonomic, a set of unconstrained generalized coordinates can be selected such that the number of generalized coordinates m equals the number of degrees of freedom n. With $m = n$, a small virtual displacement of the system will satisfy all kinematic constraints, and Lagrangian equations are the complete equations of motion for the system. Non-holonomic systems must be supplemented with additional constraint equations for a constrained set of generalized coordinates.

When implementing Lagrange's method, a function L, called the Lagrangian, is used. L is defined as

$$L = T - V$$

where T is the kinetic energy of the system, and V is its potential energy. Taking partial derivatives with respect to the generalized coordinates and velocities yields an alternative form of the Lagrangian equation:

$$\frac{d}{dt}\left(\frac{\partial L}{\partial \dot{q}_j}\right) - \frac{\partial L}{\partial q_j} = Q_j \quad \text{for} \quad j = 1, 2, \ldots m$$

By implementation of the Lagrangian and any required constraint equations, the overall equations of motion for complicated systems can be derived. System equations that would be nearly impossible to derive using the d'Alembert principle and the free body diagram analysis can be expressed with moderate effort. This is a very general methodology and is commonly adopted for solving many mechanical engineering dynamics problems. Proven methods of linearization and solution for the Lagrangian form of the equations of motion exist for a wide range of systems.

There are some limitations and disadvantages of the Lagrangian method. As is evident in the equation above, derivatives are heavily used in the formulation of the equations, which may cause numerical difficulties, and can lead to computationally intensive solutions. When generalized coordinates are used, the resulting form of the equations of motion may not be physically intuitive. This can make integration of mechanical systems formulated with the Lagrangian method difficult. Finally, the Lagrangian method leads to a very specific model formulation depending on the exact system constraints and configuration. As any new constraint or system component is introduced, the analytical solution of the Lagrangian equation for the new system must be reformulated. For example, one would have distinctive formulations of the Lagrangian equation (thus a separate set of equations to solve for simulation) for the flat-footed versus the heel-off portion of the stance phase of gait. In other words, simulations formulated with this method are not very general and not easily extended, for example, to include additional links.

Indirect Newton–Euler methods are also often employed in model formulation of rigid-link, open kinematic chains. Newton–Euler methods are based on the following two equations:

$$\text{Newton's Equation } \{F\} = m\{a\}$$

$$\text{Euler's Equation } \quad \{T\} = \{\dot{H}\}$$

where

$\{F\}$ = force vector acting on a body

m = mass of the body (scalar)

$\{a\}$ = acceleration vector of the body

$\{T\}$ = total moment of all forces, sometimes called inertial torques or free moment

$\{\dot{H}\}$ = rate of change of angular momentum

$\{H\} = [I]\{\omega\}$, angular momentum

$[I]$ = inertial matrix

$\{\omega\}$ = vector of angular velocities

One particular useful implementation of this class of methods is based on a recursive formulation. This technique can be loosely described as the simultaneous solution of the above equations so that the equations of motion for a specific linkage are numerically obtained. Although this method was originally applied as a solution to the inverse dynamics problem, it can be extended to the forward dynamics problem. An outline of how to implement this procedure is given here. Although it is noted that this formulation is valid for n-segment open kinematic chains, adaptations of this formulation have been applied to closed chains and other special cases (Pandy and Berme, 1989b; Vukobratovic, Filaretov and Korzun, 1994; Walker and Orin, 1982).

Considering the inverse dynamics equation, it has been shown that for any given open kinematic chain of n segments, the equations of motion may be presented in the form (Walker and Orin, 1982):

$$[I]\{\ddot{\vartheta}\} + \{C(\vartheta, \dot{\vartheta})\} + \{G(\vartheta)\} = \{\tau\}$$

This can be extended to include muscle force actuators as in the human system. Here the muscles are responsible for generating the resultant joint torques $\{\tau\}$. Thus the above equation becomes

$$[I(\vartheta)]\{\ddot{\vartheta}\} + \{C(\vartheta, \dot{\vartheta})\} + \{G(\vartheta)\} + \{F_m(\vartheta)\} = \{0\}$$

where

$[I(\vartheta)]$ $\quad = n \times n$ inertial matrix as a function of displacements ϑ

$\{\ddot{\vartheta}\}$ $\quad = n \times 1$ vector of angular accelerations

$\{C(\vartheta, \dot{\vartheta})\} = n \times 1$ vector of Coriolis and centrifugal terms as a function of displacements and velocities

$\{G(\vartheta)\}$ $\quad = n \times 1$ vector of gravitational terms as a function of displacements

$\{F_m(\vartheta)\}$ $\quad = n \times 1$ vector of applied muscle moments

$\{\tau\}$ $\quad = n \times 1$ vector of applied joint torques generated by the muscle forces

In the solution of the inverse problem, this method is applied by introducing the known kinematics (from measurement, gait analysis, position velocity and acceleration) and solving for the muscle moments. This is done by recurring the n-link chain. Starting at the base of the support and progressing out the chain (link 1, to link 2, ..., to link n), the inertial moment and force for each limb are computed. Recursing of the chain is then performed from n to 1 with the newly calculated force and moment applied, thus producing the required intersegmental forces and muscle moments. This is repeated for each time step in the simulation.

By modification of this technique, the equations of motion for a given model can be generated numerically for each time step. The inputs required are the geometric link parameters and the applied muscle forces–joint torques. Thus, the technique becomes applicable to a forward dynamics problem. The details of this method can be found in Pandy and Berme (1988a,b,c, 1989a) and a comparison of the computational load of this and related methods is presented in Vukobratovic, Filaretov and Korzun (1994).

The advantages of the recursive Newton–Euler formulation include its adaptability to computer/numerical solutions. The result of this method provides numerical values (coefficients) of the inertial matrix for each time step. Thus, some understanding of the physics of the system can be obtained from these values. Once it is formulated and implemented as a computer program, more links can be added without model reformulation, in contrast to the Lagrangian approach.

On the other hand, there are difficulties associated with this approach. An admitted disadvantage of using the latter method is that it can appear as a 'black box', preventing a better understanding of the system. Closed linkages present additional challenges. A computer is required to implement these techniques, because the resulting equations are not easily analytically treatable. A working knowledge of numerical methods is required to successfully simulate systems with this method.

Squat to Stand Example: Mathematical Model Formulation

Using direct inspection or the Lagrange approach will produce the following equations of motion for the system (referring to Figure 1a and Figure 1b):

$$(I_1 + m_1 d_1^2 + m_2 l_1^2 + 2m_2 l_1 d_2 \cos \theta_2 + I_2 + m_2 d_2^2)\ddot{\theta}_1 +$$
$$+ (I_2 + m_2 d_2^2 + m_2 l_1 d_2 \cos \theta_2)\ddot{\theta}_2 - m_2 l_1 d_2 \sin \theta_2 (\dot{\theta}_1 + \dot{\theta}_2)^2 +$$
$$+ m_2 l_1 d_2 \sin \theta_1 \dot{\theta}_1^2 - [m_1 d_1 \sin \theta_1 + m_2 l_1 \sin \theta_1 + m_1 d_1 \sin (\theta_1 + \theta_2)]g +$$
$$- [l_1 \cos \theta_1 + l_2 \cos (\theta_1 + \theta_2)]F_{x_3} - [l_1 \sin \theta_1 + l_2 \sin (\theta_1 + \theta_2)]F_{y_3} = \tau_1$$
$$(I_2 + m_2 d_2^2 + m_2 l_1 d_2 \cos \theta_2)\ddot{\theta}_1 + (I_2 + m_2 d_2^2)\ddot{\theta}_2 +$$
$$+ m_2 l_1 d_2 \sin \theta_2 \dot{\theta}_1^2 - m_2 d_2 \sin (\theta_1 + \theta_2)g +$$
$$- l_2 \cos (\theta_1 + \theta_2)F_{x_3} - l_2 \sin (\theta_1 + \theta_2)F_{y_3} = \tau_2$$

Applying the Newton–Euler approach yields a system of equations that appear quite different from the equations above, although they are equivalent:

$$\begin{cases} \ddot{x}_1 = -d_1 \cos \vartheta_1 \dot{\vartheta}_1^2 - d_1 \sin \vartheta_1 \ddot{\vartheta}_1 \\ \ddot{y}_1 = -d_1 \sin \vartheta_1 \dot{\vartheta}_1^2 + d_1 \cos \vartheta_1 \ddot{\vartheta}_1 \end{cases}$$

$$\begin{cases} \ddot{x}_2 = -l_1 \cos \vartheta_1 \dot{\vartheta}_1^2 - l_1 \sin \vartheta_1 \ddot{\vartheta}_1 - d_2 \cos \vartheta_2 \dot{\vartheta}_2^2 - d_2 \sin \vartheta_2 \ddot{\vartheta}_2 \\ \ddot{y}_2 = -l_1 \sin \vartheta_1 \dot{\vartheta}_1^2 + l_1 \cos \vartheta_1 \ddot{\vartheta}_1 - d_2 \sin \vartheta_2 \dot{\vartheta}_2^2 + d_2 \cos \vartheta_2 \ddot{\vartheta}_2 \end{cases}$$

$$\begin{cases} F_{x_1} = F_{x_2} + m_2 \ddot{x}_2 \\ F_{y_1} = F_{y_2} + m_2 \ddot{y}_2 + m_2 g \\ T_1 = T_2 + I_2 \ddot{\vartheta}_2 - [F_{x_1} d_2 + F_{x2}(l_2 - d_2)] \sin \vartheta_2 + [F_{y_1} d_2 + F_{y_2}(l_2 - d_2)] \cos \vartheta_2 \end{cases}$$

$$\begin{cases} F_{x_0} = F_{x_1} + m_1 \ddot{x}_1 \\ F_{y_0} = F_{y_1} + m_1 \ddot{y}_1 + m_1 g \\ T_0 = T_1 + I_1 \ddot{\vartheta}_1 - [F_{x_0} d_1 + F_{x_1}(l_1 - d_1)] \sin \vartheta_1 + [F_{y_0} d_1 + F_{y_1}(l_1 - d_1)] \cos \vartheta_1 \end{cases}$$

Please note the differences in notation and their significance by comparing the convention differences between Figure 1a and Figure 1b. Most significantly, the angular displacements are defined as the relative link angle from link to link for the direct formulation (Figure 1a, and denoted by θ), while for the Newton–Euler formulation, ϑ denotes the link angle with respect to the horizontal. Additionally, the joint numbering starts at 1 at the ankle for the direct formulation equations (Figure 1a and Figure 2), and at 0 for the Newton–Euler representation (Figures 1b and 2). Although the differences can be confusing, these distinctions were retained so that the equations could be written and treated in their classic form for each approach and to illustrate the recursive nature of the latter formulation. An equation representing the limitation in the range of motion in the knee joint (locking mechanism) can be written:

$$\dot{\vartheta}_2 = 0 \quad \text{for} \quad \vartheta_2 \geqslant \pi = \text{max. extension}$$

Even though it may seem less intuitive, the Newton–Euler approach will be used for the rest of the example. This is due to the desire to treat the more general situation where the model can be expanded to include more links for the foot or upper body. Progressing with the Lagrangian or direct approach, the model would have to be reformulated for these transitions and would become cumbersome for such multi-link models. Since one of the capabilities of the Newton–Euler approach, as implemented, is the numerical generation of the equations of motion (Pandy and Berme, 1989a,b; Vukobratovic, Filaretov and Korzun, 1994; Walker and Orin, 1982) additional links will cause no problems. Furthermore, the recursiveness of the equations makes the computer implementation very simple and thus attractive. These equations can then be used to obtain the system of differential equations representing the dynamic behaviour of the mechanical subsystem. This can be done either numerically, using the iterative Newton–Euler method, or by direct substitution and simplification of the equations.

SQUAT TO STAND EXAMPLE: A MORE PHYSIOLOGICAL MODEL

At this point, the formulation of the squat to stand simulation consists of a lumped body mass on top of an inverted double pendulum representing the lower

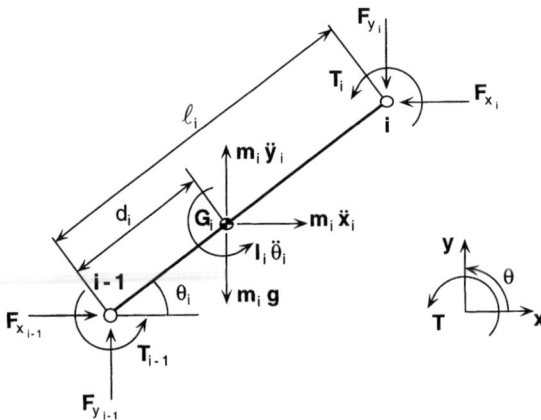

Figure 2. Newton–Euler method: applied forces and moments on a generic link

and upper legs (see Figure 1B). Note that the lumped mass for link 2 represents the mass of the limb and the upper body and can be positioned at the top of link 2 (i.e. $d = 1$ in Figure 2). The model is planar, with frictionless pin joints for the ankle and knee and the links treated as rigid. Based on the assumptions made, it is obvious that this formulation of the model will not achieve all the goals, specifically with regard to balance and recovery from perturbations. This is mainly due to the lack of a more sophisticated actuation and control model.

Squat to Stand Example: Actuator and Control Extensions

If one goal is to study human physiology by investigating the shape and function of different components of the human machine, then more accurate systems must be modelled. In the neuromuscular apparatus, there are actuators (muscle fibres), sensors (muscle spindles, Golgi apparatus), reflexes (at the spinal and central levels), signal-transmitting fibres (motoneurons, fusimotors and neurons, each one with its own transmission velocity) and a central nervous system. In order to study stability after the presentation of small perturbations in the system, many refinements can be introduced into the model. The goal of this model is to simulate not only the rising movement but any movement, with the ability to compensate for small external perturbations and to maintain a position, even an unstable one such as upright posture.

Considering the muscle, more physiological models are available. In the one presented by Dinneen and Hemami (1993), the main attribute is that it models fibre recruitment. This results in modelling the stiffness of the muscle with a parabolic function that increases linearly with the activation level. The tension force F generated by the muscle is given by

$$F = [kR + \alpha(l_a - l_0)]^2$$

where k is the amplification factor, R is the activation level and is always positive or null, α is the passive elasticity characteristic, l_a is the actual length of the fibres and l_0 is the muscle rest length. If a linear constant Hill effect, is added, the force produced becomes

$$F = [kR + \alpha(l_a - l_0)]^2 + Bi_a u(-\dot{l}_a)$$

where B is the viscous constant, $-\dot{l}_a$ is the contraction velocity and $u(x)$ is a unity step function equal to 1 for positive x and 0 for null or negative x. Note that this model can only generate tension. For this reason, it becomes necessary to define at least two such actuators for every joint, an agonistic and an antagonistic muscle group or, using different terminology, a flexor and an extensor torque.

This model supposes a force control, since modifying R affects in a direct way the force generated. However, if R is appropriately computed using position feedback, it can become a position control. Supposing the desired force–position relationship is

$$F = [k(l_a - l_d)]^2 u(l_a - l_d) + Bi_a u(-\dot{l}_a)$$

where l_d is the desired position and is always less than or equal to l_a, and comparing the two expressions for the generated tension, they are equivalent if

$$R = \left(1 - \frac{a}{k}\right)l_a + \frac{a}{k}l_0 - l_d$$

$$l_a \geqslant l_d$$

The computation of the activation level requires position information. Such information must thus be obtained from the model. One method is to introduce the muscle spindles. Considering the nonlinear moded proposed by Gielen and Houk (1987), the output of the spindle can be approximated by

$$f = q(l - th)u(l - th)\dot{l}^{\varepsilon}u(\dot{l})$$

$$l = l_a - l_0$$

where f is the sensor output, q is a positive gain constant, l is the actuator length, th is the threshold length, u is the unit step function and ε is a positive fraction. The spindle so modelled does not emit signals for a zero contraction velocity, i.e. when the muscle is not moving. In reality, the spindle is known to emit an elongation-dependent as well as a velocity-dependent signal. These two different kinds of information are generated respectively by the nuclear chain fibres and by the nuclear bag fibres present inside the spindles. The elongation signal only depends on the elongation and the activation of the spindle intrafusal muscular fibres, while the dynamic signal is a function of the velocity as well as the elongation and the activation. It is possible to slightly modify the previous model to provide two signal components: a static one, containing the position information, and a dynamic one, which is a function of the velocity of contraction as well as the elongation:

$$f_v = q(l - th)u(l - th)(\dot{l}^{\varepsilon} - 1)u(\dot{l})$$

$$f_l = q(l - th)u(l - th)$$

As for the spindles, the model of the neuromuscular control is based to some extent on what is known about the real system. There are different spinal neurons innervating the muscles: the α-motoneurons that activate the extrafusal fibres of the muscles and the γ-motoneurons activating the intrafusal fibres of the muscle spindles. At the spinal level there are different reflexes present; among them, the myotatic and the inverse myotatic are the ones included in this model. The first one is known to influence muscular tone and control stability and the second is responsible for the inhibition of the antagonistic muscle during a movement. Also, to make the model more adherent to reality, the different transmission speeds of the various afferent and efferent nervous fibres are included. In fact, the α-motoneurons are high-transmission-speed fibres (70–140 m/s), while γ-motoneurons are low-speed fibres (30–70 m/s). The spindle afferent fibres are also different: group Ia, the ones carrying the contraction velocity information, are high speed (70–140 m/s), while group II, carrying the positional information, are low speed (30–70 m/s). For simplicity, only two kinds of fibres, fast and slow, are introduced into the model. A transmission delay at every neuron is also included. The different connections and their transmission speeds for a single muscle are illustrated in Figure 3.

The web of connections for the innervation of every joint becomes quite complex, considering that every joint is actuated by two muscles, a flexor and an extensor, and that the inverse myotatic reflex is to be modelled. In Figure 4 all the exchanged signals are visible, with a different type of line depending on the signal transmission speed. The myotatic reflex has been implemented, so when a muscle is activated, its

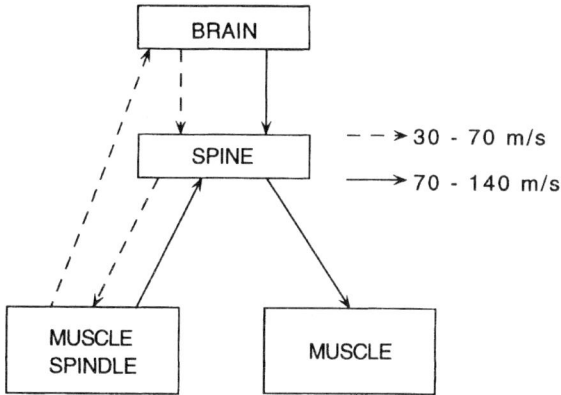

Figure 3. Muscle model: control blocks, their connections and transmission speed

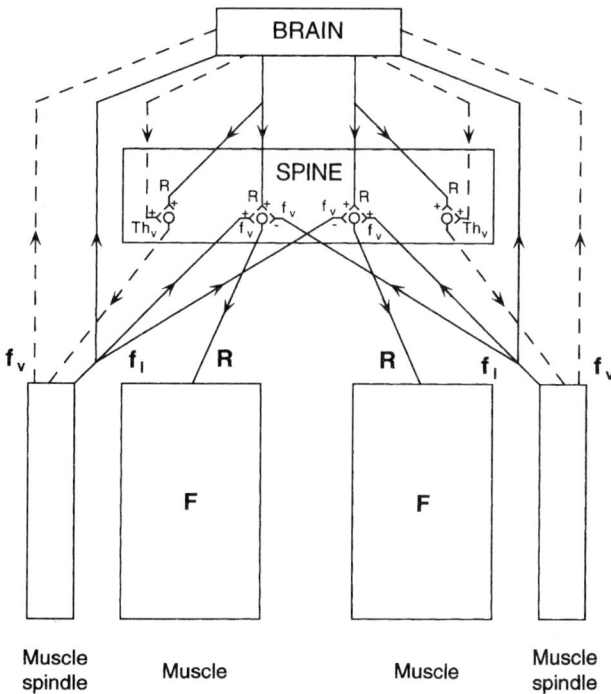

Figure 4. Joint model: actuators, control, and their connections

muscle spindle starts transmitting, and from one side it increases the muscle contraction (f_v is added to the activation signal from the brain), and on the other it reduces the activation level of the antagonistic muscle (f_v is subtracted in the synapse of the non-active muscle, reducing its activation level).

The implementation of the spinal reflexes allows for an intrinsic balancing system: when the joint is in an equilibrium position, the muscles are activated with a certain R value and the two spindles are emitting the same signal. In this condition the system is stable. As soon as a small perturbation is introduced into the system, an imbalance in the muscle spindle signals is created. This activates, through the spinal reflex, the muscle not stretched, which tries to re-establish joint equilibrium. If the perturbation goes beyond the ability of the spinal reflex to re-establish balance, the brain can intervene and change the activation level for the muscles, driving the system back to the equilibrium configuration. Having implemented the neural transmissions delays, the two balance control systems (spinal reflexes and brain) have different reaction times. It takes almost twice as long for the brain to react. This is consistent with what happens in reality. A flow chart describing the computer program used is shown in Figure 5 and Table 1.

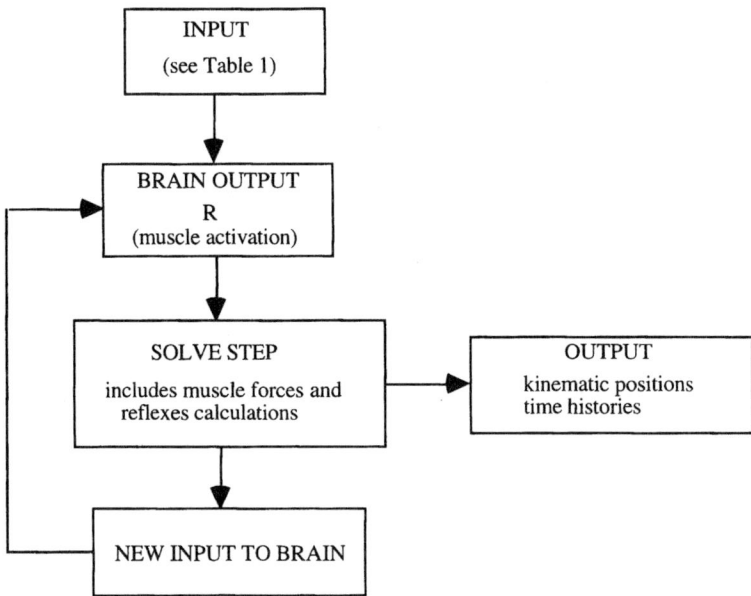

Figure 5. Computer program flow chart for solution of extended squat to stand example

Table 1. Input parameters for simulation example

Input	Variables
Number of links	[geometry (l, d, m, I), external F]
Integration parameters	[time step, t_{min}, t_{max}, integration method]
Joint constraints	[min and max θ]
Muscles	[initial l_0, k, α, β, ε, joint involved, antagonistic or agonistic]
Reflexes	[matrix]
Initial conditions	[insertion point θ and $\dot{\theta}$]

RESULTS OF SQUAT TO STAND

From preliminary results, shown in Figure 6, the effectiveness of this model has been proven: the two-links system is able to stay stable in the squatted position, even if this is not an equilibrium configuration, using a high level of muscle activation. By changing the values of the activation level and the threshold for each muscle, it is possible to make the model move towards the upright position. Furthermore, it is possible to simulate the movement with a wide range of velocity, from really slow to burst. Once the upright position is reached, the model is able to maintain stability even when a perturbation of a constant horizontal force is applied to the upper link (applied at $t = 2.0$ s). However, a small oscillation around the upright equilibrium position is present. These oscillations will explode because of accumulation of numerical error if the simulation is allowed to run for a long period of time. If an external force is applied to the model, the reflexes are able to react and to restore balance, provided that the external force is small enough to allow for position correction at the spinal level.

To highlight the usefulness of such a modular approach, using the same model some other situations can be simulated:

- introducing a nervous damage in the muscle efferent fibres, the proposed muscle model shows an elastic behaviour, similar to the real one;
- introducing nervous damage at the spine level, the reflexes are maintained;
- if the activation level of the muscles is high, the model simulates spastic behaviour due to the presence of the reflexes–therefore, the system is not able to reach a static equilibrium but oscillates around the desired position.

CONCLUDING REMARKS

The majority of this chapter has presented concepts of model formulation and associated mathematical solution techniques, including an illustrative example. Strategies for model formulation of mechanical (musculoskeletal systems), actuator (muscle models) and control (neuromuscular simulation) subsystems have been explicitly postulated. Along with the discussion of the direct or forward dynamics problem, other pertinent considerations relevant to simulation of human movement have been investigated, including balance, movement initiation, driving factors, internal forces and model complexity. The idea of goal-oriented simulation has been reinforced with a discussion about evaluation of simulation performance. Mathematical methods for obtaining the equations of motion such as direct (d'Alembert and free body diagram analysis) and indirect (Lagrangian and Newton–Euler) have been discussed with regard to their applicability. The advantages and disadvantages of each method have been

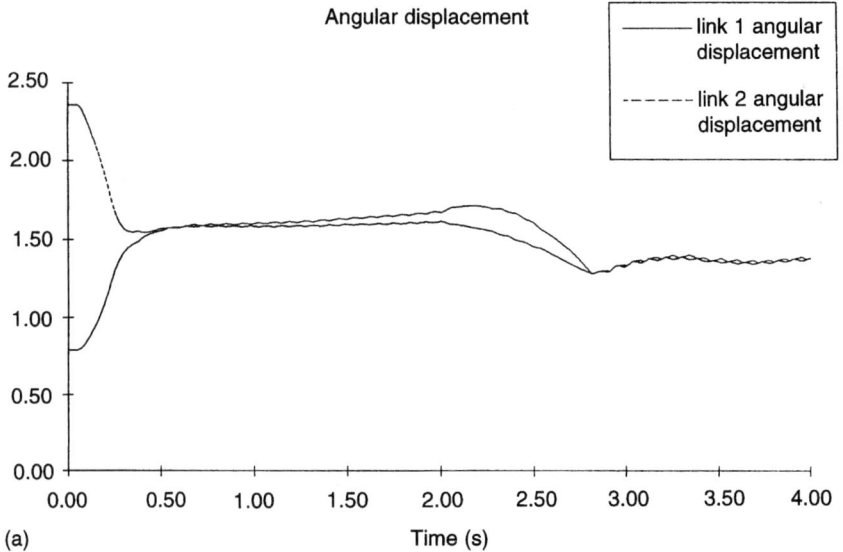

(a)

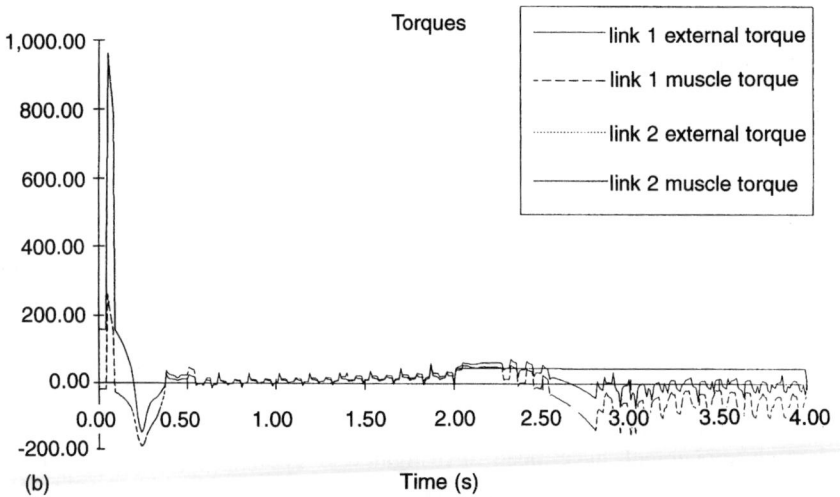

(b)

Figure 6. (a) Output kinematics: link 1 and 2 angular displacement time history. (b) Joint torques as calculated by the simulation versus time (normalized units). (c) Agonistic and antagonistic muscle spindle positions as calculated by the simulation versus time. (d) Agonistic and antagonistic muscle commands as calculated by the simulation versus time. (e) Brain or controller output of the simulation versus time

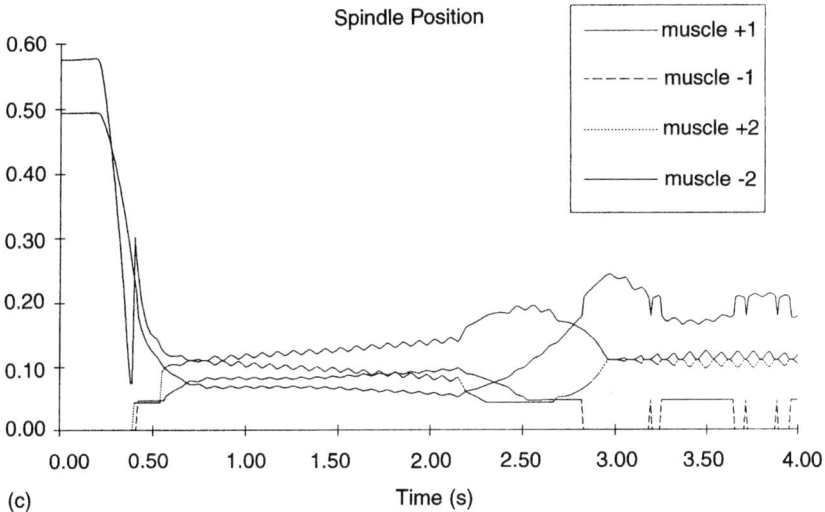

Spindle Position

(c)   Time (s)

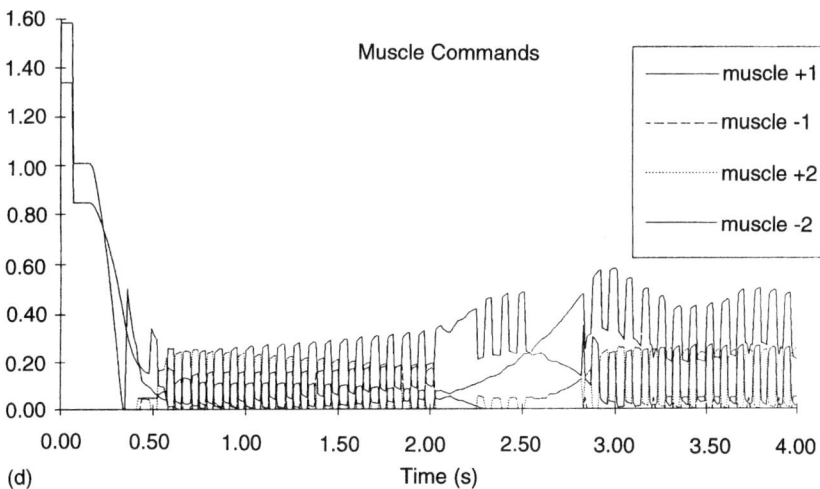

Muscle Commands

(d)   Time (s)

illustrated and enumerated. These concepts have been solidified through the presentation of an illustrative example, outlining the modelling of all the main components (mechanical, actuation or muscles, and neuromotor or control) needed for the simulation of the specific action of rising from a squatted position.

In summary, it can be reiterated that simulation must be goal-orientated to have significance. The model formulation should be as simple as possible while addressing the purpose of the simulation. The resulting formulation and solution techniques must then be chosen such that effective implementation of the model

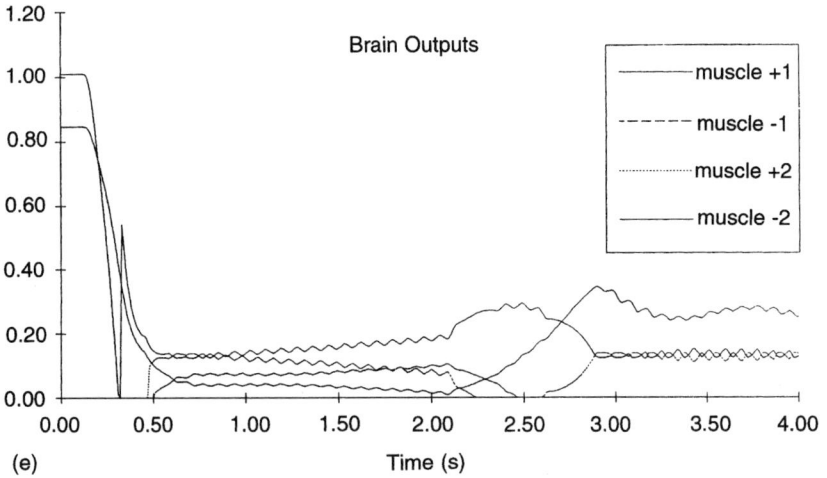

Figure 6. (*Continued*)

can be realized. Finally, the results of the simulation should be verified by comparing the model performance with that of the actual system. As modelling tools and methods evolve, more consequential simulations will be realized.

REFERENCES

American Heritage Electronic Dictionary (1992) WordStar International Inc., Novato, CA.

Aminian, K. R., Jequier, E. and Schulz, Y. (1994) Estimation of speed and incline of walking using neural network. *IEEE Instrum. Measurem. Technol. Conf.*, **1**, 110–162.

Beckett, R. and Chang, K. (1968) An evaluation of the kinematics of gait by minimum energy. *J. Biomech.*, **1**, 147–159.

Chen, B. R., Hines, M. J. and Hemami, H. (1986) Dynamic modeling for implementation of a right turn in bipedal walking. *J. Biomech.* **19**. 195–206.

Delp, S. L., Loan, J. P., Hoy, M. G., Zajac, F. E., Topp, E. L. and Rosen, J. M. (1990) An interactive graphics-based model of the lower extremity to study orthopaedic surgical procedures. *IEEE Trans. Biomed. Eng.*, **37**, 757–767.

Dinneen, J. A. and Hemami, H. (1993) Stability and movement of a one-link neuromusculoskeletal sagittal arm. *IEEE Trans. Biomed. Eng.*, **40**, 541–548.

Gasser, H. S. and Hill, A. V. (1924) The dynamics of muscular contraction. *Proc. R. Soc.*, **96-B**, 398–437.

Gielen, C. C. and Houck, J. C. (1987) A model of the motor servo: incorporating nonlinear spindle receptor and muscle mechanical properties. *Biol. Cybernet.*, **57**, 217–231.

Gubina, F., Hemami, H. and McGhee, R. B. (1974) On the dynamic stability of biped locomotion. *IEEE Trans. Biomed. Eng.* **2** 102–108.

Hatze, H. (1977a) A complete set of control equations for the human musculo-skeletal system. *J. Biomech.* **10** 799–805.

Hatze, H. (1977b) A myocybernetic control model of skeletal muscle. *Biol. Cybernet.* **25** 103–119.

Hatze, H. (1978) A general myocybernetic control model of skeletal muscle. *Biol. Cybernet.*, **28**, 143–157.

Hatze, H. (1981a) A comprehensive model for human motion simulation and its application to the take-off phase of the long jump. *J. Biomech.*, **14**, 135–142.

Hatze, H. (1981b) Analysis of stretch responses of a myocybernetic model muscle fibre. *Biol. Cybernet.*, 165–170.

Hatze, H. and Venter, A. (1981) Practical activation and retention of locomotion constraints in neuromusculoskeletal control system models. *J. Biomech.*, **14**, 873–877.

Hemami, H. (1985) Modeling, control, and simulation of human movement. *Crit. Rev. Biomed. Eng.*, **10**, 1–34.

Hemami, H. and Dinneen, J. A. (1993) A marionette-based strategy for stable movement. *IEEE Trans. Syst. Man Cybernet.*, **23**, 502–511.

Hemami, H. and Farnsworth, R. L. (1977) Postural and gait stability of the planar five link biped by simulation. *IEEE Trans. Autom. Contr.*, 452–458.

Hemami, H., Robinson, C. S. and Ceranowicz, A. Z. (1980) Stability of planar biped models by simultaneous pole assignment and decoupling. *Int. J. Syst. Sci.*, **11**, 65–75.

Hogan, N. (1984) Adaptive control of mechanical impedance by coactivation of antagonistic muscles. *IEEE Trans. Autom. Contr.*, **AC-29**, 681–690.

Hoy, M. G., Zajac, F. E. and Gordon, M. E. (1990) A musculoskeletal model of the human lower extremity: the effect of muscle, tendon, and moment arm on the moment–angle relationship of musculotendon actuators at the hip, knee, and ankle. *J. Biomech.*, **23**, 157–169.

Kuo, A. D. and Zajac, F. E. (1993a) A biomechanical analysis of muscle strength as a limiting factor in standing posture. *Biomech.*, **26** (Suppl. 1), 137–150.

Kuo, A. D. and Zajac, F. E. (1993b) Human standing posture: multi-joint movement strategies based on biomechanical constraints. *Prog. Brain Res.*, **97**, 349–358.

MacKinnon, C. D. and Winter, D. A. (1993) Control of whole body balance in the frontal plane during human walking. *J. Biomech.* **26**, 633–644.

Mochon, S. and McMahon, T. A. (1980) Ballistic walking. *J. Biomech.* **13**, 49–57.

Nigg, B. M. and Herzog, W. (1995) *Biomechanics of the Musculo-skeletal System.* John Wiley & Sons, Chichester/New York/Brisbane/Toronto/Singapore.

Onyshko, S. and Winter, D. A. (1980) A mathematical model for the dynamics of human locomotion. *J. Biomech.*, **13**, 135–142.

Pandy, M. G. and Berme, N. (1988a) An experimental and analytical study of impact forces during human jumping. *J. Biomech.*, **21**, 1061–1066.

Pandy, M. G. and Berme, N. (1988b) A numerical method for simulating the dynamics of human walking. *J. Biomech.*, **21**, 1043–1051.

Pandy, M. G. and Berme, N. (1988c) Synthesis of human walking: a planar model for single support. *J. Biomech.*, **21**, 1053–1060.

Pandy, M. G. and Berme, N. (1989a) Quantitative assessment of gait determinants during single stance via a three-dimensional model–Part 2. Pathological gait. *J. Biomech.*, **22**, 725–733.

Pandy, M. G. and Berme, N. (1989b) Quantitative assessment of gait determinants during single stance via a three-dimensional model–Part 1. Normal gait. *J. Biomech.*, **22**, 717–731.

Pandy, M. G. and Zajac, F. E. (1991) Optimal muscular coordination strategies for jumping. *J. Biomech.*, **24**, 1–10.

Pandy, M. G., Zajac, F. E., Eunsup, S. and Levine, W. S. (1990) An optimal control model for maximum-height human jumping. *J. Biomech.*, **23**, 1185–1198.

Roth, R. (1989) On constructing free-body diagrams. *Int. J. Appl. Eng. Educ.*, **5**, 565–570.

Salatian, A. W. and Zheng, Y. F. (1992) Gait synthesis for a biped robot climbing sloping surfaces using neural networks Part II: Dynamic Learning. In: *Proceedings of the IEEE International Conference on Robotics and Automation*, Nice. IEEE, Piscataway, NJ. pp. 2607–2611.

Saunders, M., Inman, V. T. and Eberhart, H. D. (1953) The major determinants in normal and pathological gait. *J. Bone Joint Surg.*, **35-A**, 543–558.

Sepulveda, F. W., Derek, M. and Vaughan, C. L. (1993) Neural network representaion of electromygraphy and joint dynamics in human gait. *J. Biomech.*, **26**, 101–109.

van Dijk, J. H. M. (1978) On the interaction between the central nervous system and the peripheral motor system. *Biol. Cybernet.*, **30**, 195–208.

Vukobratovic, M. K., Filaretov, V. F. and Korzun, A. I. (1994) A unified approach to mathematical modeling of robotic manipulator dynamics. *Robotica.* **12**, 411–420.

Walker, M. W. and Orin, D. E. (1982) Efficient dynamic computer simulation of robotic mechanisms. *J. Dynamic Syst. Measurements Control*, **104**, 205–211.

Wan, C. L. C. and Yip Pak (1991) Neural networks for 3-D motion detection from a sequence of image frames. In: *1991 IEEE International Joint Conference on Neural Networks*, IJCNN '91, Singapore. IEEE, Piscataway, pp. 2013–2018.

Winter, D. A. (1979) A new definition of mechanical work done in human movement. *J. Appl. Physiol.,* **46**, 79–83.

Winter, D. A. (1984) Biomechanics of human movement with applications to the study of human locomotion. *Crit. Rev. Biomed. Eng.*, **9**, 287–314.

Winter, D. A. (1990) *Biomechanics and Motor Control of Human Movement*, 2nd edn. John Wiley & Sons, Chichester/New York/Brisbane/Toronto/Singapore.

Yamaguchi, G. T., Moran, D. W. and Si, J. (1995) A computationally efficient method for solving the redundant problem in biomechanics. *J. Biomech.*, **28**, 999–1005.

Yamaguchi, G. T. and Zajac, F. E. (1989) Sensitivity of simulated human gait to neuromuscular control patterns. *J. Biomech.*, **22**(10), 1103.

Yamaguchi, G. T. and Zajac, F. E. (1990) Restoring unassisted natural gait to paraplegics via functional neuromuscular stimulation: a computer simulation study. *IEEE Trans. Biomed. Eng.*, **37**, 886–902.

Zajac, F. E. (1993) Muscle coordination of movement: a perspective. *J. Biomech.*, **26** (Suppl. 1); 109–124.

15

Able-bodied Gait in Men and Women

PAUL ALLARD[1,2], RÉGIS LACHANCE[1,2], RACHID AISSAOUI[1,2],
HEYDAR SADEGHI[1,2] AND MORRIS DUHAIME[2,3,4]

[1]Department of Physical Education, University of Montreal, Montreal, PQ, Canada
[2]Laboratoire d'Étude du Mouvement, Research Center, Sainte-Justine Hospital,
Montreal, PQ, Canada
[3]Orthopedic Surgery, Shriner's Hospital, Montreal, PQ, Canada
[4]McGill University, Montreal, PQ, Canada

INTRODUCTION

To some, able-bodied gait is the mean pattern of a large control group of subjects having no major orthopaedic or neurological disorders which may affect their locomotion. It is often associated with the normal gait pattern and is therefore the gold standard with which everyone else's gait pattern is compared. To others, able-bodied gait is that of a single subject rather than that of a control group. Such an able-bodied subject may exhibit large variations from the mean control group data, making comparison either difficult or misleading. Bates, Difek and Davis (1992) consider the subject to be a random response generator, making all trials independent of one another. Though many are using their own control data, little is known about the homogeneity between the groups of able-bodied subjects. Able-bodied gait is presented here to highlight the basic pattern as well as some extreme cases of normality of which we must be aware when interpreting normal gait.

Able-bodied gait has been well described in the literature. The pioneering work of Braune and Fischer (1987a,b), originally published in late 1890 and early 1900, was the first to deal with the three-dimensional (3D) aspect of human ambulation by means of two-sided chronophotography. Their work was applied

Three-dimensional Analysis of Human Locomotion. Edited by P. Allard, A. Cappozzo, A. Lundberg and C. Vaughan
© 1997 John Wiley & Sons Ltd. ISBN 0 471 96949 4

to the study of the German infantry soldier. Much later, Klopsteg (1954) and Inman, Ralston and Todd (1981) in the Biomechanics Laboratory at the University of California in San Francisco and Berkeley reported fundamental 3D data on normal and amputee locomotion for the design of artificial limbs. Since then, 3D gait analysis methods and software have been published to meet the needs of the students of human motion (Vaughan, Davis and O'Connor, 1992). Vaughan and his colleagues bring together the theory of 3D inverse dynamics applied to human gait and the tools required to perform its analysis. Allard, Stokes and Blanchi (1994) reviewed the current practices in data capture, 3D reconstruction techniques and modelling techniques in human movement as well as upcoming trends. But comprehensive, normative 3D gait data are reported for the first time in Craik and Oatis (1995) by a number of contributors. Aside from the temporal and phasic parameters, normative gait data were obtained from a single limb.

Normative data are provided here from a group of 10 male and 15 female able-bodied subjects. Their anthropometric characteristics are given in Table 1. None had presented any orthopaedic or neurological disorders which could affect their walking patterns. They were generally students in the Department of Physical Education at the University of Montreal, but none were national or provincial-level athletes.

Data for the right limb of male subjects are first presented, since it was the dominant side. Contralateral limb data are presented in the section on symmetry and, finally, data of able-bodied women are discussed before drawing conclusions.

ABLE-BODIED GAIT IN MEN

There are only a few 3D studies and fewer simultaneous bilateral evaluations of able-bodied gait. After the presentation of the spatio-temporal parameters, the results will be first discussed with respect to those obtained from the right limb of able-bodied men, since they are the most commonly available data. Then, bilateral data will be briefly presented and gait symmetry will be discussed.

Table 1. Anthropometric data for the male and female able-bodied groups

	Male	Female
Number of subjects	10	15
Number of trials	30	45
Age (years)	25.31 (σ 4.82)	20.13 (σ 2.20)
Height (m)	1.79 (σ 0.06)	1.67 (σ 0.04)
Weight (kg)	76.95 (σ 11.29)	62.03 (σ 5.61)

SPATIO-TEMPORAL FACTORS

The mean values of the spatio-temporal factors are given in Table 2. These data fall within the range reported by Craik (1995), where walking speed is between 1.2 and 1.5 m/s, with a cadence varying between 100 and 120 steps/min. Öberg, Karsznia and Öberg (1993) have performed a similar analysis for a group of 233 healthy subjects whose age varied between 10 and 79 years. For similar age groups, the walking speed was similar to that in our data but the cadence was slightly higher, at 120 steps/min. The difference may be explained in part by the methodology used to determine the spatio-temporal parameters. Öberg, Karsznia and Öberg (1993) used two photocells at a 5.5-m interval, and average values were taken from 13 walking trials. Our data are derived from a video-based system operating at 90 Hz.

Individual able-bodied gait does not necessarily follow the average pattern described by the group. The spatio-temporal factors of two male subjects who have essentially the same natural walking speed as that of our able-bodied group are also presented in Table 2. This was achieved by modulating their stride length and cadence accordingly. Their stance- and swing-phase relative durations are typical. Yet, the initial and terminal double support periods are higher, though within the limit delineated by the standard deviation of the able-bodied group.

Generally, the data of the two subjects fall within one standard deviation of those of the able-bodied group. On average, the standard deviation represents 9% of the mean value, with the smallest value being the stance- or swing-phase relative duration (4%) and the highest the double support periods (18%). Does this reflect the presence of more than one able-bodied gait pattern?

KINEMATICS

Kinematic data such as limb segment displacements, speeds and accelerations or joint range of motion are essential for estimation of dynamic joint reaction forces and muscle moments and are mostly useful in comparing able-bodied locomotion with pathological gait. Though the literature proliferates with bi-dimensional data, 3D normative data are less common. Though most studies have focused on the 3D kinematics of a single joint, only a few (Apkarian, Naumann and Cairns, 1989; Patriarco et al., 1981; Kadaba, Ramakrishnan and Wootten, 1990) have reported hip, knee and ankle ranges of motion during normal locomotion.

Wu (1995) attracts our attention to the estimation of linear acceleration obtained by direct measurement or derived from displacement data. Though the overall patterns are in good agreement throughout the gait cycle, she reports a lack of impact acceleration at heel strike when using a differentiation method. The smoothing or the filtering techniques themselves influence the results and

Table 2. Phasic parameters for an able-bodied-men group and two subjects of that group

| | Able-bodied group ($N = 30$) | | | | Individual data | |
| | Right limb | | Left limb | | | |
	Mean	σ	Mean	σ	Subject 1	Subject 2
Stance gait speed (m/s)	1.31	0.11	1.33	0.09	1.30	1.34
Cadence (step/min)	106.9	7.2	108.8	6.7	103.8	105.8
Stride length (m)	1.467	0.061	1.458	0.056	1.432	1.469
Step length (m)	0.735	0.044	0.720	0.033	0.820	0.729
Stance phase (%)	60.98	1.74	61.16	1.23	58.59	59.79
Swing phase (%)	39.02	1.74	38.84	1.23	41.41	40.21
Initial double support (%)	10.52	1.83	10.64	1.87	9.09	12.37
Terminal double support (%)	11.44	2.05	11.56	2.02	12.12	11.34
Total double support (%)	21.96		22.20		21.21	23.71

Table 3. Energies in J/kg ($\times 10^{-2}$) developed by the right limb in the sagittal plane for men

Joint	Burst	Able-bodied group Mean	Able-bodied group Standard deviation	Individual data Subject 1	Individual data Subject 2
Ankle	A1S	−18.30	4.38	−23.13	−16.97
	A2S	27.50	6.86	15.63	23.53
Knee	K1S	−5.00	3.22	−11.87	−4.80
	K2S	5.00	4.42	14.20	2.97
	K3S	−13.60	7.21	−4.73	−30.77
	K4S	−15.00	7.89	−23.57	14.07
Hip	H1S	10.50	11.07	3.73	1.50
	H2S	−14.10	12.26	−2.27	−31.77
	H3S	21.70	8.43	71.13	14.87

can render comparison difficult between different kinematic data sets (Allard et al., 1990).

Wu (1995) also reports that 18 parameters are required to describe the kinematics of a single limb segment—linear and angular displacements, velocities and accelerations. If one considers the lower and upper limbs as well as the trunk (upper and lower sections) and the head, a 15-link segment model would yield 270 parameters. The kinematics of the centre of mass of the whole body (nine parameters) and the joint ranges of motion description (72 parameters) must be added! The reader is referred to the above publications for further information as well as for normative data. Among these, Kadaba, Ramakrishnan and Wootten (1990) present average data of nine gait cycles from each of the 40 subjects ($N = 360$).

MOMENTS, MUSCLE POWERS AND MECHANICAL ENERGIES

Although ground reaction forces and muscle moments are useful parameters with which to describe human locomotion (Herzog et al., 1989) and to quantify gait disorders, muscle powers provide a good indication of the person's ability to control and propel the lower limb. Muscle power had been defined by Winter (1990) as the rate of work done by the muscles. It is expressed as the product of the net muscle moment and angular velocity and can be either positive or negative. When the signs or the polarities of the moment and the angular velocity are the same, their product is positive. This represents a power generation and is associated with a concentric muscle contraction. When the polarities are of opposite signs, the power is negative and this corresponds to an eccentric muscle contraction.

Muscle powers and their corresponding moments are given in Figures 1 to 3. The muscle power bursts have been labelled according to Eng and Winter (1995). These were obtained from 10 able-bodied men and represent 30 simultaneous bilateral trials. Right limb data were normalized with respect to body weight and expressed in percentage of the duration of the gait cycle (GC). The standard deviation calculated at each percentage of the GC is overlaid on the right limb mean data. Data of the left limb are also shown on the same figure. The left limb began the swing phase at about 10% of the GC of the right limb, and toe-off occurred at about 112% of GC. This representation of bilateral data has the advantage of keeping the temporal relationship between the limbs, which is important for taking into account asymmetries in pathological gaits.

A word of caution is necessary before examining the power curves. The sagittal moments are generally comparable; however, the moments calculated in the frontal and transverse planes vary widely from one study to another. For example, Crowinshield et al. (1978), Bowsher and Vaughan (1995) and Kadaba et al. (1989) reported an internal hip rotation moment between 5% and 25% GC followed by an external moment, while Apkarian et al. (1989) and Eng and Winter (1995) found the opposite. At the ankle, Apkarian, Naumann and Cairns (1989) documented an inversion moment, while Kadaba et al. (1989) reported an eversion moment. In the Eng and Winter (1995) study, the subjects developed an everted moment at the ankle which was followed by an inverted moment. We assume that this variability in the results is due to either different normal walking patterns, adaptations to control the lower limb or a combination of both.

Recently, 3D muscle power data have been used to describe both normal (Allard et al., 1996; Ounpuu, Gage and Davis, 1991; Eng and Winter, 1995) and pathological gait (Allard et al., 1995; Czerniecki, Gitter and Munro, 1991; Loizeau et al., 1995). The reader is referred to these publications for a detailed explanation of the powers absorbed and generated during gait and should keep in mind that powers and energies are scalar terms and were partitioned into three arbitrary planes to facilitate data interpretation.

The mechanical energy values obtained by time integration of the power curves are reported in Tables 3 to 5 for the 30 trials as well as for two individuals who formed a part of the able-bodied group. The individual data followed the energy absorption and generation pattern of the able-bodied group, with the exception of K4S burst. Subject 1 had a stronger energy absorption than that of the control group, while subject 2 generated 14.07 J/kg ($\times 10^{-2}$). Large variations were also noted between the two individuals. For example, the H2S energy absorption was 14 times higher in subject 2, but the H3S generation was nearly five times higher in subject 1. This shows that individuals may not comply completely with a group mean but may have different but still quite normal patterns.

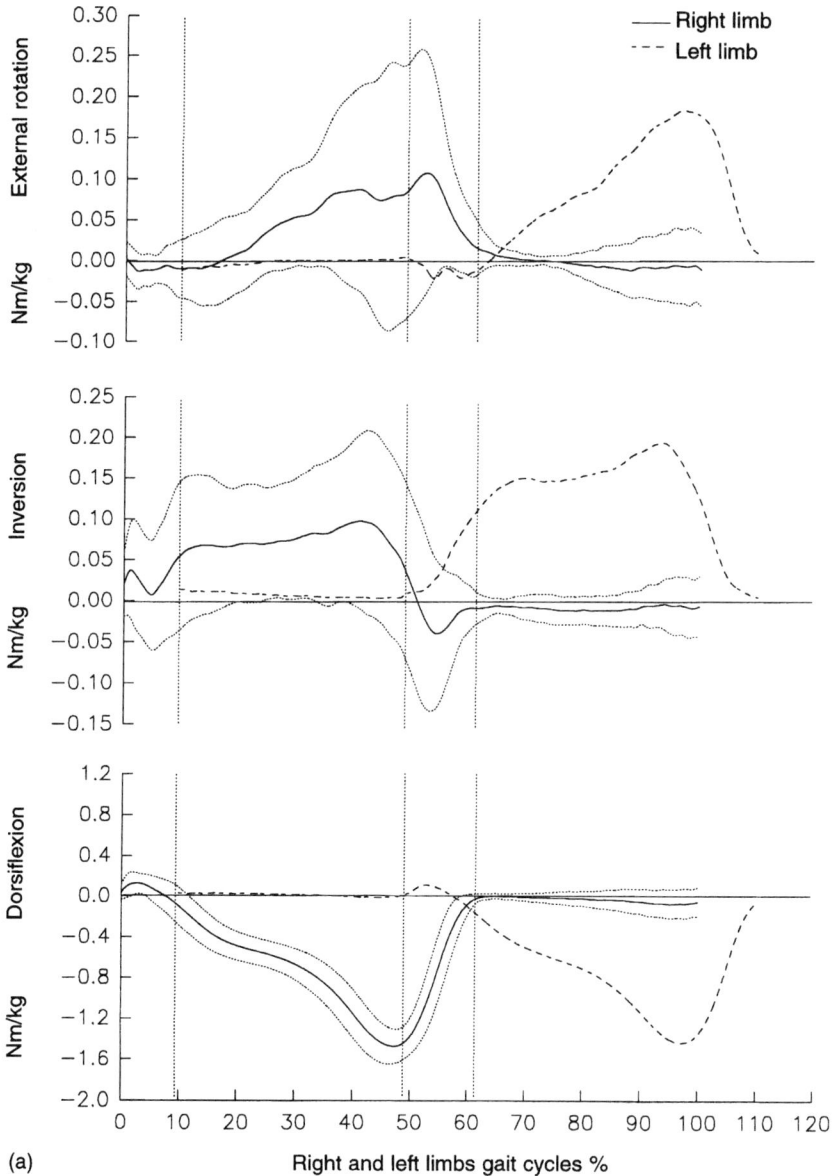

Figure 1. Joint (a) moments and (b) muscle powers developed at the ankle in 10 able-bodied men. The solid line corresponds to the right limb data and the dashed line to the left limb data. The standard deviation is overlaid onto the right limb data. The dotted vertical lines at about 10%, 50% and 60% of the gait cycle indicate left limb toe-off, left-limb heel strike and right limb toe-off, respectively. Thus the 0–10% interval corresponds to the initial double support, while the 50–60% interval is associated with the terminal double support of the right limb

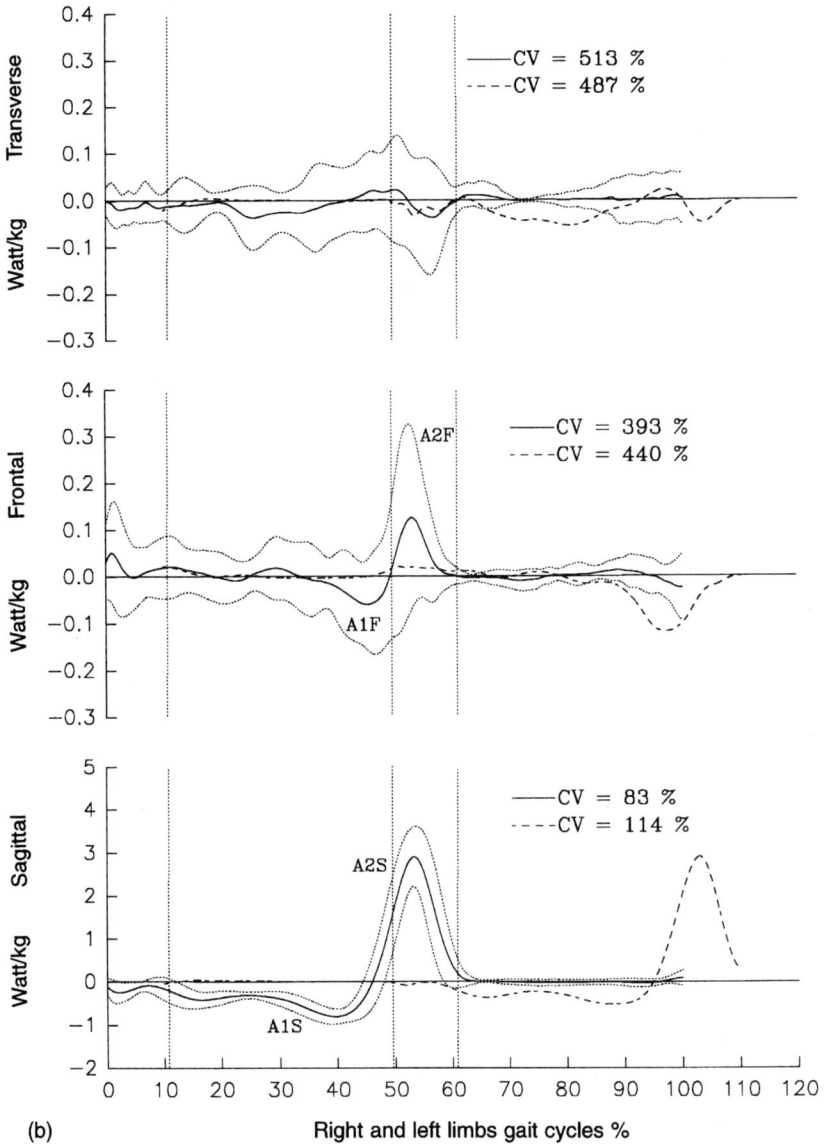

Figure 1. (*continued*)

GAIT SYMMETRY

The main task of the lower extremities during gait is to keep the body upright, and to displace it in an orderly and stable manner. Since the gait patterns of the right and left lower limbs were assumed to be symmetrical, numerous

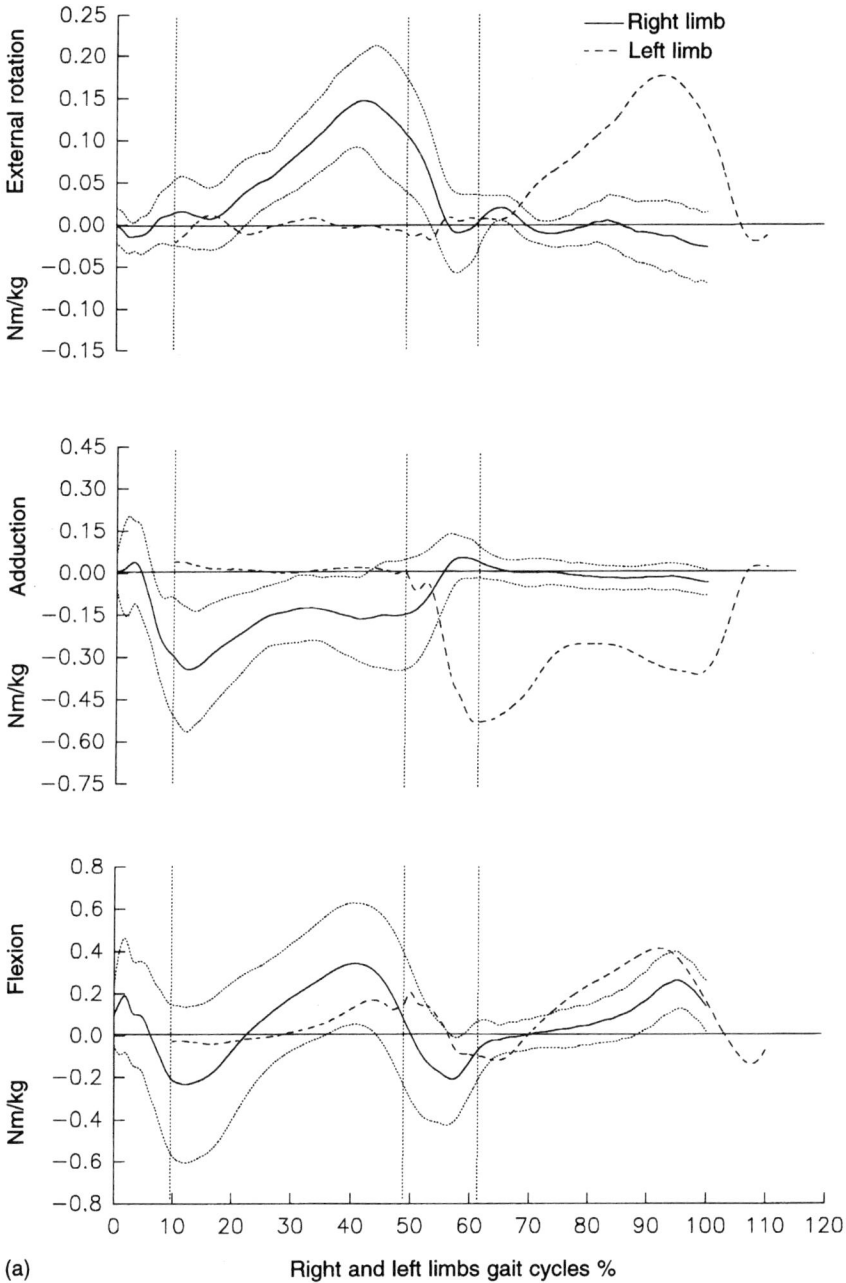

Figure 2. Joint (a) moments and (b) muscle powers developed at the knee in 10 able-bodied men. See Figure 1 caption

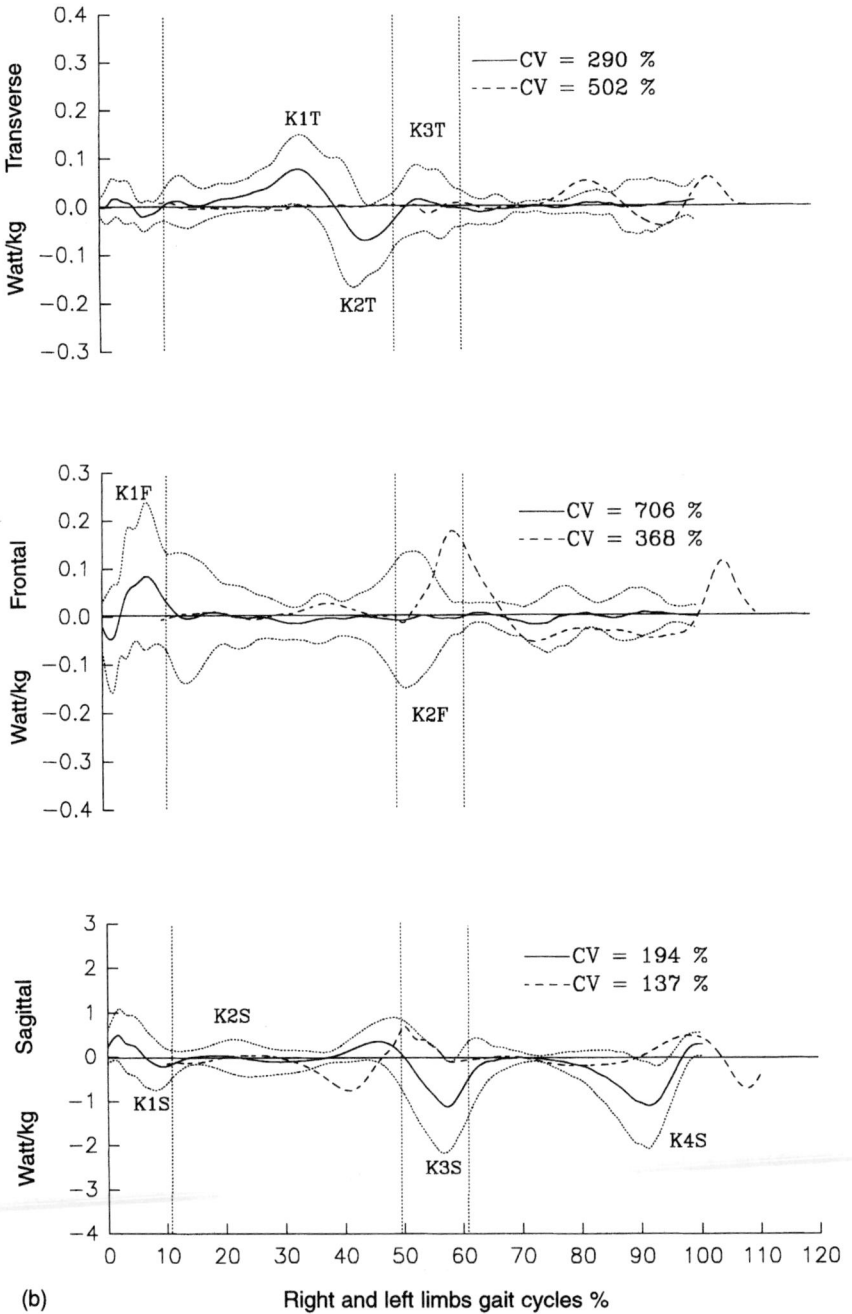

Figure 2. (*continued*)

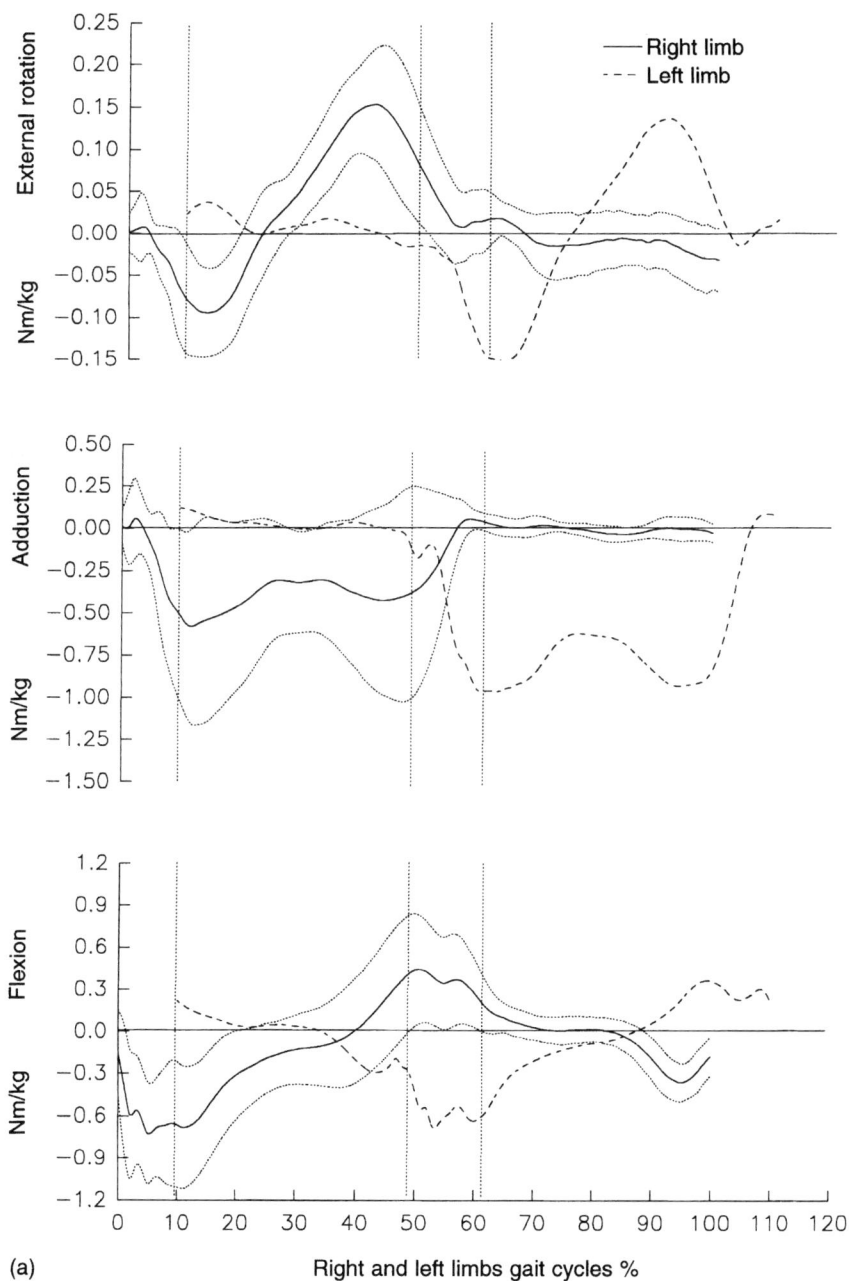

Figure 3. Joint (a) moments and (b) muscle powers developed at the hip in 10 able-bodied men. See Figure 1 caption

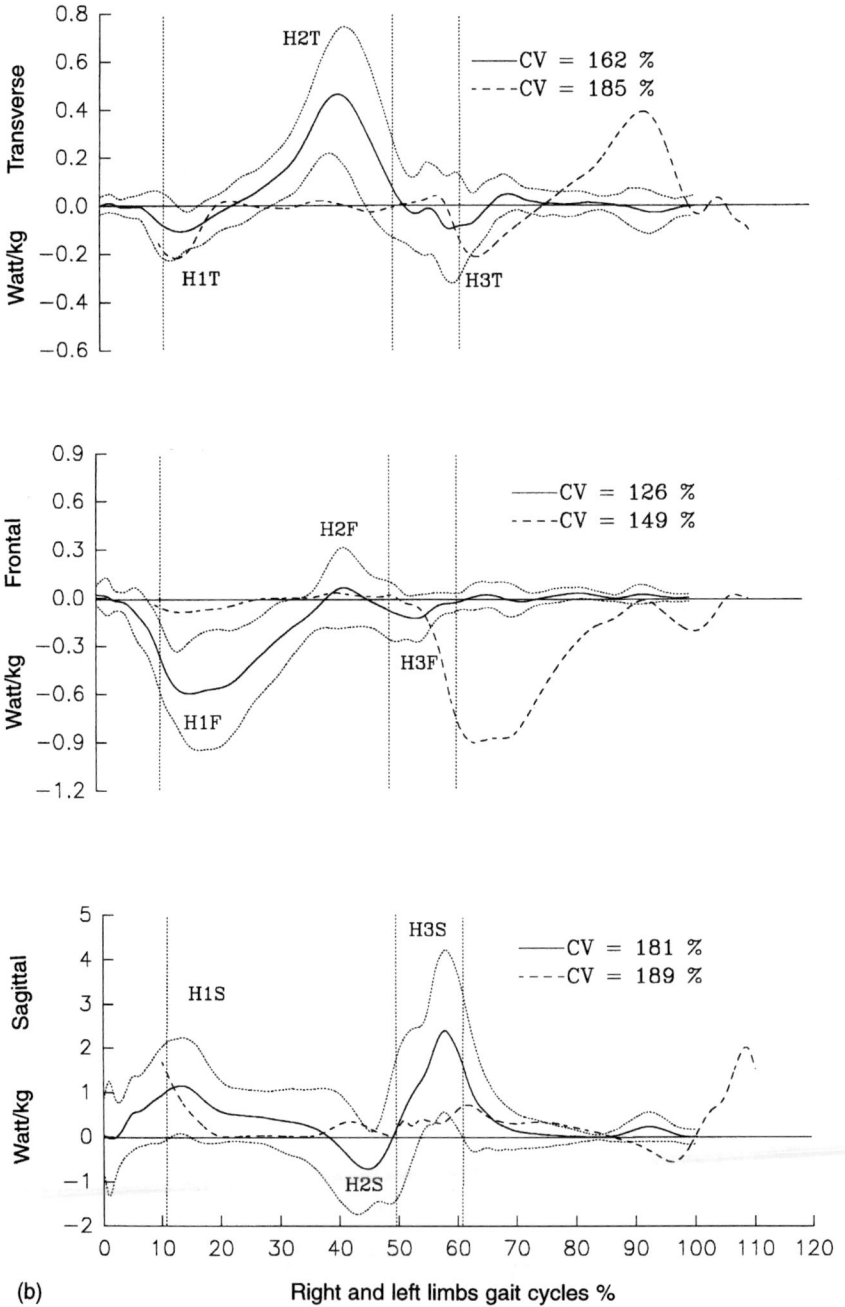

(b)

Right and left limbs gait cycles %

Figure 3. (*continued*)

Table 4. Energies in J/kg ($\times 10^{-2}$) developed by the right limb in the frontal plane for men

Joint	Burst	Able-bodied group		Individual data	
		Mean	Standard deviation	Mean	Standard deviation
Ankle	A1F	−2.10	1.84	Nil	−0.90
	A2F	1.60	0.94	2.00	Nil
Knee	K1F	0.50	1.01	0.77	0.82
	K2F	−1.90	2.89	−1.61	−2.75
Hip	H1F	−16.40	8.25	−11.80	−21.47
	H2F	1.10	1.40	1.37	1.50
	H3F	−1.40	1.18	0.00	−2.37

Table 5. Energies in J/kg ($\times 10^{-2}$) developed by the right limb in the transverse plane for men

Joint	Burst	Able-bodied group		Individual data	
		Mean	Standard deviation	Mean	Standard deviation
Knee	K1T	1.20	0.76	1.67	0.53
	K2T	−0.90	0.68	−1.83	−1.03
	K3T	0.30	0.34	0.00	1.47
Hip	H1T	−1.60	1.57	−1.60	−2.10
	H2T	4.10	2.12	1.07	0.90
	H3T	−1.00	1.09	−1.00	−1.57

able-bodied gait studies have relied on unilateral data (Apkarian, Naumann and Cairns, 1989; Kadaba et al., 1989; Eng and Winter, 1995; Sutherland, 1984; Vaughan, Davis and O'Connor, 1992). There have been a number of bilateral gait studies but often limited to the assessment of a single limb at a time (Loizeau et al., 1995). Ounpuu, Gage and Davis (1991) obtained 3D bilateral simultaneous data on 31 children but reported combined right and left limb values. Using bilateral triaxial electrogoniometers attached to the hips and the knees, Hannah, Morrison and Chapman (1984) concluded that their 12 able-bodied subjects walked with reasonable symmetry. Recently, Allard et al. (1996) presented temporal and 3D bilateral kinematic data obtained from 19 able-bodied male subjects. They reported statistically different mechanical energies for the right and left limbs, even though both limbs had the same walking speed.

Anthropometric factors, dominance and orthopaedic disorders may interfere with gait symmetry. Chhibber and Singh (1970) have found that the muscle weight of the right limb was significantly greater on the dominant side, though no correlation has been found between the upper and lower limb dominance. After assessing lower limb dominance by means of the mobility and stability tasks, Gundersen et al. (1989) performed a bilateral planar kinematic analysis of 14 able-bodied subjects. They concluded that symmetry cannot be assumed but should be examined in relation to the subject's own variability and could be predicted by the dominance.

The literature is divided in its support of gait symmetry. There are numerous studies which document gait symmetry in terms of temporal parameters (Hamill, Bates and Knutzen, 1984; Vanden der Straaten and Scholton, 1978), kinematics (Hannah, Morrison and Chapman, 1984; Rosenrot, 1980) and ground reaction forces (Hamill, Bates and Knutzen, 1984). Others reported gait asymmetry in spatio-temporal parameters (Brandstater et al., 1983), kinematics (Henrichs, 1992), kinetics (Herzog et al., 1989) and electromyographic activity (Arsenault, Winter and Marteniuk, 1986; Ounpuu and Winter, 1989). The presence of gait asymmetry is definitely confirmed, though some subjects may exhibit less than others. Some questions still remain to be answered: which gait parameters are the most significant and why asymmetry is present and what causes it.

To determine gait asymmetry, different methods have been used, namely statistical and symmetry indices. Gundersen et al. (1989) have used a two-way multivariate analysis of variance (MANOVA) with repeated measures because the temporal and kinematic variables were not independent of each other. A Newman–Keul *post hoc* test was performed for variables showing a significant within-subject difference between limbs. They also used a Pearson product–moment correlation to determine the relationships among the 12 measured gait variables. Hannah, Morrison and Chapman (1984) passed a best-fit straight line between the right and left data sets. Perfect symmetry was expressed by a coefficient of value one. They observed that in the frequency domain, the detection rate of asymmetries is markedly improved.

The symmetry index (SI) proposed by Robinson, Herzog and Nigg (1987) was used by Herzog et al. (1989) to determine asymmetries in ground reaction force patterns in normal gait. It is expressed as

$$\text{SI} = \frac{(X_R - X_L)}{0.5(X_R + X_L)} \times 100\%$$

where X_R and X_L were the values of the gait variable measured for the right and left limb respectively. A zero SI index indicates perfect symmetry. Because the difference between the right and left limb values is reported against their average value, this index has two limitations. Where a large asymmetry is present, the average value does not correctly reflect the performance of either limb. Furthermore, parameters having large values but relatively small interlimb differences will tend to lower the index and reflect symmetry. Using this index, Becker et al. (1995) were able to show that the successful surgical treatment of ankle fractures in young adults resulted in an improvement of gait symmetry in terms of plantar pressure distribution.

The ratio between the value for one limb and that of the other was used by Ganguli, Mukherji and Boss (1974) to investigate peak velocity during the gait cycle of below-knee amputees and by Andres and Stimmel (1990) in assessing lower limb prosthetic alignment. It was also applied by Wall and Turnbull (1986) to determine temporal gait asymmetries in 25 patients with residual hemiplegia. A ratio of one was indicative of a reciprocal gait pattern, while higher or lower values reflected asymmetries. A similar approach involving the ratio of the standard deviations was considered by Klajic, Bajd and Stanic (1975) to represent the perfection of the gait pattern.

Vagenas and Hoshizaki (1992) developed a new ratio:

$$I_a = \frac{(L - R)}{\max(L,\ R)} \times 100\%$$

where R and L stand for right and left limb values respectively. The scores were recorded to reflect trichotomized values of 1, 0 and -1. Asymmetries larger than 1% were given a value of ± 1, and in all other cases a value of 0 was given to represent symmetry.

Though these indices have not been used here to determine the presence of gait asymmetry, a paired t-test was applied between right and left limb data. The spatio-temporal parameters did not reveal any significant differences between the limbs. However, differences were noted between the mechanical energies, as shown in Tables 6 to 8. Most of the differences were observed in the sagittal plane and were generally associated with energy absorption. This was related to the greater absorption of the right limb, while both limbs maintained the same speed of locomotion (Allard et al., 1996).

•

ABLE-BODIED GAIT IN WOMEN

There have been relatively few studies of able-bodied gait in women (Murray, Kory and Sepic, 1970; Finley, Cody and Finizie, 1969). These were limited to the description of the spatio-temporal and kinematic parameters.

SPATIO-TEMPORAL FACTORS

Table 9 presents the spatio-temporal parameters obtained from a group of 15 able-bodied women. Their average speed and cadence are respectively about 10% faster and 9.5% slower than those of a comparable age group in Öberg, Karsznia and Öberg (1993) but still well within their 95% confidence limit. No significant difference was noted between the right and left limb values. The

Table 6. Energies in J/kg ($\times 10^{-2}$) developed by the able-bodied-men group in the sagittal plane by both limbs

		Able-bodied group ($N = 30$)			
		Right limb		Left limb	
Joint	Burst	Mean	Standard deviation	Mean	Standard deviation
Ankle	A1	−18.30*	4.38	−16.40*	4.35
	A2	27.60	6.86	28.70	5.79
Knee	K1	−5.00	3.22	−4.10	3.81
	K2	5.00	4.42	4.50	4.82
	K3	−13.60*	7.21	−5.10*	3.19
	K4	−15.00*	7.89	−8.50*	2.27
Hip	H1	10.00*	11.07	21.70*	19.81
	H2	−14.10*	12.26	−3.90*	6.47
	H3	21.70*	8.43	16.00*	4.61
Total limb generated		64.70	21.15	70.90	19.94
Total limb absorbed		−66.00*	14.84	−37.90*	9.52

* $p < 0.05$.

Table 7. Energies in J/kg ($\times 10^{-2}$) developed by the able-bodied-men group in the frontal plane by both limbs

		Able-bodied group ($N = 30$)			
		Right limb		Left limb	
Joint	Burst	Mean	Standard deviation	Mean	Standard deviation
Ankle	A1F	−2.10	1.84	−1.80	0.83
	A2F	1.60*	0.94	0.70*	0.88
	K1F	0.50*	1.01	1.80*	1.61
	K2F	−1.90	2.89	−1.00	1.32
Hip	H1F	−16.40	8.25	−18.70	10.21
	H2F	1.10	1.40	0.70	1.19
	H3F	−1.40	1.18	−1.30	1.48
Total limb generated		3.20	2.04	3.20	2.66
Total limb absorbed		−21.80	11.34	−22.80	10.57

* $p < 0.05$.

Table 8. Energies in J/kg ($\times 10^{-2}$) developed by the able-bodied-men group in the transverse plane by both limbs

| | | Able-bodied group ($N = 30$) | | | |
| | | Right limb | | Left limb | |
Joint	Burst	Mean	Standard deviation	Mean	Standard deviation
Ankle	K1T	1.20	0.76	0.90	0.94
	K2T	-0.90^*	0.68	-0.60^*	0.57
	K3T	0.30	0.34	0.20	0.44
Hip	H1T	-1.60^*	1.57	-0.90^*	1.37
	H2T	4.10	2.12	3.90	2.70
	H3T	-1.00	1.09	-0.60	0.69
Total limb generated		5.60	2.06	5.00	3.38
Total limb absorbed		-3.50^*	1.60	-2.10^*	1.83

$^* p < 0.05.$

Table 9. Phasic parameters for an able-bodied-women group

| | Able-bodied group ($N = 45$) | | | |
| | Right limb | | Left limb | |
	Mean	σ	Mean	σ
Stance gait speed (m/s)	1.37	1.10	1.37	0.12
Cadence (step/min)	113.5	6.6	112.8	7.7
Stride length (m)	1.438	0.058	1.445	0.058
Step length (m)	0.725	0.037	0.719	0.032
Stance phase (%)	59.30	2.04	59.15	2.06
Swing phase (%)	40.70	2.04	40.85	2.06
Initial double support (%)	9.52	2.08	9.54	2.08
Terminal double support (%)	9.21	2.13	9.54	2.18
Total double support (%)	18.73		18.78	

women had a faster walking speed than the able-bodied-men group, which can be explained in part by a higher cadence (Yamasaki, Sasaki and Masafumi, 1991).

MUSCLE POWER AND MECHANICAL ENERGY

The muscle powers and the joint moments for the ankles, hips and knees are presented in Figures 4 to 6. The moments calculated at each joint were similar to those reported for the able-bodied-men group of our study. Furthermore, the

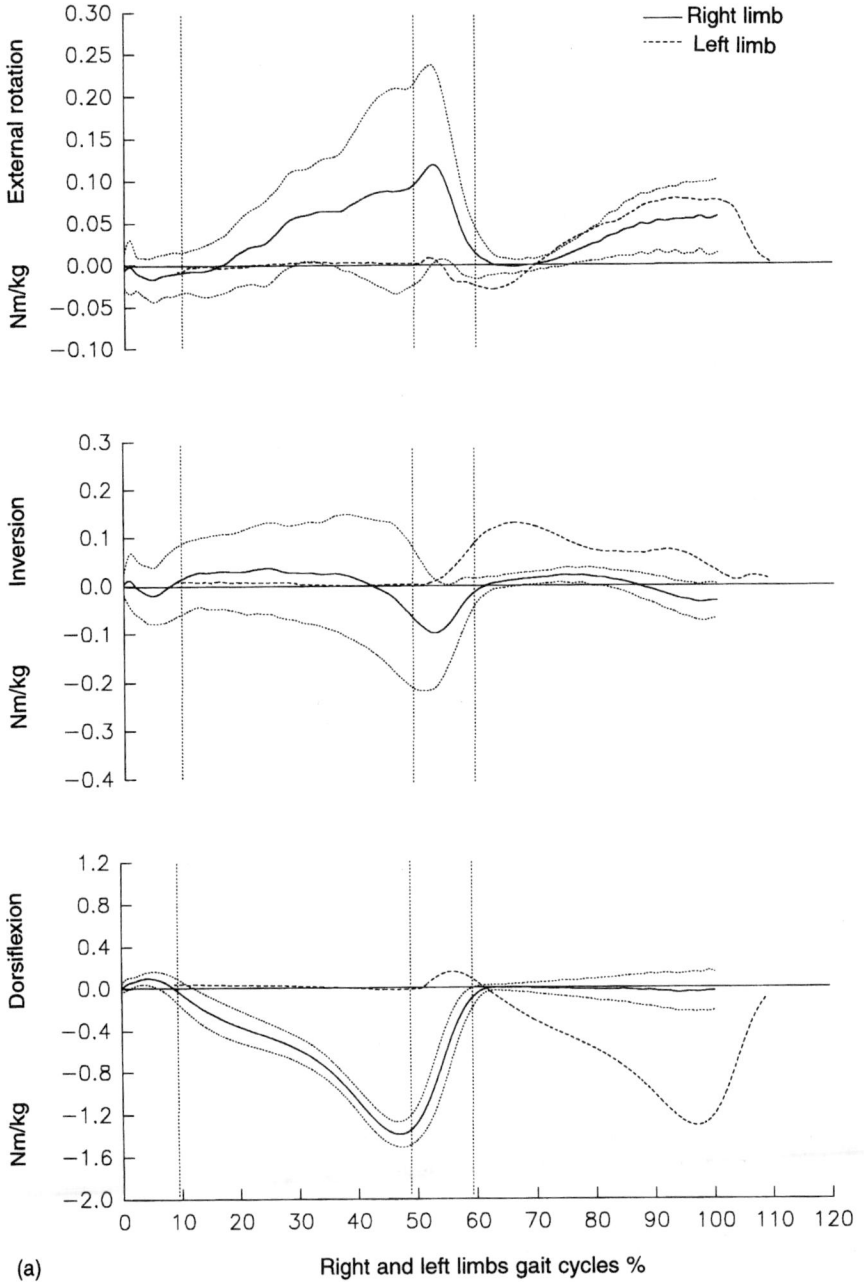

Figure 4. Joint (a) moments and (b) muscle powers developed at the ankle in 15 able-bodied women. See Figure 1 caption

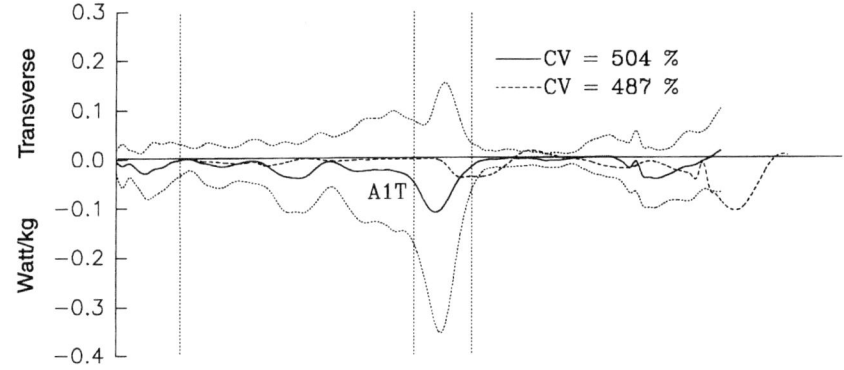

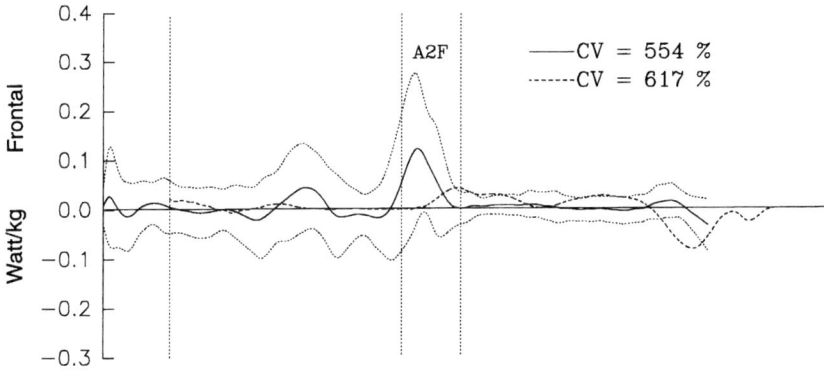

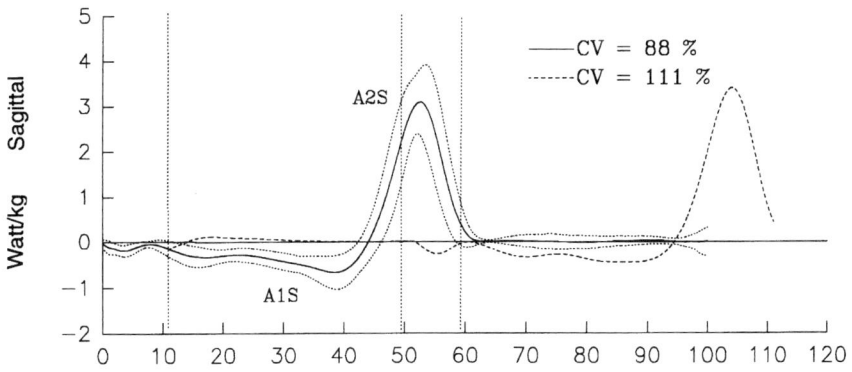

(b) Right and left limbs gait cycles %

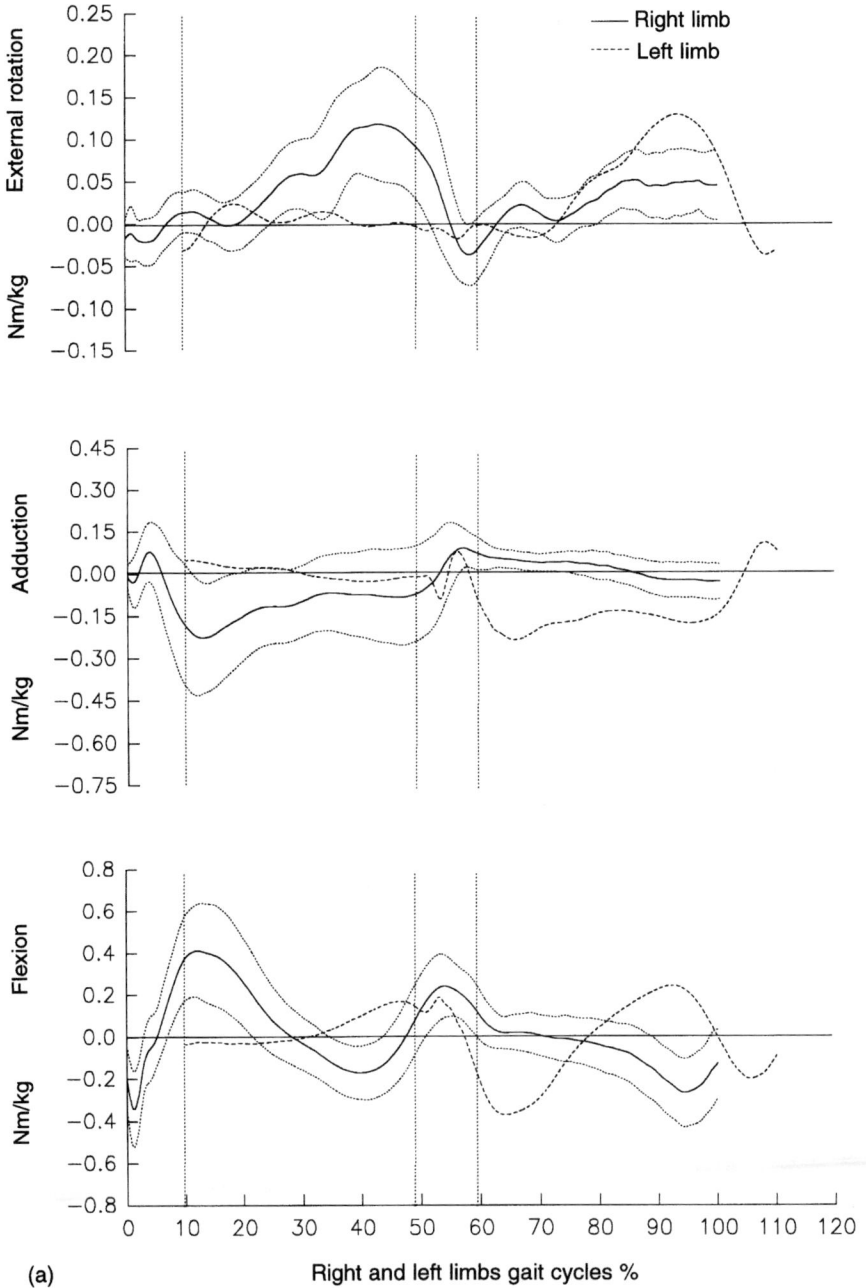

Figure 5. Joint (a) moments and (b) muscle powers developed at the knee in 15 able-bodied women. See Figure 1 caption

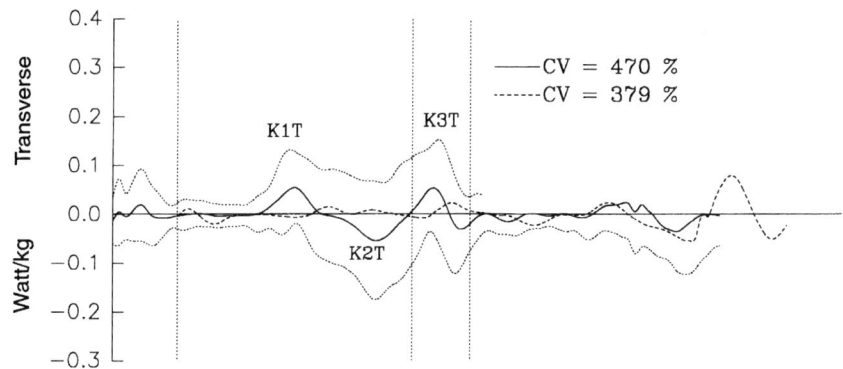

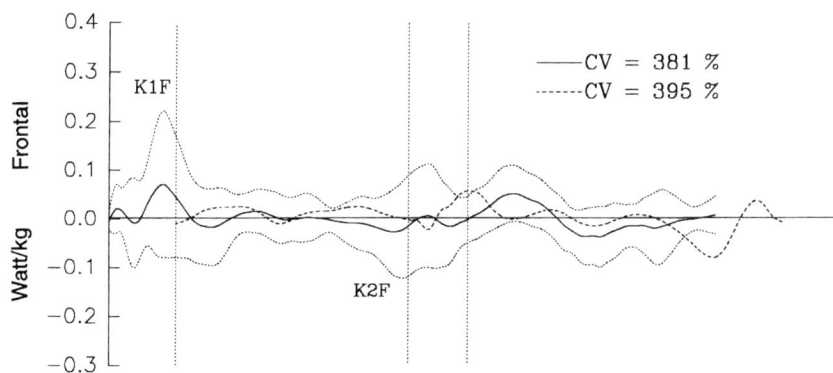

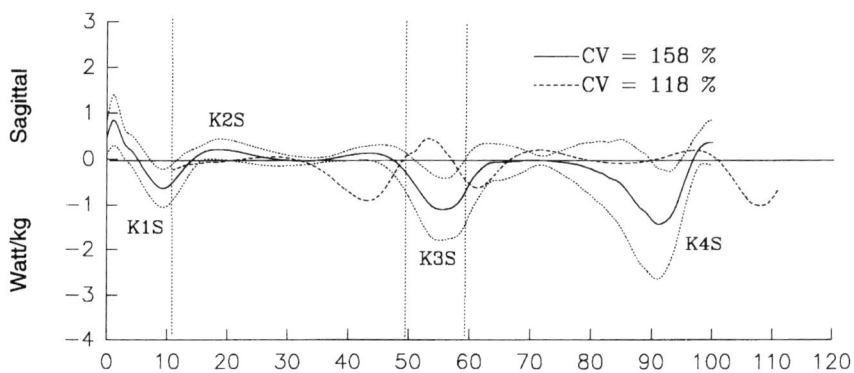

(b) Right and left limbs gait cycles %

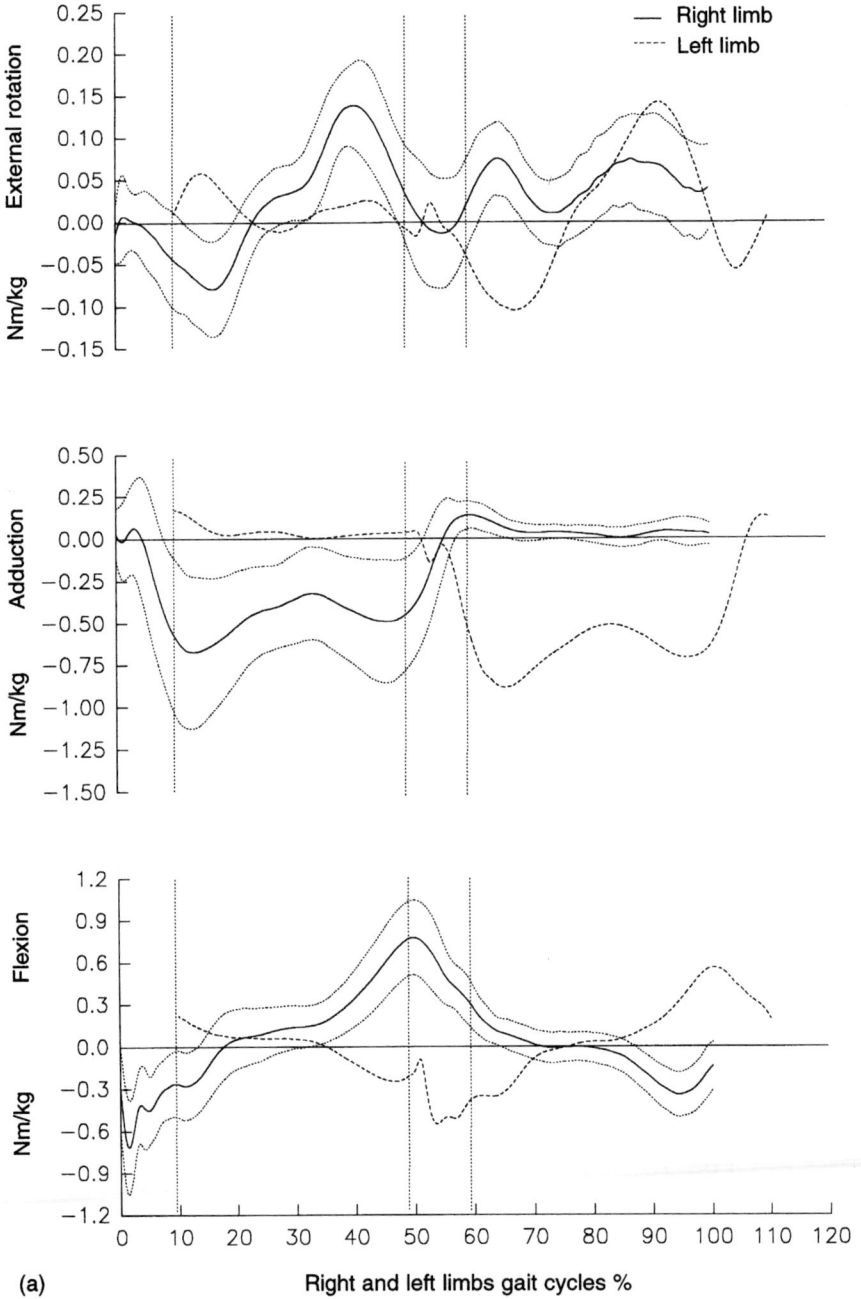

Figure 6. Joint (a) moments and (b) muscle powers developed at the hip in 15 able-bodied women. See Figure 1 caption

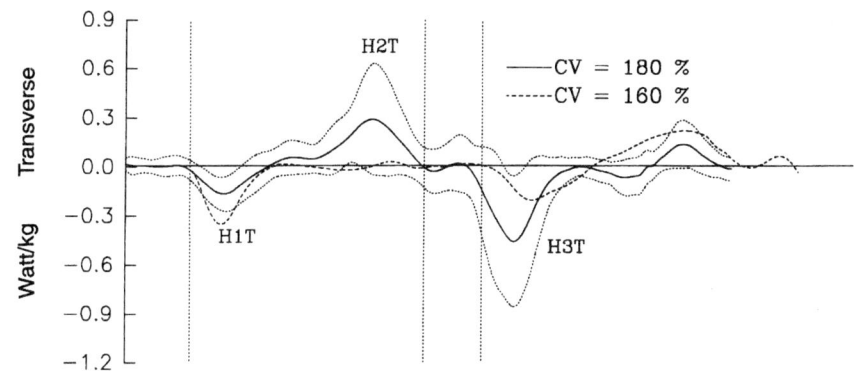

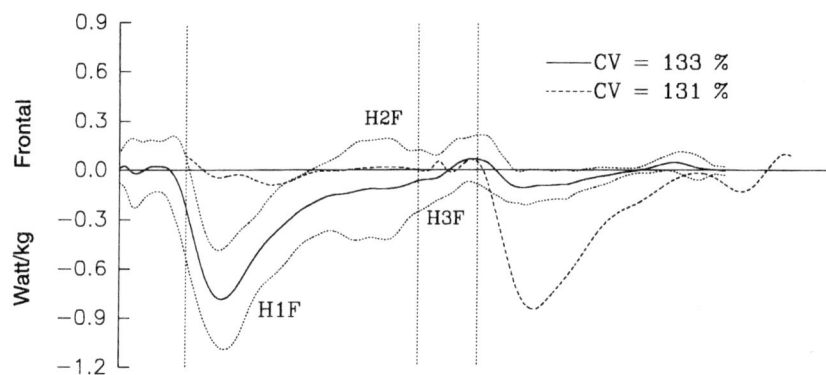

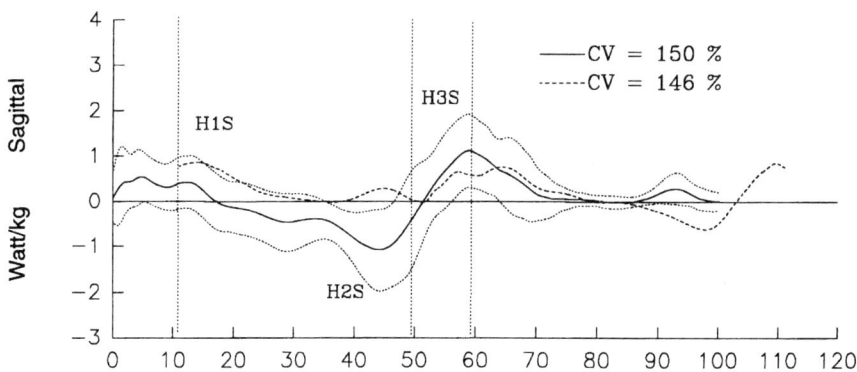

(b) Right and left limbs gait cycles %

right and left limb moments were generally quite symmetrical. Additionally, the powers developed at each joint and in each plane were also symmetrical and similar to those of the able-bodied-men group.

Mechanical energies absorbed and generated are reported in Tables 10 to 12 for the sagittal, the frontal and transverse planes respectively. Once normalized with respect to body mass, these values are close to but usually smaller than those reported here for the able-bodied men. Again, the right limb absorbed more than the left limb, even though both limbs have the same walking speed and similar energy generation values. The sagittal plane accounted for most of the significant differences, and out of the four parameters which were different in both the women and men groups, three were related to energy absorption.

TRENDS

Data from the literature, as well as those presented in this chapter, show a large variability in the results obtained for able-bodied groups. Even within a group, gait data of some able-bodied subjects varied from the group's mean. This issue of data variability is further complicated when considering simultaneous bilateral information.

Differences between the right and left limbs may reflect a natural limb

Table 10. Energies in J/kg ($\times 10^{-2}$) developed by the able-bodied-women group in the sagittal plane by both limbs

| Joint | Burst | Able-bodied group ($N = 45$) | | | |
| | | Right limb | | Left limb | |
		Mean	Standard deviation	Mean	Standard deviation
Ankle	A1S	-13.60^*	4.94	-11.10^*	3.92
	A2S	28.40	5.65	31.50	9.65
Knee	K1S	-4.40	1.82	-4.70	3.41
	K2S	3.50^*	2.53	2.60^*	2.09
	K3S	-13.20^*	7.45	-8.70^*	4.29
	K4S	-13.20	16.79	-9.10	2.20
Hip	H1S	8.90^*	8.59	12.10^*	12.57
	H2S	-22.40^*	18.86	-10.60^*	7.59
	H3S	17.20	9.45	15.10	10.04
Total limb generated		58.00	16.82	71.40	19.61
Total limb absorbed		-66.80^*	22.28	-37.90^*	12.53

$^* p < 0.05$.

Table 11. Energies in J/kg ($\times 10^{-2}$) developed by the able-bodied-women group in the frontal plane by both limbs

Joint	Burst	Right limb		Left limb	
		Mean	Standard deviation	Mean	Standard deviation
Angle	A1F	−0.90	0.60	−2.10	1.80
	A2F	2.20	3.46	1.70	1.37
Knee	K1F	−0.90	0.73	−1.30	1.92
	K2F	−1.50*	0.81	0.80*	0.66
Hip	H1F	−13.30*	6.86	−16.90*	8.83
	H2F	0.90	1.07	1.10	1.81
	H3F	−1.50	1.92	−1.30	1.60
Total limb generated		4.60	3.35	3.60	2.41
Total limb absorbed		−16.00	7.54	−21.60	10.18

*$p < 0.05$.

Table 12. Energies in J/kg ($\times 10^{-2}$) developed by the able-bodied-women group in the transverse plane by both limbs

Joint	Burst	Right limb		Left limb	
		Mean	Standard deviation	Mean	Standard deviation
Angle	A1T	−1.40	1.74	−1.40	1.63
Knee	K1T	0.60*	0.51	0.20*	0.46
	K2T	−0.90*	0.75	−0.50*	0.87
	K3T	0.80*	0.72	0.40*	0.72
Hip	H1T	−1.90	1.97	−2.30	1.75
	H2T	3.80	3.10	3.70	2.89
	H3F	−3.40	2.69	−1.90	3.17
Total limb generated		5.20	3.39	4.90	5.11
Total limb absorbed		−7.60	4.80	−6.10	4.53

*$p < 0.05$.

adjustment pattern to maintain a relatively constant walking speed rather than gait asymmetry, but normal gait asymmetry may also be present, at least to some extent. Different methods have been proposed to quantify gait differences associated with asymmetry, but consensus has yet to be achieved.

Different methods and approaches have been used to characterize normal and pathological gait by one or several typical patterns of relevant gait parameters. Olney, Griffin and McBride (1994) have used multiple linear regression to determine the relationship of temporal, kinematic and kinetic variables to walking speed in patients with hemiplegia. A stepwise regression identified the most useful parameters in predicting speed. Using a principal component analysis approach, Mah et al. (1994) were able to reduce the number of kinematic variables that needed to be analysed. They then modelled the shape pattern of these variables by a distortion analysis to resolve small distributed changes in the gait patterns within subjects. Loslever, Laassel and Angue (1994) have developed a combined statistical method to study gait patterns which involved both principal component analysis and multiple correspondence analysis. Holzreiter and Köhle (1993) applied artificial neural network modelling techniques to distinguish between able-bodied and pathological gait. These advanced statistical methods are providing new and promising directions in the characterization of normal and pathological gait.

REFERENCES

Allard, P., Stokes, I. A. F. and Blanchi, J.-P. (1994) *Three-dimensional Analysis of Human Movement*. Human Kinetics, Champaign, IL.

Allard, P., Blanchi, J.-P., Gautier, G. and Aissaoui, R. (1990) Techniques de lissage et de filtrage de données biomécaniques. *Sci. Sport*, **5**, 27–38.

Allard, P., Trudeau, F., Prince, F., Dansereau, J., Labelle, H. and Duhaime, M. (1995) Modeling and gait evaluation of an asymmetrical keel foot prosthesis. *Med. Biol. Eng. Comput.*, **33**, 2–7.

Allard, P., Lachance, R., Aissaoui, R. and Duhaime, M. (1996) Simultaneous bilateral 3-D gait analysis. *Hum. Movem. Sci.*, **15**, 327–346.

Andres, P. O. and Stimmel, S. K. (1990) Prosthetic alignment effects on gait symmetry: a case study. *Clin. Biomech.*, **5**, 88–96.

Apkarian, J., Naumann, S. and Cairns, B. (1989) A three-dimensional kinematic and dynamic model of the lower limb. *J. Biomech.*, **22**, 143–155.

Arsenault, A. B., Winter, D. A. and Marteniuk, R. G. (1986) Bilaterism of EMG profiles in human locomotion. *Am. J. Phys. Med.*, **65**, 1–16.

Bates, B. T., Difek, J. S. and Davis, H. P. (1992) The effect of trial size on statistical power. *Med. Sci. Sports Exer.*, **24**, 1059–1068.

Becker, H. P., Rosenbaum, D., Kriese, T., Gerngross, H. and Claes, L. (1995) Gait asymmetry following successful surgical treatment of ankle fractures in young adults. *Clin. Orthop. Relat. Res.*, **311**, 262–269.

Brandstater, M. E., de Bruin, H., Gowland, C. and Clark, B. H. (1983) Hemiplegic gait: analysis of temporal variables. *Arch. Phys. Med. Rehabil.*, **64**, 583–587.

Bowsher, K. A. and Vaughan, C. L. (1995) Effect of foot-progression angle on the hip joint moments during gait. *J. Biomech.*, **28**, 759–762.

Braune, W. and Fischer, O. (1987a) *The Human Gait* (translated by Maquet, P. and Furlong, R.). Springer-Verlag, Berlin.

Braune, W. and Fischer, O. (1987b) *On the Center of Gravity of the Human Body as*

Related to the Equipment of the German Infantry Soldier (translated by Maquet, P. and Furlong, R.). Springer-Verlag, Berlin.

Chhibber, S. R. and Singh, I. (1970) Asymmetry in muscle weight and one-sided dominance in the human lower limbs. *J. Anat.*, **106**, 553–556.

Craik, R. B. L. (1995) Spatial and temporal characteristics of foot fall patterns. In: *Gait Analysis: Theory and Application* (ed. R. B. L. Craik and C. A. Oatis), Mosby, St Louis, MN, pp. 143–158.

Craik, R. B. L. and Oatis, C. A. (1995) *Gait Analysis: Theory and Application*. Mosby, St Louis, MN.

Crowninshield, R. D., Johnston, R. C., Andrews, J. G. and Brand, R. A. (1978) A biomechanical investigation of the human hip. *J. Biomech.*, **11**, 75–85.

Czerniecki, J. M., Gitter, A. and Munro, C. (1991) Joint moment and muscle power output characteristics of below-knee amputees during running: the influence of energy storing prosthetic feet. *J. Biomech.*, **24**, 63–75.

Eng, J. J. and Winter, D. A. (1995) Kinetic analysis of the lower limbs during walking: what information can be gained from a three-dimensional model? *J. Biomech.*, **28**, 753–758.

Finley, F. R., Cody, K. A. and Finizie, R. V. (1969) Locomotion patterns in elderly women. *Arch. Phys. Med. Rehabil.*, **50**, 140–147.

Ganguli, S., Mukherji, P. and Boss, K. S. (1974) Gait evaluation of unilateral below-knee amputees fitted with patellar-tendon-bearing prostheses. *J. Indian Med. Assoc.*, **63**, 256–259.

Gundersen, L. A., Valle, D. R., Barr, A. E., Danoff, J. V., Stanhope, S. J. and Snyder-Mackler, L. (1989) Bilateral analysis of the knee and ankle during gait: an examination of the relationship between lateral dominance and symmetry. *Phys. Ther.*, **69**, 640–650.

Hamill, J., Bates, B. T. and Knutzen, K. M. (1984) Ground reaction force symmetry during walking and running. *Res. Q. Exer. Sport*, **55**, 288–294.

Hannah, R. E., Morrison, J. B. and Chapman, A. E. (1984) Kinetic symmetry of the lower limbs. *Arch. Phys. Med. Rehabil.*, **65**, 155–158.

Henrichs, R. N. (1992) Case studies of asymmetrical arm action in running. *Int. J. Sport Biomech.*, **8**, 111–128.

Herzog, W., Nigg, B. M., Read, J. and Olsson, E. (1989) Asymmetries in ground reaction force patterns in normal human gait. *Med. Sci. Sports Exer.*, **21**, 110–144.

Holzreiter, S. H. and Köhle, M. E. (1993) Assessment of gait patterns using neural networks. *J. Biomech.*, **26**, 645–651.

Inman, V. T., Ralston, H. J. and Todd, F. (1981) *Human Walking*. Williams and Wilkins, Baltimore MD.

Kadaba, M. P., Ramakrishnan, H. K. and Wootten, M. E. (1990) Measurement of lower extremity kinematics during level walking. *J. Orthop. Res.*, **8**, 383–392.

Kadaba, M. P., Ramakrishnan, H. K., Wootten, M. E., Gainey, J., Gorton, G. and Cochran, G. V. B. (1989) Repeatability of kinematic, kinetic and electromyographic data in normal adult gait. *J. Orthop. Res.*, **7**, 849–860.

Klajic, M., Bajd, T. and Stanic, U. (1975) Quantitative gait evaluation of hemiplegic patients using electrical stimulation orthoses. *IEEE Trans. Biomed. Eng.*, **22**, 438–441.

Klopsteg, P. E. (1954) *Human Limbs and their Substitutes: Presenting Results of Engineering and Medical Studies of the Human Extremities and Application of the Data to the Design and Fitting of Artificial Limbs and Training of Amputees*. McGraw-Hill, New York.

Loizeau, J., Allard, P., Landjerit, B. and Duhaime, M. (1995) Bilateral gait patterns in subjects fitted with a total hip prosthesis. *Arch. Phys. Med. Rehabil.*, **76**, 552–557.

Loslever, P., Laassel, E. L. and Angue, J.-C. (1994) Combined statistical study of joint angles and ground reaction forces using component and multiple correspondence analysis. *IEEE Trans. Biomed. Eng.*, **41**, 1160–1167.

Mah, C. D., Hulliger, M., Lee, R. G. and O'Callaghan, I. S. (1994) Quantitative analysis of human movement synergies: constructive pattern analysis for gait. *J. Motor Behav.*, **26**, 83–102.

Murray, M. P., Kory, R. C. and Sepic, S. B. (1970) Walking pattern of normal women. *Arch. Phys. Med. Rehabil.*, **51**, 637–650.

Öberg, T., Karsznia, A. and Öberg, K. (1993) Basic gait parameters: reference data for normal subjects 10–79 years of age. *J. Rehabil. Res. Dev.*, **30**, 210–223.

Olney, S. J., Griffin, M. and McBride, I. D. (1994) Temporal, kinematic and kinetic variables related to gait speed in subjects with hemiplegia: a regression approach. *Phys. Ther.*, **74**, 872–885.

Ounpuu, M. S., Gage, J. R. and Davis, R. B. (1991) Three-dimensional lower extremity joint kinetics in normal children. *J. Pediatr. Orthop.*, **11**, 341–349.

Ounpuu, S. and Winter, D. (1989) Bilateral electromyographical analysis of the lower limbs during walking in normal adults. *Electroencephalogr. Clin. Neurophysiol.*, **72**, 429–438.

Patriarco, A. G., Mann, R. W., Simon, S. R. and Mansour, J. M. (1981) An evaluation of the approaches of optimization models in the prediction of muscle forces during human gait. *J. Biomech.*, **14**, 513–525.

Robinson, R. O., Herzog, W. and Nigg. B. M. (1987) The use of force platform variables to quantify the effects of chiropractic manipulation on gait symmetry. *J. Manip. Physiol. Ther.*, **10**, 172–176.

Rosenrot, P. (1980) Asymmetry of gait and the relationship to lower limb dominance. *Human Locomot.*, **1**, 26–27.

Sutherland, D. H. (1984) *Gait Disorders in Childhood and Adolescence*. Williams and Wilkins, Baltimore, MD.

Vagenas, G. and Hoshizaki, B. (1992) A multivariable analysis of lower extremity kinematic asymmetry in running. *Int. J. Sports Biomech.*, **8**, 11–29.

Vanden der Straaten, J. H. M. and Scholton, P. M. J. (1978) Symmetry and periodicity in gait patterns of normal and hemiplegic children. In: *Biomechanics VI* (ed. R. Easmussen and R. Jorgense) University Park Press, Champaign, IL, pp. 287–292.

Vaughan, C. L., Davis, B. L. and O'Connor, J. C. (1992) *Dynamics of Human Gait*. Human Kinetics, Champaign, IL.

Wall, J. C. and Turnbull, G. I. (1986) Gait asymmetries in residual hemiplegia. *Arch. Phys. Med. Rehabil.*, **67**, 550–553.

Winter, D. A. (1990) *Biomechanics and Motor Control of Human Movement*. Wiley-Interscience publication, John Wiley and Sons Inc, New York.

Wu, G. (1995) A review of body, segmental displacement, velocity and acceleration in human gait. In: (ed. R. B. L. Craik and C. A. Oatis), Mosby, St Louis, MN, pp. 205–222.

Yamasaki, M., Sasaki, T. and Masafumi, T. (1991) Sex difference in the pattern of lower limb movement during treadmill walking. *Eur. J. Appl. Physiol.*, **62**, 99–103.

16

Gait of Normal Children and Those with Cerebral Palsy

CHRISTOPHER L. VAUGHAN[1], DIANE L. DAMIANO[2]
AND MARK F. ABEL[2]

[1]University of Cape Town, Cape Town, South Africa
[2]University of Virginia, Charlottesville, VA, USA

NORMAL CHILDREN

INTRODUCTION

It is generally accepted that an infant will acquire the ability to sit independently at approximately 6 months, to walk without support between a year and 15 months, and to run at 18 months (Sutherland et al., 1980). During this crucial period, and for the subsequent few years, the central nervous system (CNS) will mature in parallel with musculoskeletal growth. The child's characteristic gait pattern will be influenced by these two factors. As the child grows older, the primary gait variables—temporal distance parameters, angular kinematics and joint kinetics—will change, and it is natural to ask the question: how much of the change can be attributed to maturation and how much to growth?

This is a complex question to answer but it is nevertheless one with which most gait laboratories must grapple. Many of their patients will be children with some form of neurological deficit and, in assessing their gait, the laboratory personnel require normative data for the purpose of comparison. Is it possible to standardize gait parameters so that a normal 5-year-old can be compared to a normal 12-year-old? Can a 6-year-old child with cerebral palsy be compared to a 6-year-old with no disabilities? We believe that such comparisons are indeed feasible and we will provide some supportive evidence in this first section of the chapter.

Three-dimensional Analysis of Human Locomotion. Edited by P. Allard, A. Cappozzo, A. Lundberg and C. Vaughan
© 1997 John Wiley & Sons Ltd. ISBN 0 471 96949 4

Our approach is to scale the gait parameters to body size by adopting the approach suggested by Hof (1996). The premise on which his method is based is the simple assumption that taller people tend to walk with longer steps and lower step rates than shorter people, while heavier individuals will exert greater forces and moments than those who are lighter. When normalizing a gait parameter it is thus necessary to account for both body length and mass. Table 1 is a summary of the primary gait parameters and the formulae for creating dimensionless numbers.

By rendering the gait parameters dimensionless, we are able to factor out the effects of growth. Any differences that exist, say between normal 5- and 12-year-olds, can be attributed to CNS maturation. Similarly, when comparing patients with normals, the differences will be primarily a function of the pathology. While Sutherland (1996), who has been a pioneer in the field of paediatric gait (Sutherland, 1978 and Sutherland et al., 1980, 1988), supports the approach of Hof (1996), he believes that it is applicable only after age 4, when CNS maturation has occurred.

TEMPORAL-DISTANCE PARAMETERS

The two fundamental temporal-distance parameters in human gait are step length l (measured in metres) and cadence or frequency f (measured in steps per second). They combine to form another parameter, velocity:

$$v = lf \tag{1}$$

Note that some authors (Beck et al., 1981) prefer to use stride length (which is simply two consecutive steps, left plus right) and measure cadence in steps/min. However, equation (1) is in a simple form and the units (m/s for velocity) are consistent. Furthermore, it is fairly straightforward to transform temporal-distance data between these two conventions.

Table 1. Gait parameters can be normalized and rendered dimensionless according to the approach suggested by Hof (1996)

Quantity	Symbol	Dimension	Dimensionless number
Mass	m	M	$m' = m/m_0$
Length	l	L	$l' = l/l_0$
Time	t	T	$t' = t/\sqrt{(l_0 g)}$
Frequency	f	T^{-1}	$f' = f/\sqrt{(g/l_0)}$
Velocity	v	LT^{-1}	$v' = v/\sqrt{(g/l_0)}$
Force	F	MLT^{-2}	$F' = F/m_0 g$
Moment	M	ML^2T^2	$M' = M/m_0 g l_0$
Angle	ϕ		Already dimensionless

Note: m_0 = body mass; l_0 = leg length (distance from greater trochanter to floor); and g = acceleration due to gravity (9.81 m/s^2).

As demonstrated by research groups based in San Diego (Sutherland et al., 1980, 1988) and in Chicago (Beck et al., 1981), step length increases steadily with age. Beck et al. (1981) suggested normalizing this parameter by dividing by the child's height and expressing as a percentage. Hof (1996) has argued that it makes more sense to divide by leg length. This is the approach that we used in our laboratory at the University of Virginia in a study of 75 normal children between 2 and 13 years of age. Figure 1a shows the actual stride lengths (in metres) expressed as a function of age, and the increase over the 10 years is quite evident. When these data are normalized by leg length, there is no difference across the age range (Figure 1b).

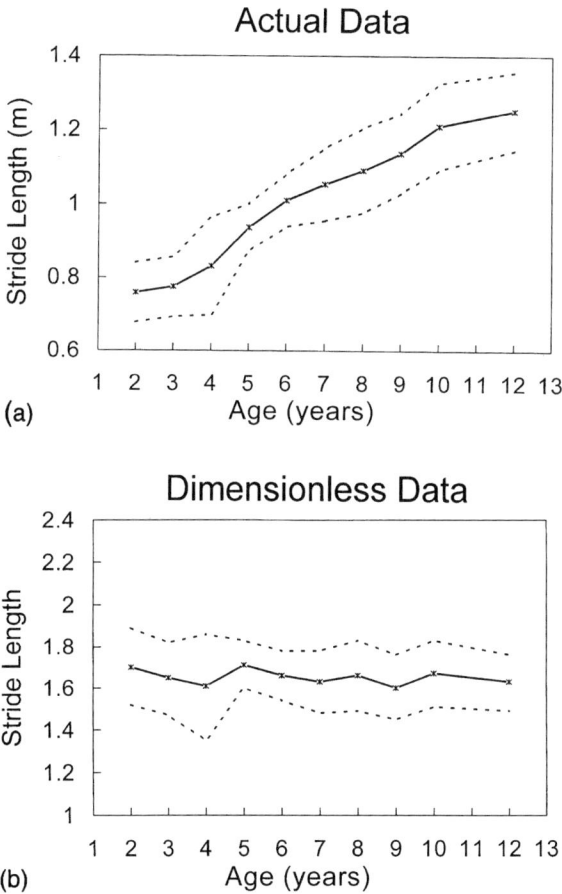

Actual Data

(a)

Dimensionless Data

(b)

Figure 1. Stride length plotted as a function of age: (a) actual data in metres; (b) dimensionless data normalized according to Table 1. These data are based on a study of 75 normal children performed at the University of Virginia

In contrast to stride length, cadence decreases with age (Figure 2a). However, when this parameter is normalized as described in Table 1, Figure 2b shows that there is again no difference across the age range (Zijlstra, Prokop and Berger, 1996).

Our data suggest that when children are asked to walk at their naturally selected speed, there is no statistical difference in their stride length and frequency when these parameters are normalized according to Hof (1996). O'Malley (1996), in re-analysing the data of Sutherland et al. (1988), has used a least squares correlation technique to normalize step length and frequency with respect to both age and height. He concluded that this statistical approach was more flexible than the simple normalization approach proposed by Beck et al. (1981) and Hof (1996), and it had the added benefit of retaining the original units. As might be expected, walking velocity will also increase with age (Figure

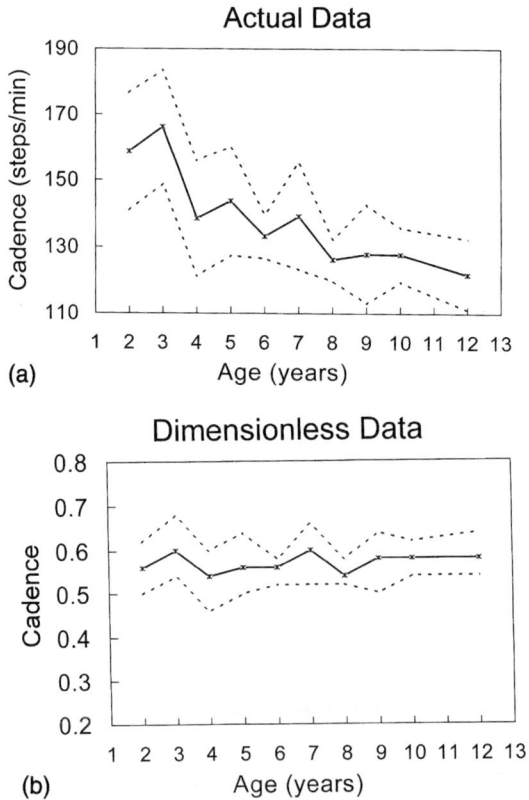

(a)

(b)

Figure 2. Cadence plotted as a function of age: (a) actual data in steps/min; (b) dimensionless data normalized according to Table 1. These data are based on a study of 75 normal children performed at the University of Virginia

3a), but, when normalized appropriately (Table 1 and equation 1), it too has a constant value. Note that this parameter is truly dimensionless, in contrast to some reports, which have expressed velocity in statures/s. The data in Figures 1 to 3 suggest that, as far as these temporal-distance parameters are concerned, children have established a mature gait pattern by age 2.

Sutherland et al. (1980) have argued that maturation occurs only by age 4, and their thesis is bolstered by another temporal-distance parameter, the duration of single limb stance (expressed as a percentage of the gait cycle). We have reproduced their data in Figure 4, where the steady increase in this parameter from age 1 (31%) to age 4 (36%) is clearly evident. These data are in stark contrast to our own data, also plotted in Figure 4, where the single limb stance time is approximately constant at 39%. It is unclear why there is this discrepancy, but it may be related to the fact that our method for determining the gait events

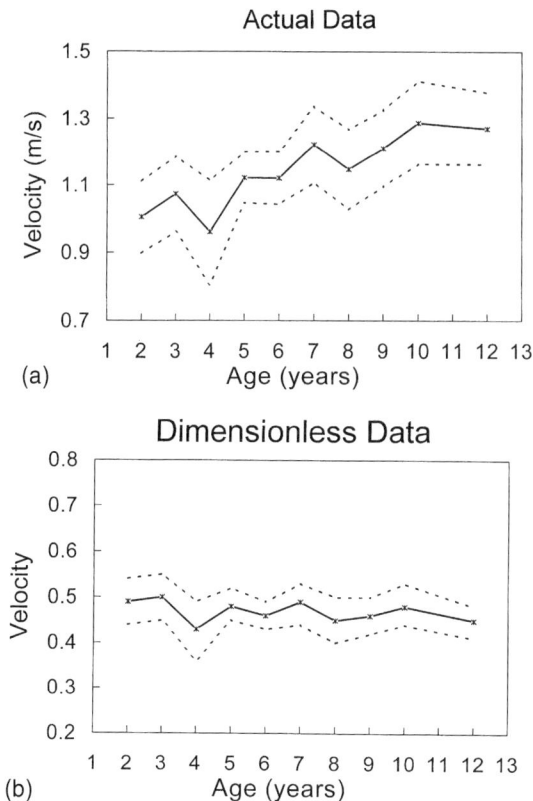

Figure 3. Velocity plotted as a function of age: (a) actual data in metres/s; (b) dimensionless data normalized according to Table 1. These data are based on a study of 75 normal children performed at the University of Virginia

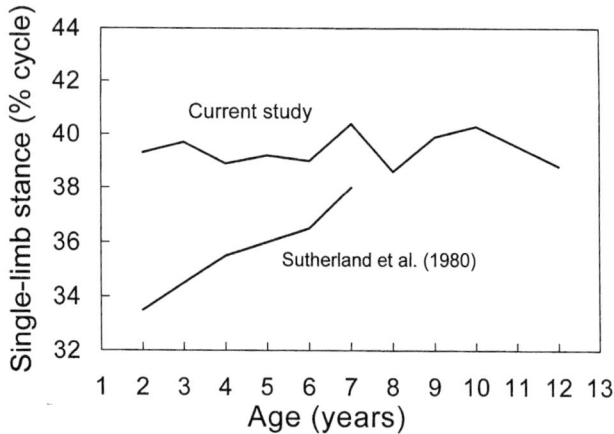

Figure 4. Duration of single limb stance (expressed as a percentage of the gait cycle) plotted as a function of age. The data of Sutherland et al. (1980) have been contrasted with our own study of 75 normal children performed at the University of Virginia

was based on force plate and video records, while Sutherland et al. (1980) used 16 mm movie films.

ANGULAR KINEMATICS

Joint angles, particularly those in the sagittal plane for the hip, knee and ankle, are very repeatable and readily identified gait parameters. Fortunately, they are already dimensionless and so do not need to be normalized (Table 1). Sutherland et al. (1988), in their detailed anlaysis of over 400 children, showed that the joint kinematics were very similar for 1-, 2- and 7-year-olds, although there were some subtle differences. We also studied the three-dimensional (3D) kinematics of our 75 children and have reproduced the ensemble averages for 2-, 3- and 12-year-olds in Figure 5.

A brief examination of Figure 5 reveals that the curves for these three age groups are very similar. However, a more careful study shows that 2-year-olds have some subtle differences, whereas the 3- and 12-year-olds are virtually indistinguishable. For the hip joint angle (Figure 5a), the 2-year-olds have about 5–10° more flexion, exhibiting a slightly exaggerated anterior pelvic tilt. For the knee joint angle (Figure 5b), the 2-year-olds have more flexion at heel strike (0% and 100%), midstance (40%) and midswing (70%). The lack of extension at heel strike and midstance may be attributed to the 2-year-old child lacking some stability. The ankle joint angle (Figure 5c) shows the 2-year-olds with slightly more dorsiflexion in stance and less plantarflexion in swing. The data in Figure 5 are consistent with those of Sutherland et al. (1980, 1988), confirming

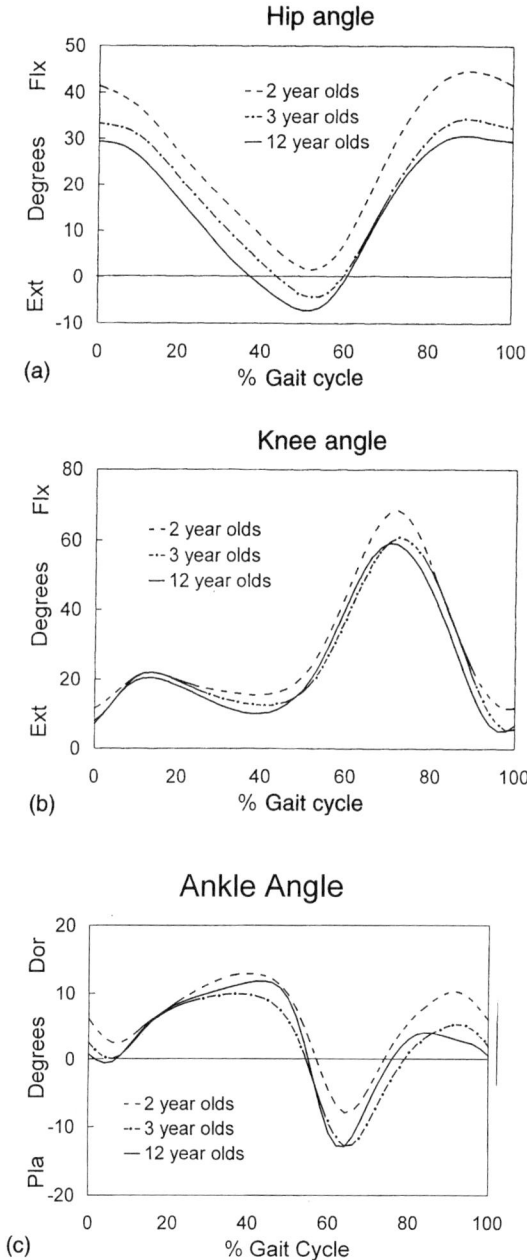

Figure 5. Angular kinematics for 2-, 3- and 12-year-olds plotted as a function of the gait cycle: (a) hip joint; (b) knee joint; and (c) ankle joint. These are ensemble curves based on a study of 75 normal children performed at the University of Virginia

that 2-year-olds have joint angle patterns that are similar to older children. By age 3 the patterns are well established and essentially identical to those of teenagers and, indeed, adults (Ounpuu, Gage and Davis, 1991).

JOINT KINETICS

The resultant moments (or torques) at the hip, knee and ankle, also referred to as the joint kinetics, play an important causative role in human gait (Vaughan, Davis and O'Connor, 1992). Despite their importance, there have been very few reports in the archival literature on the joint moments for normal children, with the study by Ounpuu, Gage and Davis (1991) being a notable exception.

Ounpuu, Gage and Davis (1991) studied 31 normal children (19 females and 12 males) with no previous history of musculoskeletal problems and ranging in age from 5 to 14 (mean age of 9.6). They presented ensemble average data for electromyography, kinematics, joint moments and joint powers. We have digitized their data for the ankle joint moment and reproduced the curve in Figure 6. Note that Ounpuu, Gage and Davis (1991) have normalized the joint moment by dividing by body mass in kilograms.

The standard deviations in Figure 6, represented by dashed lines, suggest that there is not much variability in the ankle joint moment across the age range of 5–14 years. However, as we (Bowsher and Vaughan, 1995) and others (Ramakrishnan et al., 1987; Hof, 1996) have argued, when normalizing for body size, you should account for weight *and* segment length differences. Support for this

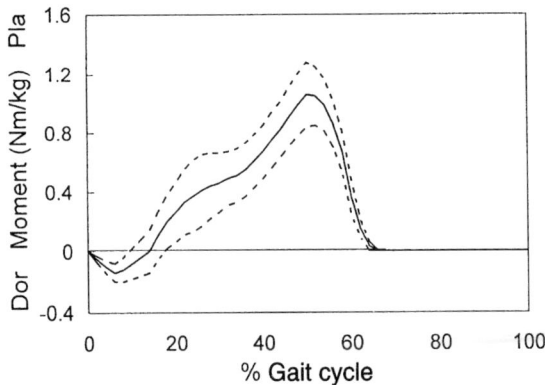

Figure 6. Ankle moment plotted as a function of the gait cycle. The curves are based on data in Ounpuu, Gage and Davis (1991) for 31 normal children (average age 9.6 years). Note: the moments have been normalized by dividing by body mass in kilograms; the solid line is the ensemble average, while the dashed lines indicate plus and minus one standard deviation

argument is provided in Figure 7, where the ankle moments for our 2- and 12-year-olds have been compared. In Figure 7a, the data are presented in actual units of Nm, in Figure 7b the units of Nm/kg are used (Ounpuu, Gage and Davis, 1991), while in Figure 7c we have normalized by body weight and leg length (Table 1).

In Figure 7a the difference between the 2- and 12-year-olds is quite obvious. When the ankle joint moment is normalized by body mass (Figure 7b), the difference is much less dramatic. However, a careful examination of Figures 7b and 6 reveals that our 2-year-olds have a moment pattern that is similar to the lower standard deviation of Ounpuu, Gage and Davis (1991), while the pattern of our 12-year-olds is similar to their upper standard deviation. The dimensionless ankle moment (Figure 7c) demonstrates that 2- and 12-year-olds have very similar patterns. Our dimensionless data for the hip and knee joint moments also show that infants and teenagers (and indeed adults) exhibit almost identical profiles, providing further evidence that mature gait may be established by as early as 2 years of age.

CHILDREN WITH CEREBRAL PALSY

CHARACTERIZATION IN CEREBRAL PALSY GAIT

Cerebral palsy (CP) is a broad diagnostic category, the common underlying abnormality being an injury to the CNS within the first 2 years of life. While the precise neurophysiological causes are not known, in two-thirds of cases prematurity, anoxia, or other perinatal complications are related factors (Torfs, van der Berg and Oechsli, 1990). Although many patients may have associated cognitive or sensory deficits, the hallmark of this disorder is motor dysfunction. The vast majority of patients exhibit spasticity, which is a sign of pyramidal involvement, while a small percentage exhibit athetosis, ataxia or dystonia, which are extrapyramidal signs. The anatomical classification is based on body regions affected, such as hemiplegia, where the upper and lower extremity on one side of the body are involved, quadriplegia, where all four extremities are affected, and diplegia, where both lower extremities are involved with minimal if any deficits in the upper extremities.

Disorders of balance, tone and selective muscle control all contribute to the motor deficits seen in this population (Gage, 1991). However, virtually all children with diplegia and hemiplegia, and a small proportion of those with quadriplegia, will eventually walk with some difficulty (Molnar, 1991). The ability to walk is a primary concern for parents of a child with CP in early childhood (Bleck, 1990). Consequently, the majority of interventions provided to these children are designed to facilitate, improve or maintain this skill. Although other disabilities may take precedence in adulthood (Bleck, 1984), physical

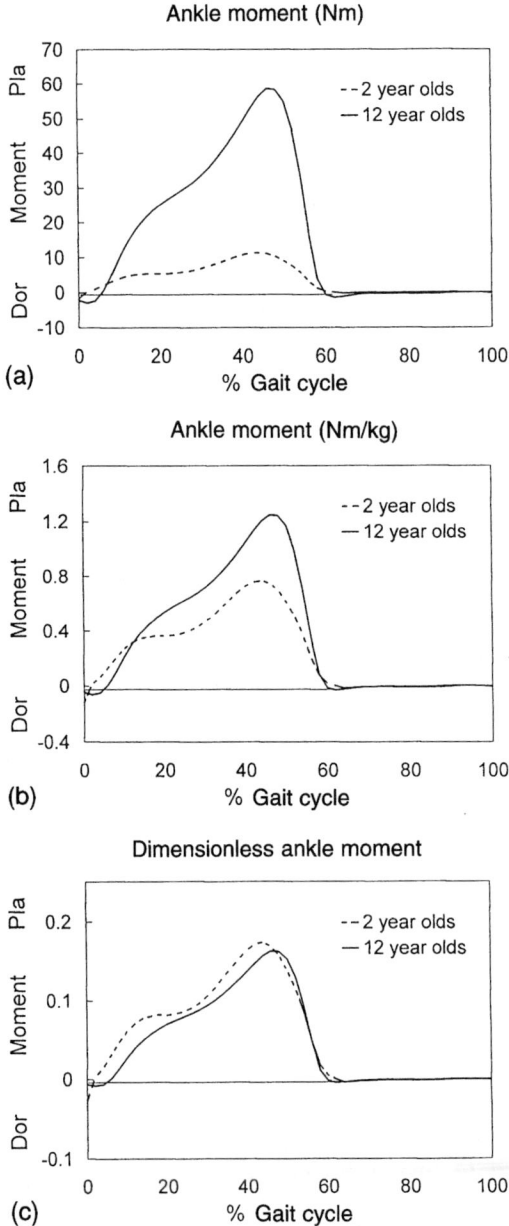

Figure 7. Ankle moment plotted as a function of the gait cycle for 2- and 12-year-olds: (a) actual data in Nm; (b) data normalized by dividing by body mass in kilograms; and (c) dimensionless data normalized according to Table 1. These are ensemble curves based on a study of 75 normal children performed at the University of Virginia

limitations are still a major concern for these individuals, and the desire to maintain ambulation persists (Harris, Abel and Damiano, 1997).

When evaluating the gait of a child or an adult with CP, normal gait parameters are used as the standard for comparison, and therapeutic interventions are often designed to minimize the differences between normal and CP gait. However, the primary goal should not be to create a gait pattern that appears normal but rather to improve function. Persons with CP typically walk slower (Norlin and Odenrick, 1986; Abel and Damiano, 1996) and with a greater energy cost (Rose et al., 1989; Harris, Abel and Damiano, 1997), than an age- and size-matched comparison group. They also tend to be less stable, spending proportionately greater time in single or double limb stance (Abel and Damiano, 1996). Not only are children with CP generally slower, but they also have a diminished capacity to increase their walking velocity compared to their peers and, as illustrated in Figure 8, they tend to rely on an increase in cadence to produce a greater speed (Abel and Damiano, 1996). An additional finding is that in contrast to paediatric normative values, velocity appears to decrease with age in CP (Norlin and Odenrick, 1986) in the absence of surgical intervention (Johnson, Abel and Damiano, 1997), suggesting a deterioration in gait function over time.

The functional limitations that occur in CP are the result of more specific kinematic, kinetic and EMG abnormalities that may vary considerably across patients, even those within the same diagnostic category. A major dilemma for clinicians and researchers who assess outcomes of treatments for CP has been

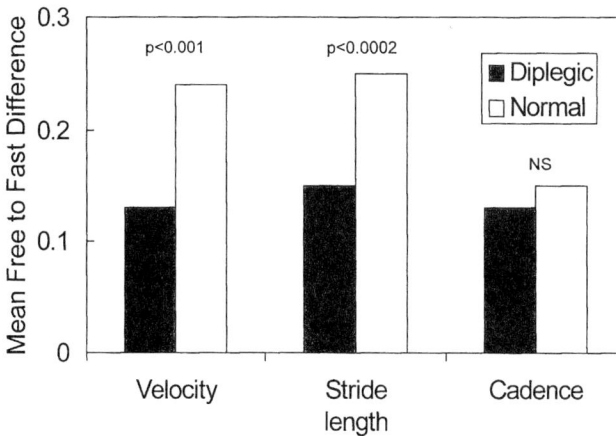

Figure 8. Histogram comparing the strategy by which normal children and those with spastic diplegia increase their walking velocity (Abel and Damiano, 1996). The data have been normalized according to the method described in Table 1

the wide heterogeneity of patients carrying the diagnosis, and the differential responses to interventions in this population. Gait analysis has been shown to be very useful in providing a quantitative and comprehensive way to document this diversity (Damiano and Abel, 1996) and to categorize these individuals based on their ambulatory patterns (Winters, Gage and Hicks, 1987; O'Malley, Abel and Damiano, 1995). This information could ideally be used to examine the effects of interventions within and across more homogeneous patient subgroups.

Nearly all children with spastic diplegia have deficits at the foot and ankle and, as severity increases, the ability to exert control at the knee and hip joint is compromised. These patients typically demonstrate a toe-strike pattern with ankle equinus, a stiff, crouched position at the knees with limited stance-phase extension and reduced swing-phase flexion; there is also exaggerated flexion, adduction and internal rotation at the hips. In the involved lower extremities of children with hemiplegia, distal muscles are also primarily affected, with proximal involvement seen as severity increases. These children tend to have a slight reduction in velocity, with reduced step length and stance time on the affected side. Patterns of motion include the lack of heel-strike with frequent equinovarus deformities, slight knee crouch during stance, a tendency for the hip to be positioned in more flexion and adduction (as compared to the opposite extremity), with elevation and retraction of the hemi-pelvis on the hemiplegic side.

Although some advances have been made in understanding the pathophysiology of CP, as well as identification and prevention of some of the predisposing factors, no 'cure' exists for the primary brain lesion. All currently available therapeutic interventions therefore aim to alleviate the peripheral manifestations of this CNS disorder, and at best serve only to modify the course of this chronic disease. The most common treatments for ambulatory persons with CP include: physiotherapy; the application of orthoses; spasticity-reducing medications; neurosurgery, selective dorsal rhizotomy; and orthopaedic surgery. The advent and increasing utilization of gait analysis technology has altered clinical practice for those with access to these facilities, and has led to a dramatic increase in published research reports on the motor dysfunction seen in ambulatory CP. Gait analysis offers additional insights into the complex, 3D walking disorders in CP that are undetectable with mere visual observation (Sutherland, 1989). It is an evaluative rather than a diagnostic tool which is used in combination with other clinical measures to monitor progress over time, to help plan interventions, and to assess the outcomes of those interventions (Abel, 1995).

EFFECTIVENESS OF PHYSIOTHERAPY IN IMPROVING GAIT FUNCTION IN CEREBRAL PALSY

With rare exceptions, virtually all children diagnosed with CP receive physiotherapy services. Despite its prevalence, studies examining the efficacy of traditional therapy approaches have failed to demonstrate consistent beneficial effects (Campbell, 1990). Surprisingly few studies utilizing gait analysis as an outcome measure for assessing physical therapy interventions in CP have been published. Motion analysis methodology has been used to assess the effectiveness of physiotherapy treatments such as inhibitory casting (Bertoti, 1986; Hinderer et al., 1988) and neurodevelopmental therapy (NDT) and ankle–foot orthoses (Embrey, Yates and Mott, 1990). In a novel experimental design, the change in the qualitative performance of a reaching task due to an NDT programme was evaluated with this technology, by quantifying the directness of the movement and the pattern of acceleration before and after the intervention (Kluzik, Fetters and Coryell, 1990). Gait analysis was also instrumental in documenting the positive effects of posterior walkers as compared to anterior ones (Logan, Byers-Hinkley and Ciccone, 1990). The posterior design encouraged a more upright posture, which also led to improvement in joint alignment and excursion in gait.

The disappointing functional outcomes from traditional therapy approaches in CP have spurred the profession to explore alternative clinical strategies. One approach to treating motor deficits in CP that is gaining popularity is transcutaneous electrical stimulation (Pape et al., 1993). The theoretical basis underlying this treatment strategy is to apply a low-intensity stimulation to a non-spastic antagonist over a prolonged period of time in an attempt to reduce spasticity and thereby enhance motor function. While preliminary reports document positive motor outcomes, including a measurable improvement in walking ability, using standard developmental scales, published articles documenting the precise effects on gait function have not yet been forthcoming.

Another approach that was virtually ignored during the predominance of NDT, but is now experiencing a resurgence, is the use of muscle strengthening. Although weakness is recognized clinically in spastic CP (Guiliani, 1991), strength training has not been a standard practice in the physiotherapy treatment approach with these children. No evidence exists to support the current clinical prejudice against the use of resistance exercise in persons with CNS involvement; in fact, research findings are accumulating to the contrary. Not only are children with CP weak (Damiano, Kelly and Vaughan, 1995), but the degree of weakness has been shown to be directly related to motor function (Damiano, Vaughan and Abel, 1995; Kramer and MacPhail, 1994). The underlying pathology does not appear to inhibit the ability of these children to increase strength (Damiano, Kelly and Vaughan, 1995), although it is still unknown how increased peripheral strength affects force production during motor tasks. Positive outcomes from strengthening in CP include increased rate of torque production (McCubbin and Shasby, 1985), improved range of motion (Horvat, 1987) and positive gait changes such as increased energy efficiency (MacPhail and Kramer, 1995), decreased crouch and increased stride length (Damiano, Vaughan and Abel, 1995). Further impetus for the use of strengthening is that it is an additive approach to motor dysfunction, in contrast to orthopaedic surgery procedures, which improve joint excursion by reducing strength, an approach that could prove problematic in children who are already weak.

The CNS damage in spastic diplegia produces a common gait pattern characterized by excessive knee flexion in stance phase. With time, and often despite treatment, the crouched posture may result in permanent contracture of the hamstring muscles. This is manifested by an inability to extend the knee completely. The hypothesis guiding this research was that muscle imbalance at the knee was in part responsible for crouch gait, and quadriceps strengthening would alleviate the crouch by enabling the quadriceps to extend the knee. Results were anticipated to be similar to those found after surgery to lengthen the hamstring tendons, which address the muscle imbalance by decreasing strength. We therefore initiated a project to examine the degree of weakness in the muscles surrounding the knee and to determine the effect of quadriceps strengthening on both isometric force production and gait function (Damiano, Kelly and Vaughan, 1995; Damiano, Vaughan and Abel, 1995).

First, quadriceps and hamstring strength was measured, and expressed as percentage maximum muscle contraction, in 14 children with spastic diplegia and 25 normal children. All values were normalized by body weight to allow for comparisons across subjects. Then, the children with diplegia were entered into a 6-week quadriceps-strengthening programme. None of the children who participated in the programme had a static knee flexion contracture. These studies revealed that both the hamstrings and quadriceps muscles in the children with diplegia were weaker than in normal subjects. Furthermore, significant increases in muscle strength, with reduction in crouch gait, were found as a result of the strength-training programme. Increased stride length was an additional positive change found after the 6-week training programme (Table 2). Specifically, these children showed greater knee extension in terminal swing, which elongated their stride and pre-positioned the knee in more extension prior to initial contact. No improvement in knee extension was noted in midstance, which can be explained

Table 2. Kinematic gait parameters at freely selected and fastest speeds for 14 children who participated in a strength-training programme compared to normative data for a 9-year-old child

	Before		After		Normal (9 year)
	Free	Fast	Free	Fast	Free
Velocity (m/s)	0.76	1.20	0.82	1.30	1.23
Stride length (m)	**0.74**	**0.83**	**0.84**	**0.97**	1.14
Cadence (steps/min)	119	173	120	162	128
Knee flexion at IC	**32.0**	33.9	**26.6**	30.2	3.0
Knee flexion in MS	12.7	14.4	10.7	12.2	0.0

IC = initial contact; MS = midstance. Knee flexion in degrees.
Bold type indicates significant differences between conditions ($p < 0.05$).

by the fact that knee motion at this point in the cycle in controlled primarily by the plantarflexion–knee extension force couple (Gage, 1991).

While this study clearly demonstrated that strengthening is a viable treatment option in CP (Damiano, Kelly and Vaughan, 1995; Damiano, Vaughan and Abel, 1995), it only addressed one aspect of the motor dysfunction seen in this population. A comprehensive rehabilitation programme must consider the other components of the motor impairment in CP as well, such as the primary symptoms of spasticity and centrally mediated muscle imbalance, and the secondary peripheral effects, which may include limitations in passive range of motion and bony deformities. Furthermore, while gait analysis technology appears to be an ideal, objective tool for evaluating gross motor function in ambulatory individuals, it remains underutilized by the physiotherapy profession for documentation of the functional effects of therapeutic approaches in CP.

EFFECTIVENESS OF ORTHOSES IN MANAGING GAIT DISORDERS IN CEREBRAL PALSY

Approximately 70% of the children diagnosed with CP will eventually become ambulatory, and nearly all of them will receive one, and more likely several, ankle–foot orthoses (AFOs) during their lifespan. The current standard of practice is to use orthoses as a first-line treatment for those without significant ankle joint contractures, to improve ankle–foot patterns during walking. The major goals of orthotic intervention, regardless of pathology, are: (1) to prevent deformity; (2) to support normal joint alignment and mechanics; (3) to provide variable ranges of motion; and (4) to facilitate function (Knutson and Clark, 1991). Although an orthosis may only have a direct effect on foot and ankle joint alignment, this distal control is believed to exert a positive effect on more proximal joints as well (Condie and Meadows, 1993).

Metal bracing had been employed in some patients with CP for several decades, but the use of AFOs as we know them today did not really become prevalent in this population until the 1970s. Their rapid proliferation was spurred on by the clinical popularity of inhibitory casts in the treatment of tone and alignment abnormalities in CP, and the development of mouldable plastics for in-shoe correction of dynamic deformities. This technology has improved considerably with the development of stronger, lighter and more flexible orthoses. New designs such as hinged and supramalleolar orthoses have emerged, along with the incorporation of more sophisticated design features, as exemplified in the more recent introduction of the 'dynamic' AFOs. However, scientific evidence for the effectiveness of orthoses for improving motor function in CP, or the relative efficacy of different designs, has not kept pace with technology.

The effects of AFOs on walking function have been studied by using both simple and more sophisticated gait analysis techniques. Using video technology alone, Powell, Silva and Grindeland (1989) showed no significant effect of AFOs on cadence, velocity or stride length. Thomas, Mazur and Wright (1989) used kinematic and EMG information to evaluate 17 children with CP walking barefoot and in AFOs. They demonstrated an improvement in ankle motion for all children, and also an increase in the hip and knee motions for at least 80% of the patients. Perhaps their most interesting finding was a significant reduction in co-contraction of some muscle groups, and more phasic patterns in the tibialis anterior, gastrocnemius, vastus lateralis and medial hamstrings. They therefore speculated that these changes in muscle function decreased the energy consumption required for walking, a conclusion also reached by Mossberg, Linton and Friske (1990).

Condie and Meadows (1993) compared the dynamics of normal paediatric gait with that of children who have spastic diplegia with and without AFOs. They demonstrated that the use of an appropriate AFO in CP reduced the high impact forces seen in early stance with CP, while the vertical reaction forces in late stance were increased, suggesting an improved ability to support body weight and to generate push-off.

In our laboratory, we performed a study to quantify the effects of fixed AFOs on equinus gait and crouch gait in diplegic CP (Abel et al., 1997). A cohort of 35 subjects with spastic diplegia was evaluated to assess the effectiveness of AFOs in controlling crouch gait ($n = 17$) and equinus gait ($n = 18$). Subjects were tested barefoot and in the orthoses on the same day. Our primary hypothesis was that elimination of abnormal foot contact pattern would produce more normal joint moments, enhance stability and increase stride length and velocity. Physical examination and computerized 3D motion analysis were used to determine bracing indications and effect of bracing on gait. The equinus group was characterized by equinus at foot strike and midstance with an increased plantarflexion moment in early stance and a typically reduced moment in late stance. The crouch group demonstrated increased excessive knee flexion and an exaggerated knee extensor moment in midstance which were associated with planovalgus foot deformities.

For the patient group as a whole, a significant increase in velocity and stride length was seen using the AFOs compared with bare feet ($p < 0.003$). This increase was accomplished by an increase in single limb stance time and increased excursion at the pelvis, hip and knee while anke excursion was decreased (Table 3). The effect of bracing on velocity and stride length was similar for both indications (equinus and crouch). In conclusion, compared with barefoot walking, AFOs provided better stability, improved joint alignment and thus increased walking velocity for both dynamic equinus and crouch gait patterns.

A second project examined the biomechanical effects of bilateral rigid AFOs compared to dynamic supramalleolar orthoses (SMOs) in 11 children with spastic diplegia, ranging from 4 to 11 years of age. A four-camera system was used to obtain 3D kinematic data, while two force plates provided ground

Table 3. Comparison of mean values for temporal-distance and kinematic gait parameters in bare feet and AFOs for 35 children with CP
(Abel et al., 1997)

	Barefoot	AFO
Velocity (m/s)	**0.72**	**0.82**
Stride length (m)	**0.69**	**0.79**
Cadence (steps/min)	123.5	121.2
Single stance (%)	**32.2**	**35.0**
Double stance (%)	**33.9**	**30.0**
Stance (%)	**67.0**	**65.0**
Pelvic excursion (degrees)	7.1	8.1
Hip excursion (degrees)	**40.8**	**45.2**
Knee excursion (degrees)	**36.8**	**42.0**
Ankle excursion (degrees)	**25.8**	**13.7**
Ankle position at contact (degrees)	1.26	6.91
Ankle maximum (degrees)	14.3	15.7
Ankle minimum (degrees)	**−9.1**	**2.6**

Bold type indicates significant differences between conditions ($p < 0.05$).

reaction force data. An inverse dynamics approach, combining anthropometric, kinematic and force plate data, was employed to calculate joint moments and powers (Vaughan, 1996). Each child was studied in both orthoses, with order of use randomly assigned, and in shoes alone. In summary, on the positive side, the AFOs functioned successfully by limiting the range of motion at the ankle, positioning the foot appropriately prior to initial foot contact, absorbing less power following initial foot contact, and generating a larger ankle moment during push-off. From a negative point of view, the AFOs did not decrease the undesirable plantarflexion moment peak at 20% of the cycle, and their use led to a reduction in the ankle power generated during push-off (Vaughan, 1996). In contrast to the AFOs, which altered the ankle joint mechanics quite noticeably, the SMOs appeared to have elicited almost no changes at all (cf. Figure 9). This is an important finding, since SMOs are widely prescribed and, because of their cosmetic appeal, are well tolerated by patients and their families (Carlson et al., 1995).

While research evidence suggests that orthoses are beneficial for ankle–foot deformities in CP, not all patients benefit to the same degree; some may show equivocal results, and still others may be hampered by their use. We need to be able to predict the patient and orthotic factors that determine success prior to prescription, and also to verify improvement for an individual patient once prescribed. One limitation of a standard gait assessment is that the foot is assumed to be a rigid segment, and all motion occurring within the foot is therefore recorded as ankle motion. Dynamic plantar pressure distribution measurement systems provide a more accurate and precise evaluation of ankle–foot deformities and the biomechanical effects of an orthosis. The clinical importance

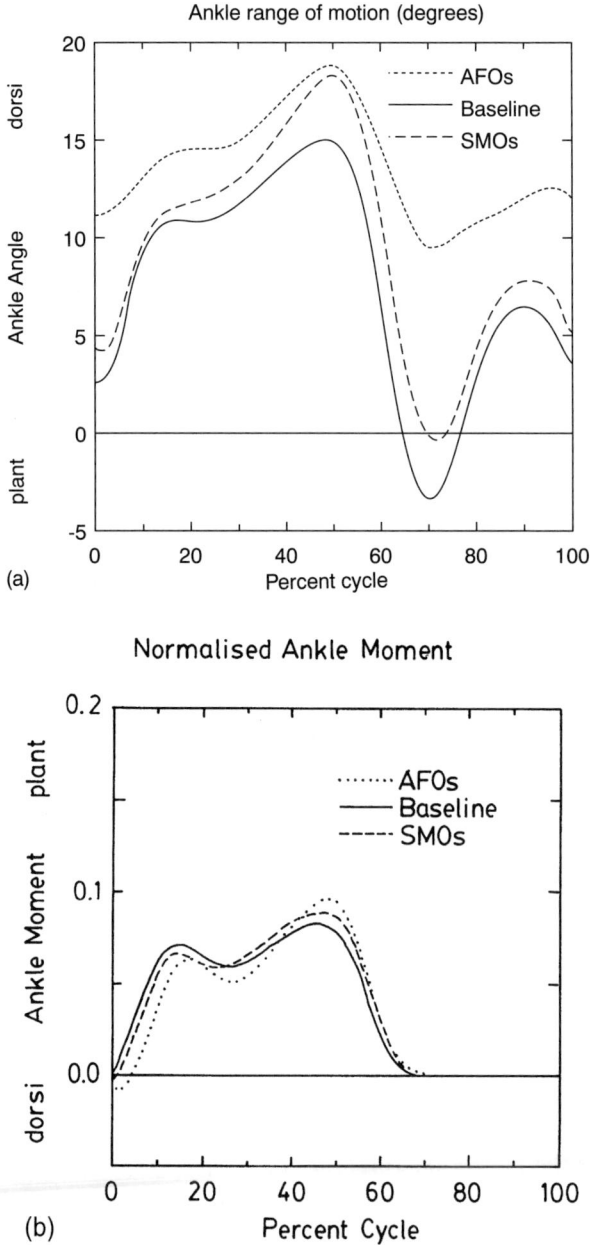

Figure 9. Ensemble averages, based on 11 children with spastic diplegia, for (a) sagittal plane ankle angle; and (b) sagittal plane ankle moment normalized according to Table 1. The three conditions illustrated are: AFOs; SMOs; and baseline, where the same shoes, but no orthoses, were worn (Vaughan, 1996)

of analysing specific areas of pressure under the plantar surface of the foot has been well documented in diabetes (Cavanagh and Ulbrecht, 1994), and in many surgical reconstructive and revascularization procedures for the foot and ankle (Cooper and Dietz, 1995; Widhe and Bergren, 1994). Surprisingly, this technology has not been applied extensively to the CP population, where virtually all individuals have distal motor involvement that leads frequently to ankle and foot deformities.

GAIT CHANGES AFTER PHARMACOLOGICAL AND NEURO-SURGICAL REDUCTION OF SPASTICITY

Pharmacological interventions directly aimed at reducing spasticity in ambulatory patients with CP are not widespread and, with the exception of intrathecal baclofen, have been shown to be minimally effective at best (Park and Owen, 1992). Botulinum-toxin A (Botox), injected intramuscularly, has been used in patients with CP and, while its most direct effect is the production of muscle weakness, it has also been shown to be capable of reducing spasticity (Koman et al., 1993). Reports have also documented gait improvements as a result of injections into the gastrocnemius and hamstring muscle groups in children with spastic CP (Koman et al., 1993; Cosgrove, Corry and Graham, 1994). However, Botox has a short-lived positive effect (3–6 months) and may therefore only be a temporary solution for these patients.

For children who are ambulatory or pre-ambulatory, selective dorsal rhizotomy is a neurosurgical procedure that offers a permanent reduction in spasticity with the primary objective of facilitating or improving gait function. Ideal candidates are those who are limited in their functional mobility by spasticity, yet have sufficient underlying muscle strength and voluntary control. Other good prognostic indicators of success from rhizotomy include: diagnosis of spastic diplegia secondary to prematurity; intelligence within or close to the normal range; a highly motivated patient and family; and no evidence of extrapyramidal signs (Oppenheim, Staudt and Peacock, 1991). This procedure is typically performed on children less than 12 years of age, although it has also been performed in adolescence and adulthood, with one report documenting positive functional outcomes in 77% of the patients studied (Peter and Arens, 1994).

Virtually all patients show an immediate reduction in spasticity postoperatively, as typically measured with the modified Ashworth Scale (Bohannon and Smith, 1987). Interestingly, improvement in muscle tone and function in parts of the body not directly affected by the surgery have been reported, with an improvement in sitting balance being especially evident (Lazareff, Mata-Acosta and Garcia Mendez, 1990). Several gait analysis studies (Cahan et al., 1989; Boscarino et al., 1993; Thomas et al., 1996; Vaughan et al., 1988; Vaughan, Berman and Peacock, 1991) have shown consistent improvements in free walking velocity, stride length, and hip and knee excursion, with a reduction in cadence. Our long-term results have only recently become available (Subramanian et al., 1997) and suggest that motor function continues to improve in these patients with age (cf. Figure 10). However, many of those studied continued to receive other interventions such as physiotherapy and orthopaedic surgery after the rhizotomy surgery.

In summary, rhizotomy effectively alleviates spasticity, which is one of the dominant negative symptoms in patients with CP. It does not create, but rather

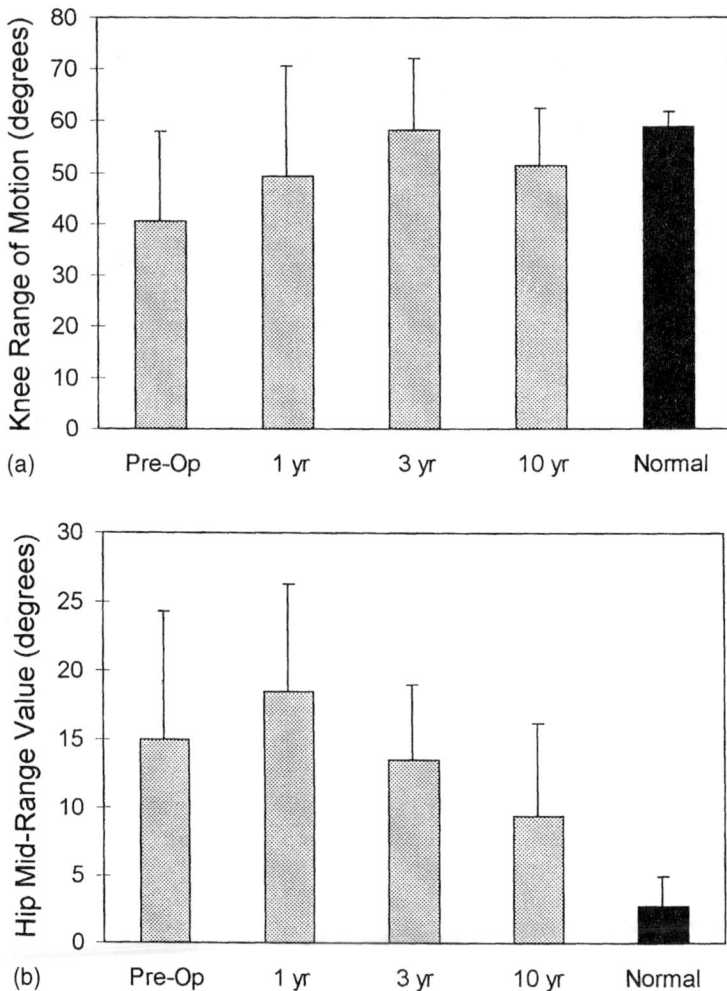

Figure 10. Gait analysis data: (a) knee range of motion; and (b) hip mid-range value for a long-term study of 11 children with CP who underwent selective posterior rhizotomy surgery (Vaughan et al., 1988; Vaughan, Berman and Peacock, 1991; Subramanian et al., 1997)

unmasks, weakness (Guiliani, 1991). While it has been shown to improve both active and passive range of motion in the lower extremities, orthopaedic surgery is still often required to address residual or recurrent static and dynamic contractures or correct bony deformities as well to maximize outcomes. Additionally, an intensive physiotherapy programme focusing on muscle strengthening postoperatively has been shown to enhance the functional results of this surgery and may be responsible for some of the positive outcomes seen (McLaughlin et al., 1994).

EFFECT OF ORTHOPAEDIC SURGERY ON GAIT IN CEREBRAL PALSY

Traditionally, orthopaedic procedures in the form of tendon lengthening or bone realignment have played a central role in the treatment of deformity caused by CP (Bleck, 1987a; Park and Owen, 1992). The strong belief that surgery is beneficial underlies the fact that most children with CP in the USA and elsewhere undergo some form of orthopaedic surgery during childhood. Indications for orthopaedic surgery in spastic CP include dynamic or static joint contractures which interfere with ambulation or lead to joint malalignment. The overall goal of these interventions is to improve mobility and functional independence. Surgery is invasive and involves a permanent change in the structure of muscles and/or bones, and is recommended only when the deformity is not amenable to treatment by more conservative means such as physiotherapy or bracing (Hoffer et al., 1988; Bleck, 1987a,b).

Three-dimensional gait laboratories have been acknowledged as important tools for assessment of neuromuscular disorders (Skrotsky, 1983; DeLuca, 1991; Sutherland, 1989) and we have shown that gait performance is representative of generalized gross motor function in CP (Damiano and Abel, 1996). Gait analysis provides data which are used to differentiate primary gait deviations from compensatory patterns of motion to aid in surgical decision-making. The postoperative outcome can also be quantified with this technology (Gage et al., 1984; Sutherland, 1989; Thometz, Simon and Rosenthal, 1989; Lee, Goh and Bose, 1992; Rose et al., 1993; Etnyre et al., 1993). Gage et al. (1984) used quantitative gait analysis to evaluate the outcome of multiple orthopaedic surgeries performed at one sitting in 20 patients. By summation of various gait parameters, they concluded that 13 patients were improved, six were unchanged and one was worse. The patients who remained unchanged postoperatively showed less severe involvement than the improved group, although no description of preoperative motor involvement was provided.

In a similar study, we have prospectively analysed 25 consecutive children with spastic diplegia undergoing multiple muscle–tendon lengthening to improve walking. All patients were assessed in our motion analysis laboratory preoperatively and at 3, 6 and 9 months postoperatively in order to monitor recovery and to minimize the confounding effect of improvements in gait due solely to limb growth. In addition, the gross motor function measure (GMFM) and a comprehensive physical examination consisting of lower extremity passive range of motion were performed. The GMFM is a validated evaluation tool measuring a variety of gross motor skills ranging from lying to running and jumping (Russell et al., 1991). We hypothesized that the improved alignment and joint excursion following surgery should be reflected by increased walking speed and stride length. Furthermore, if the weakening effect of muscle–tendon lengthening were detrimental, a deterioration of gait variables and other motor skills would be expected.

The mean number of muscle–tendon units treated on the 25 subjects was 7.0 (range 2–13) or 3.5 units per limb. As predicted, passive range of motion improved at the joints directly affected by the surgical procedure. In examining the functional outcomes (Table 4), we found significant increases in walking velocity and stride length at 9 months postoperatively, with no change in cadence. This suggests that the gait was not only faster, but more efficient postoperatively. The increased stride length resulted from different combinations of increased pelvic, hip and knee motion across subjects, and the knee position was more extended during stance, indicating better joint alignment and a more upright posture. Improvements at the individual joints were more apparent when the patients were subgrouped by procedures. For example, in the subjects who underwent gastrocnemius lengthenings, ankle motion shifted to a more normal range, with a marked reduction in the equinus positioning of the ankle that was evident preoperatively. Finally, significant improvements in the standing and walking domains of the GMFM were seen.

In conclusion, the data on surgical treatment to improve ambulation in CP indicate that the technical outcome in terms of greater passive joint range of motion is quite predictably achieved ($> 90\%$). When examining active motion or positive functional changes in gait, it is apparent that, while most patients do

Table 4. Comparison of preoperative and postoperative gait and gross motor function measure results for 25 patients who underwent orthopaedic surgery at the University of Virginia

	Preop.	9 Month postop.	ANOVA p-value
Velocity (m/s)	**0.69**	**0.80**	**0.017**
Cadence (steps/min)	123	125	0.735
Stride length (m)	**0.66**	**0.75**	**0.001**
Double support time (%)	33.2	32.3	0.992
Stance time (%)	66.4	66.7	0.809
GMFM 5	**46.87**	**49.62**	**0.042**
GMFM total	**79.41**	**80.81**	**0.037**

Bold type indicates significant difference at $p < 0.05$.

seem to improve in these measures, some clearly do not change, and a few may even be worse as a consequence of the weakness imposed by surgery (Figure 11). Gait analysis provides a more detailed biomechanical assessment of motion, but surgical recommendations are still dependent on the clinical skill, judgement and biases of the surgeons who utilize these data in decision-making. The resultant recommendations may span the range from no surgery, to an operation at a single joint, to simultaneous bony and soft-tissue procedures at multiple joints. Gage (1991, 1994) contends that gait analysis is essential in the orthopaedic treatment of a child with CP, and its use has led to the development of new procedures as well as the elimination of less effective and even injurious ones. Furthermore, the gait record is far more objective than passive range of motion measurements and can be used to reappraise decisions (Abel, 1995). DeLuca (1991) and others (Lee, Goh and Bose, 1992) found that the additional information provided by gait analysis has led to the performance of more effective and perhaps fewer orthopaedic procedures.

ACKNOWLEDGEMENTS

This chapter resulted from a collaboration between the three authors when they were affiliated with the University of Virginia. We gratefully acknowledge the financial support of the National Institutes of Health (RO1 HD30134), the US Department of Education

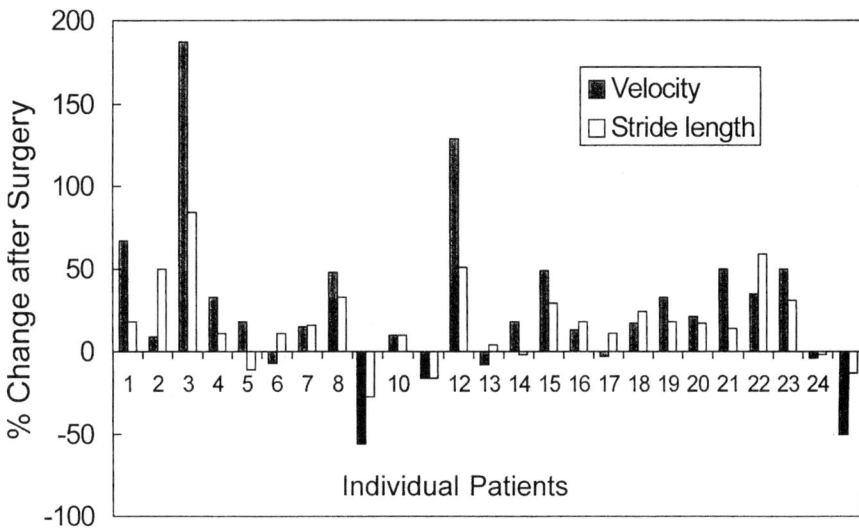

Figure 11. The percentage change in velocity and stride length for 25 patients who underwent orthopaedic surgery, comparing preoperative and 9-month postoperative assessments

(Rehabilitation Engineering Training Grant H133P10006), and the Orthopaedic Research and Education Foundation. In particular, we would like to thank our students who assisted in gathering and analysing the data: Scott Colby, Gary Brooking, Robert Abramczyk, Jerome Watson, David Johnson, Jeff Bush, Tim Harris, Mary Beth Wiley, Kristen Bowsher, John Paul Romano, Warren Carlson, Greg Juhl and Michael Pannunzio.

REFERENCES

Abel, M. F. (1995) Letter to Editor in response to 'Gait laboratory analysis for preoperative decision making in spastic cerebral palsy: is it all that it is cracked up to be?' *J. Pediatr. Orthop.*, **15**, 698–699.

Abel, M. F. and Damiano, D. L. (1996) Strategies for increasing walking speed in diplegic cerebral palsy. *J. Pediatr. Orthop.*, **16**, 753–758.

Abel, M. F., Juhl, G. A., Vaughan, C. L. and Damiano, D. L. (1997) Functional assessment of ankle foot orthoses in children with spastic diplegia. *Arch. Phys. Med. Rehab. (Br.)* (in press).

Beck, R. J., Andriacchi, T. P., Kuo, K. N., Fermier, R. W. and Galante, J. O. (1981) Changes in the gait patterns of growing children. *J. Bone Joint Surg. (Am.)*, **63**, 1452–1456.

Bertoti, D. B. (1986) Effect of short leg casting on ambulation in children with cerebral palsy. *Phys. Ther.*, **66**(10), 1522–1529.

Bleck, E. E. (1984) Where have all the CP children gone?—the needs of adults. *Dev. Med. Child Neurol.*, **26**, 669–676.

Bleck, E. E. (1990) Management of the lower extremities in children who have cerebral palsy. *J. Bone Joint Surg. (Am.)*, **72**, 140–144.

Bleck, E. E. (1987a) Special assessments and investigations. In: *Orthopaedic Management in Cerebral Palsy* (ed. E. E. Bleck), MacKeith Press, Philadelphia, PA, pp. 65–105.

Bleck, E. E. (1987b) Goals, treatment and management. In: *Orthopaedic Management in Cerebral Palsy*, (ed. E. E. Bleck), MacKeith Press, Philadelphia, PA, pp. 142–212.

Bohannon, R. W. and Smith M. B. (1987) Interrater reliability of a modified Ashworth scale of muscle spasticity. *Phys. Ther.*, **67**, 206–207.

Boscarino, L. F., Ounpuu, S., Davis, R. B., Gage, J. R. and DeLuca, P. A. (1993) Effects of selective posterior rhizotomy on gait in children with cerebral palsy. *J. Pediatr. Orthop.*, **13**, 174–179.

Bowsher, K. A. and Vaughan, C. L. (1995) Effect of foot-progression angle on hip joint moments during gait. *J. Biomech.*, **28**, 759–762.

Cahan, L. D., Adams, J. M., Perry, J. and Beeler, L. M. (1990) Instrumented gait analysis after selective dorsal rhizotomy. *Dev. Med. Child Neurol.*, **32**, 1037–1043.

Campbell, S. K. (1990) Consensus conference on the efficacy of physical therapy in the management of cerebral palsy. *Pediatr. Phys. Ther.*, **2**, 123–124.

Carlson, W. E., Damiano, D. L., Abel, M. F. and Vaughan, C. L. (1995) Biomechanics of orthotic management of gait in spastic diplegia. *Gait Posture*, **3**, 102.

Cavanagh, P. R. and Ulbrecht, J. S. (1994) Clinical plantar pressure measurement in diabetes: rationale and methodology. *The Foot*, **4**, 123–135.

Condie, D. N. and Meadows, C. B. (1993) Ankle–foot orthoses. In: *Biomechanical Basis of Orthotic Management* (ed P. Bowker, D. Condie, D. Bader and D. Pratt), Butterworth-Heinemann, Oxford, pp. 99–123.

Cooper, D. M. and Dietz, F. R. (1995) Treatment of idiopathic clubfeet: a thirty year follow-up note. *J. Bone Joint Surg. (Am.)*, **77**(10), 1477–1489.

Cosgrove, A. P., Corry, I. S. and Graham, H. K. (1994) Botulinum toxin in the management of the lower limb in cerebral palsy. *Dev. Med. Child Neurol.*, **36**, 386–396.

Damiano, D. L. and Abel, M. F. (1996) Relationship of gait analysis to gross motor function in spastic cerebral palsy. *Dev. Med. Child Neurol.*, **38**, 389–396.

Damiano, D. L., Kelly, L. E. and Vaughan, C. L. (1995) Effects of quadriceps femoris muscle strengthening on crouch gait in children with spastic diplegia. *Phys. Ther.*, **75**(8), 658–667.

Damiano, D. L., Vaughan, C. L. and Abel, M. F. (1995) Muscle response to heavy resistance exercise in children with spastic cerebral palsy. *Dev. Med. Child Neurol.*, **37**, 731–39.

DeLuca, P. A. (1991) Gait analysis in the treatment of the ambulatory child with cerebral palsy. *Clin. Orthop. Relat. Res.*, **64**, 65–75.

Embrey, D. G., Yates, L. and Mott, D. H. (1990) Effect of neurodevelopmental treatment and orthoses on knee flexion during gait: a single subject design. *Phys. Ther.*, **70**, 626–637.

Etnyre, B., Chambers, C. S., Scarborough, N. H. and Cain, T. E. (1993) Preoperative and postoperative assessment of surgical intervention for equinus gait in children with cerebral palsy. *J. Pediatr. Orthop.*, **13**, 24–31.

Gage, J. R. (1991) *Gait Analysis in Cerebral Palsy.* MacKeith Press, New York, Cambridge.

Gage, J. R. (1994) The role of gait analysis in the treatment of cerebral palsy. *J. Pediatr. Orthop.*, **14**, 701–702.

Gage, J. R., Fabian, D., Hicks, R. and Tashman, S. (1984) Pre- and postoperative gait analysis in patients with spastic diplegia: a preliminary report. *J. Pediatr. Orthop.*, **4**, 715–725.

Guiliani, C. A. (1991) Dorsal rhizotomy for children with cerebral palsy: support for concepts of motor control. *Phys. Ther.*, **71**(3), 248–259.

Harris, T., Abel, M. F. and Damiano, D. L. (1997) The effect of spastic cerebral palsy on gait patterns and joint function in adults. *Dev. Med. and Child Neur.*, **39**(9), (in press).

Hinderer, K. A., Harris, S. R., Purdy, A. H., Chew, D. E., Staheli, L. T., McLaughlin, J. F. and Jaffe, K. M. (1988) Effects of 'tone-reducing' vs. standard plaster casts on gait improvement of children with cerebral palsy. *Dev. Med. Child Neurol.*, **30**, 370–377.

Hof, A. L. (1996) Scaling gait data to body size. *Gait Posture*, **4**, 222–223.

Hoffer, M. M., Rinsky, L., Root, L., Simon, S. R. and Sutherland, D. H. (1988) Symposium: management of cerebral palsy in the lower extremity. *Contemp. Orthop.*, **16**, 79–111.

Horvat, M. (1987) Effects of a progressive resistance training program on an individual with spastic cerebral palsy. *Am. Correct. Ther. J.*, **41**(1), 7–11.

Johnson, D., Abel, M. F. and Damiano, D. L. (1997) The evolution of gait in childhood and adolescent cerebral palsy. *J. Pediatr. Orthop.*, **17**(3), 392–396.

Kluzik, J., Fetters, L. and Coryell, J. (1990) Quantification of control: a preliminary study of the effects of neurodevelopmental treatment on reaching in children with spastic cerebral palsy. *Phys. Ther.*, **7**, 65–78.

Knutson, L. M. and Clark, D. E. (1991) Orthotic devices for ambulation in children with cerebral palsy and myelomeningocele. *Phys. Ther.*, **71**, 947–960.

Koman, L. A., Mooney, J. F. 3rd, Smith, B., Goodman, A. and Mulvaney, T. (1993) Management of cerebral palsy with botulinum-A toxin: preliminary investigation. *J. Pediatr. Orthop.*, **13**(4), 489–495.

Kramer, J. F. and MacPhail, H. E. A. (1994) Relationships among measures of walking efficiency, gross motor ability and isokinetic strength in adolescents with cerebral palsy. *Pediatr. Phys. Ther.*, **6**, 3–8.

Lazareff, J. A., Mata-Acosta, A. M. and Garcia Mendez, M. A. (1990) Limited posterior rhizotomy for the treatment of spasticity secondary to infantile cerebral palsy: a preliminary report. *Neurosurgery*, **27**(4), 535–538.

Lee, E. H., Goh, J. C. H. and Bose, K. (1992) Value of gait analysis in the assessment of surgery in cerebral palsy. *Arch. Phys. Med. Rehabil.*, **73**, 642–646.

Logan, L., Byers-Hinkley, K. and Ciccone, C. D. (1990) Anterior versus posterior walkers: a gait analysis study. *Dev. Med. Child Neurol.*, **32**, 1044–1048.

MacPhail, H. E. A. and Kramer, J. F. (1995) Effect of isokinetic strength-training on functional ability and walking efficiency in adolescents with cerebral palsy. *Dev. Med. Child Neurol.*, **37**, 763–775.

McCubbin, J. A. and Shasby, G. B. (1985) Effects of isokinetic exercise on adolescents with cerebral palsy. *Adapt. Phys. Activ. Q.*, **2**, 56–64.

McLaughlin, J. F., Bjornson, K. F., Astley, S. J., Hays, R. M., Hoffinger, S. A., Armantrout, E. A. and Roberts, T. S. (1994) The role of selective dorsal rhizotomy in cerebral palsy: critical evaluation of a prospective clinical series. *Dev. Med. Child Neurol.*, **36**, 755–769.

Molnar, G. (1991) Rehabilitation in cerebral palsy. *Western J. Med.*, **154**(5), 569–572.

Mossberg, K. A., Linton, K. A. and Friske, K. (1990) Ankle–foot orthoses: effect on energy expenditure of gait in spastic diplegic children. *Arch. Phys. Med. Rehabil.*, **71**, 490–494.

Norlin, R. and Odenrick, P. (1986) Development of gait in spastic children with cerebral palsy. *J. Pediatr. Orthop.*, **6**, 674–680.

O'Malley, M. J. (1996) Normalization of temporal-distance parameters in pediatric gait. *J. Biomech.*, **29**(5), 619–625.

O'Malley, M. J., Abel, M. F. and Damiano, D. L. (1995) Fuzzy clustering of temporal-distance and kinematic data for children with cerebral palsy. *Gait Posture*, **3**, 92.

Oppenheim, W. A., Staudt, L. A. and Peacock, W. J. (1991) The rationale for rhizotomy. In: *The Diplegic Child* (ed. M. D. Sussman), American Academy of Orthopaedic Surgery Press, Rosemont, IL, pp. 271–285.

Ounpuu, S., Gage, J. R. and Davis, R. B. (1991) Three-dimensional lower extremity joint kinetics in normal pediatric gait. *J. Pediatr. Orthop.*, **11**(3), 341–349.

Pape, K. E., Kirsch, S. E., Galil, A., Boulton, J. E., White, A. and Chipman, M. (1993) Neuromuscular approach to the motor deficits of cerebral palsy: a pilot study. *J. Pediatr. Orthop.* **13**(5), 628–633.

Park, T. S. and Owen, J. H. (1992) Surgical management of spastic diplegia in cerebral palsy. *N. Eng. J. Med.*, **326**, 745–749.

Peter, J. C. and Arens, L. J. (1994) Selective posterior rhizotomy in teenagers and young adults with spastic cerebral palsy. *Br. J. Neurosurg.*, **8**, 135–139.

Powell, M. M., Silva, P. D. and Grindeland, T. (1989) Effects of two types of ankle–foot orthoses on the gait of children with spastic diplegia. *Dev. Med. Child Neurol. (Suppl. 59)*, **31**(5), 8–9.

Ramakrishnan, H. K., Kadaba, M. P. and Wootten, M. E. (1987) Lower extremity moments and ground reaction torque in adult gait. In: *Biomechanics of Normal and Prosthetic Gait* (ed. J. L. Stein), BED- Vol. 4, American Society of Mechanical Engineers, New York, pp. 87–92.

Rose, J., Gamble, J. G., Medeiros, J., Burgos, A. and Haskell, W. L. (1989) Energy cost of walking in normal children and in those with cerebral palsy: comparison of heart rate and oxygen consumption. *J. Pediatr. Orthop.*, **9**, 276–279.

Rose, S. A., DeLuca, P. A., Davis, R.B., Ounpuu, S. and Gage, J. R. (1993) Kinematic and kinetic evaluation of the ankle after lengthening of the gastrocnemius fascia in children with cerebral palsy. *J. Pediatr. Orthop.*, **13**, 727–732.

Russell, D., Rosenbaum, P., Cadman, D., Gowland, C., Hardy, S. and Jarvis, S. (1991) The gross motor function measure: a means to evaluate the effects of physical therapy. *Dev. Med. Child Neurol.*, **31**, 351–352.

Skrotsky, K. (1983) Gait analysis in cerebral palsied and nonhandicapped children. *Arch. Phys. Med. Rehabil.*, **64**, 291–295.

Subramanian, N., Vaughan, C. L., Peter, J. C. and Arens, L. J. (1997) Gait before and ten years after rhizotomy in children with cerebral palsy spasicity. *J. Neurosurgery* (in press).

Sutherland, D. H. (1978) Gait analysis in cerebral palsy. *Dev. Med. Child Neurol.*, **20**, 807–813.

Sutherland, D. H. (1984) *Gait Disorders in Childhood and Adolescence*. Williams & Wilkins, Baltimore, MD.

Sutherland, D. H. (1989) Gait analysis in neuromuscular diseases. In: *Instructional Course Lectures*, **39**, pp. 333–341. American Academy of Orthopaedic Surgeons, Rosemont, IL.

Sutherland, D. H. (1996) Dimensionless gait measurements and gait maturity. *Gait Posture*, **4**, 209–211.

Sutherland, D. H., Olshen, R., Cooper, L. and Woo, S. L. Y. (1980) The development of mature gait. *J. Bone Joint Surg. (Am.)*, **62-A**, 336–353.

Sutherland, D. H., Olshen, R. A., Biden, E. N. and Wyatt, M. P. (1988) *The Development of Mature Walking*. MacKeith Press, London.

Thomas, S. S., Mazur, J. M. and Wright, N. (1989) Quantitative assessment of AFOs for children with cerebral palsy. *Dev. Med. Child Neurol. (Suppl. 59)*, **31**(5), 7.

Thomas, S. S., Aiona, M. D., Pierce, R. and Piatt, J. H. (1996) Gait changes in children with spastic diplegia after selective dorsal rhizotomy. *J. Pediatr. Orthop.*, **16**, 747–752.

Thometz, J., Simon, S. and Rosenthal, R. (1989) The effect on gait of lengthening of the medial hamstrings in cerebral palsy. *J. Bone Joint Surg. (Am.)*, **71**, 345–353.

Torfs, C. P., van der Berg, B. J. and Oechsli, F. W. (1990) Prenatal and perinatal factors in the etiology of cerebral palsy. *J. Pediatr.*, **116**, 615–619.

Vaughan, C. L. (1996) Are joint torques the Holy Grail of human gait analysis? *Human Movem Sci.*, **15**, 423–443.

Vaughan, C. L., Berman, B., Staudt, L. A. and Peacock, W. J. (1988) Gait analysis of cerebral palsy before and after rhizotomy. *Paediatr. Neurosci.*, **14**, 297–300.

Vaughan, C. L., Berman, B. and Peacock, W. J. (1991) Cerebral palsy and rhizotomy. A three year follow up evaluation with gait analysis. *J. Neurosurg.*, **74**, 178–184.

Vaughan, C. L., Davis, B. L. and O'Connor, J. C. (1992) *Dynamics of Human Gait*. Human Kinetics, Champaign, IL.

Vaughan, C. L., Carlson, W. E., Damiano, D. L. and Abel, M. F. (1995) Biomechanics of orthotic management of gait in spastic diplegia. Report of a Consensus Conference on the Lower Limb Orthotic Management of Cerebral Palsy (ed. D. N. Condie), International Society for Prosthetics and Orthotics, Copenhagen, pp. 181–191.

Widhe, T. and Bergren, L. (1994) Gait analysis and dynamic foot pressure in the assessment of untreated clubfoot. *Foot Ankle Int.*, **15**(4), 186–190.

Winters, T. F., Gage, J. R. and Hicks, R. (1987) Gait patterns in spastic hemiplegia in children and young adults. *J. Bone Joint Surg. (Am.)*, **69**, 437–441.

Zijlstra, W., Prokop, T. and Berger, W. (1996) Adaptability of leg movements during normal treadmill walking and split-belt walking in children. *Gait Posture*, **4**, 212–221.

17

Locomotion in Healthy Older Adults

MARK D. GRABINER

Department of Biomedical Engineering, Lerner Research Institute,
The Cleveland Clinic Foundation, Cleveland, Ohio, USA

The normal ageing process is associated with significant changes to the nervous, muscular, and skeletal systems that influence physical capability (Grabiner and Enoka, 1995). These changes occur with significant within- and between-subject variability relative to the onset, rate and absolute magnitude of change. Some of these changes are inevitable and irreversible. Others, however, demonstrate sensitivity to various interventions. Thus, from the standpoint of basic motor control, changes in locomotion that accompany the ageing process could provide insight into how the control of a complex system is adapted to the rather substantial changes to central and peripheral neuromuscular mechanisms.

From the standpoint of an orthopaedist, physical therapist or engineer, knowledge of the forces and moments acting on musculoskeletal tissues during locomotion and other mobility-related activities may assist in formulating reconstructive surgical procedures, rehabilitation protocols, and the selection of materials and design of an implant. From yet another standpoint, studying the manner in which gait changes with age may lend itself to explanations underlying the nonlinear, age-related increase in mobility-related falls after age 75. Characterizing the underlying causes could be invaluable to those seeking to reduce the morbidity and mortality in older adults due to falls. It is obvious that relative to the general population, the rate of growth of older adults has significantly increased and, furthermore, the rate of growth is continuing to increase. Clearly, as the population of very old people increases, the impact of

Three-dimensional Analysis of Human Locomotion. Edited by P. Allard, A. Cappozzo, A. Lundberg and C. Vaughan
© 1997 John Wiley & Sons Ltd. ISBN 0 471 96949 4

their health-related problems contributes a disproportionately large burden to an already troubled health-care system.

In addition to the age-related changes to the neuromuscular and musculoskeletal systems, it is evident that the motion patterns of older adults can be substantially different from those of young adults (Grabiner and Enoka, 1995). What is the influence of the ageing process on gait? Why does the ageing process exert these influences on gait? Can gait maladaptations occurring independently of disease be reversed? The prevalence of mobility problems in older adults and the socio-economic and medical ramifications of these problems are likely determinants of the increased interest in the biomechanics of mobility in this population. Indeed, a Medline search revealed that during the periods 1984–1990 and 1990–1995 the number of published manuscripts concerning gait in older adults was 242 and 333, respectively. However, much of this literature was not biomechanical in nature, and the biomechanics research to date has been limited to two-dimensional (2D) analysis. This is not surprising, given that walking is an activity dominated by motion in the plane of progression. However, falls occurring during locomotion are not limited to this plane.

A comprehensive review of the literature concerning the gait of older adults is beyond the scope of this chapter. Rather, this chapter will review some of the existing data concerning gait in older adults within the context of what the literature suggests to us about locomotion-related falls in healthy older adults. An emphasis has been placed on work that has included both young and older adults in the same study. The existing data are essentially descriptive. With regard to interpretations and extrapolations to falling behaviour, and to productive directions concerning reducing falls in older adults, the existing literature sheds little direct light. However, the question that may be asked is 'What can 3D analysis of locomotion in older adults offer in terms of further understanding and controlling the incidence of falling behaviour, and its sequela?'

BIOMECHANICAL DISTINCTIONS BETWEEN YOUNG AND OLDER ADULT GAIT

Locomotion in older adults has been investigated, in detail, for nearly 30 years. Perhaps the most robust observation regarding the gait of older adults is that the preferred, or freely chosen, speed of walking is slower than that of young adults (Wang and Olney, 1994; Ferrandez, Pailhous and Durap, 1990; Williams and Bird, 1992; Elble et al., 1991; Winter et al., 1990; Kaneko et al., 1991; Finley, Cody and Finizie, 1969; Hageman and Blanke, 1986; Bendall, Bassey and Pearson, 1989; Chen et al., 1991, 1994; Leiper and Craik, 1991; Blanke and Hageman, 1989; Himann et al., 1988; Martin, Rothstein and Larish, 1992; Blin, Ferrandez and Serratrice, 1990). However, after three decades of investigation, a solution to the question 'Why do older adults tend to walk slower?' has not emerged. Figure 1 presents walking velocity data as a function of age compiled from 21 studies. A linear regression analysis limited to the data for adults older than 50 years of age accounted for only 36.1% (adjusted) of the shared variance. This interstudy comparison is somewhat consistent with other intrastudy analyses. For example, the regression by Dobbs et al. (1993) of age and preferred walking velocity in 144 healthy subjects (age: 30–89 years) accounted for 40% (unadjusted) of the shared variance. Figure 1 illustrates the variability of the relationship between age and preferred walking velocity. Some of the variability may reflect male–female differences, differences in data collection procedures, including the laboratory environment and instruction set given to the subjects, and differences in the data analysis methods. A large source of variability could be differences in the health status and exercise habits of the subjects, i.e. the effects of chronological age versus those of biological age.

Figure 2 presents data compiled from studies reporting walking velocity as a function of age when the general instruction to the subjects was to walk as fast as possible. The influence of age seems more pronounced in these data. The regression equation accounts for 68% (adjusted) of the variability. Notably, when specifically requested to walk quickly, healthy older adults up to about 75 years of age appear capable of achieving mean walking velocities (1.41 ± 0.28 m/s) that are similar to the mean preferred walking velocity of young adults, which is

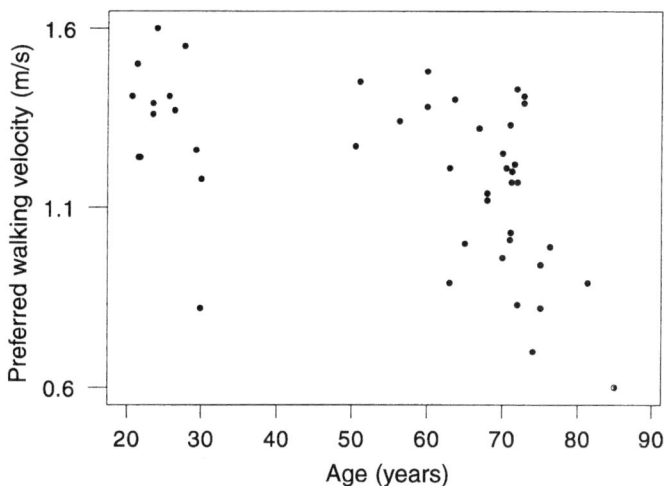

Figure 1. Compilation of data from studies reporting the effect of age on preferred walking velocity. A regression of data greater than or equal to 50 years of age generated an equation

$$\text{Preferred walking velocity} = 2.42 - (0.0188 \times \text{age})$$

The regression accounted for an adjusted 36.1% of the shared variance. The standard error of measurement was 0.1848

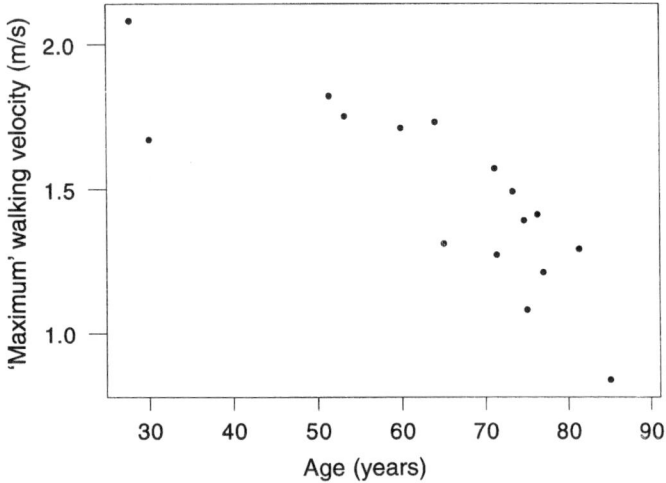

Figure 2. Compilation of data from studies reporting the effect of age on walking velocity under the general instruction to walk as fast as possible. A regression of data greater than or equal to 50 years age generated an equation

$$\text{Preferred walking velocity} = 3.07 - (0.0236 \times \text{age})$$

The regression accounted for an adjusted 68.0% of the shared variance. The standard error of measurement was 0.1588.

1.33 ± 0.20 m/s. Age-related decreases in preferred walking velocity are not necessarily obligatory. For example, Blanke and Hageman (1989) reported that the preferred walking velocity of older male adults (63.6 years) was similar to that of the young adults (24.5 years), 1.39 and 1.31 m/s, respectively. Martin, Rothstein and Larish (1992) reported that the preferred walking velocity of a group of active older adults (1.43 m/s) was similar to that of a group of sedentary young adults (1.41 m/s). These data underscore the effect of health and/or exercise status of subjects.

Stride length and stride frequency are, of course, the basic determinants of walking speed, and the effect of ageing on these variables has been reported. Differences in the way in which stride length is calculated do not generally affect the conclusion that, with age, stride length generally decreases in both older women and men. For example, Hirasaki et al. (1993) reported a 21% decrease in non-normalized step length data with age (0.62 m in young adults versus 0.49 m in the older adults). The age-related decrease in step length normalized to body height was much smaller, 11.4%. The effects of age on stride frequency, on the other hand, seem to be more variable. The data of Ferrandez, Pailhous and Durap (1990), Kaneko et al. (1991) and Himann et al. (1988) support a general age-related decrease in stride frequency. Himann et al. reported slightly faster stride frequencies in young adult women than in young adult men but not in older adults, and a larger age-related decrease in stride frequency with age in women. They also reported that whereas older men increased walking speed by increasing step length, older women increased walking velocity by modulating both stride length and stride frequency.

The stride length and frequency data generally explain, with slightly greater detail, why older adults walk slower than young adults. However, these variables offer little insight into the falling behaviour problem of older adults. The published literature regarding joint kinematics presents the next level of investigation. Finley, Cody and Finizie (1969) compared kinematics of a group of 23 older women (mean age: 74.4 years) and 12 young women (mean age: 29.9 years) walking at their preferred velocity (0.82 and 0.70 m/s, respectively). The patterns of sagittal and frontal plane ankle, knee and hip time–displacement curves were qualitatively similar, which led the authors to conclude that chronological age did not affect the measured joint kinematics. Williams and Bird (1992) also reported a slower walking velocity in older women compared to young women that accompanied somewhat gross measures in interlimb coordination.

Older women tend to demonstrate a significantly smaller maximum range of ankle joint plantarflexion–dorsiflexion motion than young women (Hageman and Blanke, 1986; Kaneko et al., 1991; Hirasaki et al., 1993). Finley, Cody and Finizie (1969) qualitatively reported older women as having a smaller ankle joint plantarflexion angle at toe-off compared to the young women (95° versus 100°, respectively), although comparisons at other instants during the gait cycle were similar. Kaneko et al. (1991) also reported smaller ankle joint plantarflexion angle at toe-off in older women. With regard to this observation, Winter et al. (1990) had previously reported that older adults generate less plantarflexor-related work during late stance. The smaller plantarflexor work correlated with the shorter stride length in older adults, but one may question whether the step length is smaller because of reduced plantarflexor work, or whether the reduced plantarflexor work simply reflects the mechanical requirement of the reduced step length. Bendall, Bassey and Pearson (1989) reported that plantarflexion strength was significantly related to preferred walking velocity in older men and women. In contrast to this is the work of Martin et al. (1993), who reported the effects of a 16-week lower extremity strength-training programme on the preferred walking velocity of young and older adults. Despite a significant increase in plantarflexion strength (as well as knee extension, hip flexion and hip extension strength), there was no change in the preferred walking velocity from pre-training (1.36 m/s) to post-training (1.38 m/s).

Finley, Cody and Finizie (1969) reported a difference between young and older adults in knee joint flexion angle at toe-off, 55° versus 49°, respectively. The older women also had a smaller knee flexion angle at midswing than the young women, 38° versus 43°, respectively. Elble et al. (1991) and Hirasaki et al. (1993) reported a larger knee range of motion in young adults, 61.7° and

$60.5° \pm 7.0°$, respectively, compared to the $54.2°$ and $50.6° \pm 4.8°$, respectively, for the older adults. The data of Kaneko et al. (1991) also reveal decreased knee flexion angle at toe-off as well as an age-related decrease in maximum knee joint angular displacement. With regard to the hip, older and young adults display similar maximum ranges of hip joint flexion–extension (Finley, Cody and Finizie, 1969; Elble et al., 1991; Hirasaki et al., 1993). From the above data, one may generalize that kinematically, if differences exist between the gaits of young and older adults, they are subtle. The above-mentioned studies shed little light on the mechanisms underlying falling behaviour in older adults. From a methods standpoint, these studies collectively tend to disregard the sensitivity of these types of variables to walking velocity which may actually serve to decrease, in a statistical sense, the observed differences (Kirtley, Whittle and Jefferson, 1985; Martin and Marsh, 1992).

Joint kinetics provide an estimate of how much muscle force and/or power is required to perform a specific task. With regard to joint kinetics in older adults, one can turn to the orthopaedic literature associated with lower extremity arthroplasty to define the kinetic envelope within which older adults normally perform. For example, Andriacchi, Galante and Fermier (1982) used a control group of 14 older male and female subjects in a study describing knee joint kinetics during level walking and stair climbing. Table 1 presents the maximum knee joint moments observed during level walking in these subjects (mean age: 62.4 years). The table also includes the data of Brugioni, Andriacchi and Galante (1990), who used a control group of 15 older male and female subjects (age range: 41–75 years) in a study of walking and stair climbing. Application of some average older male weight (757 N) and height (1.78 m) values allows comparison of these kinetics to the general capabilities of older subjects. A difficulty in this exercise, however, is the selection of a representative value for maximum voluntary knee extension strength.

The literature reveals significant differences in strength depending on whether the measures were made isometrically or isokinetically. In the case of the latter,

Table 1. Maximum knee joint moments[a] observed during level walking at 1.0 m/s and ascending stairs

	Andriacchi, Galante and Fermier (1982) (walk)	Brugioni, Andriacchi and Galante (1990) (walk)	Brugioni, Andriacchi and Galante (1990) (stairs)
Extension	2.29 ± 1.10	2.64 ± 1.35	2.89 ± 1.04
Flexion	2.74 ± 0.98	3.69 ± 6.76	11.12 ± 2.47
Adduction	3.57 ± 0.90	3.23 ± 1.02	2.57 ± 1.51
Abduction	0.25 ± 0.25	0.39 ± 0.32	1.23 ± 1.05
External rotation	0.54 ± 0.14	0.18 ± 0.12	0.4 ± 0.20
Internal rotation	0.41 ± 0.19	0.86 ± 0.53	1.2 ± 0.28

[a]Normalized to, and expressed as, a percentage of (body weight × body height).

the speed of isokinetic movement will substantially influence the maximum voluntary strength value. Additionally, these types of measurement are most often made on isolated joints, which are quite different from the multiple joint system that is characteristic of gait. Based upon data from Owings and Grabiner (1995) and Poulin et al. (1992), a generous mean value of maximum concentric isokinetic knee extension strength may be 150 Nm. Using this value results in an estimate that walking in older men requires approximately 36.6% of the maximum available knee extension strength. A similar comparison using the data of Brugioni, Andriacchi and Galante (1990) for the maximum knee flexion moment observed during stair ascent ($11.12 \pm 2.47\%$ body weight $\times$ body height) suggests that this task would require approximately 99% of the maximum available knee extension strength. This pointedly reveals weaknesses in conclusions drawn from comparisons between measurements of strength and calculations of joint moments required to perform various tasks. It certainly underscores the contention of Schultz (1992) that the implicit belief in a functional relationship between mobility impairments in older adults and muscular strength as presently evaluated 'warrants careful consideration'. However, because the importance of such comparisons is not diminished, continued and more comprehensive efforts to derive meaningful relationships are needed.

THE WHY ...

So the originally posed, and possibly related, questions 'Why do older people walk slower?', 'Is the slower walking velocity of older people a maladaptation reflective of a failing neuromuscular system?', and 'Does the gait of older people predispose them to falling behaviour?' remain unanswered. The slower gait of older adults has been hypothesized to be related to issues of energy costs, stability, and improved adaptability to changing environmental conditions. With regard to energy costs, there is a specific gait speed that results in maximum economy and that reflects the amount of oxygen used to walk a specified distance. It has been suggested that humans, as well as animals, subscribe to a maximum economy principle in selecting gait speed (Larish, Martin and Mungiole, 1988). Larish, Martin and Mungiole and later Martin, Rothstein and Larish (1992) found that older adults demonstrated lower economy than young adults when walking at similar speeds. However, the older people were found to demonstrate their highest economy within the same range of walking velocities as the young adults. Further, the preferred walking velocity of the active and sedentary older adults was not the speed at which maximum economy was observed. Thus, a relationship between age-related changes in walking speed and energy costs may be questioned.

The issues of increased stability and adaptability would seem to be related, in so far as the purpose of locomotion is to move safely from point A to point B. It seems intuitive that safe arrival may take precedence over early arrival. Adaptability may, in part, reflect something of a 'safety factor' and relate to the speed with which environmentally derived data such as surface slope, surface friction, obstacles, etc., may be processed via cognitive and non-cognitive pathways. Likely to be aware of this type of change in ability, a person might voluntarily reduce walking velocity to provide a larger time envelope in which to perceive and respond to unexpected events.

The issue of quantifying stability during gait has proven to be difficult. Stability during the 'static' conditions of upright posture can be described in terms of the relationship between the location of the centre of gravity and the base of support. Although rather simple to describe, stability of the human body during locomotion has proven less simple to define mathematically and implement as in the static condition. The task of maintaining stability during locomotion is considerably more difficult for the central nervous system. The problem is that the centre of gravity falls outside the boundaries of the base of support for a dominant portion of the single support phase of gait. In the static condition, once the centre of gravity falls outside the base of support, the body is by definition unstable and is falling. By the same definition, during locomotion the body is unstable and falling most of the time. The development of a valid mathematical definition of stability that can be applied to locomotor activities may represent one of the more challenging problems presently facing biomechanists in the area of locomotion.

There have been a number of experimental approaches to differentiating stable from unstable gait. Yack and Berger (1993) used Fourier analysis of trunk accelerations during preferred walking velocity to derive an 'index of smoothness'. Older adult subjects were retrospectively categorized as stable and unstable based upon self-reported falling history. The index of smoothness was reported to have statistically distinguished between these two groups. The authors' conclusion that individuals having stability problems may have diminished control of the trunk during locomotion was reasonably consistent with the findings of Krebs et al. (1992) and MacKinnon and Winter (1993). This contention has been indirectly supported by Grabiner et al. (1993, 1996), who have reported that control of the trunk is a factor contributing to successful recovery from anteriorly directed trips. Eng and Winter (1995) have documented the importance of three-dimensional kinetic analysis, particularly in relation to the hip joint, where they have reported a large amount of frontal plane work performed to control the pelvis and trunk.

In the area of upper body control during gait, Prince et al. (1994) reported evidence supporting an anticipatory control mechanism through which stabilization of the head and trunk occurred in a top-down fashion. Hirasaki et al. (1993) had previously reported on Fourier analysis of sagittal plane head motions as a function of age. Given that two of the three systems to which postural control is subservient reside in the head (visual and vestibular systems), it is reasonable that one aspect of successful gait is to provide a stable platform for these systems. Notably, the mean frequency of vertical head acceleration (taken from the Fourier analysis of acceleration data) for the older adults was significantly higher than that of the young adults.

Stride width is a variable that has been suggested as being age-sensitive and related to side-to-side stability during gait. However, the existing data do not reveal a consistent sensitivity of this variable to age. Hageman and Blanke (1986) reported that older women had significantly larger stride width values (10.0 cm) compared to young adult women (8.3 cm). However, Blanke and Hageman (1989) reported that the stride width of young male subjects (10.8 cm) was significantly larger than that of older adult men (8.25 cm). Further, Elble et al. (1991) reported similar step widths for both young and older adults (10.9 cm).

Another factor related to stability during locomotion is foot trajectory. The control of foot trajectory, according to Winter (1989), comprises toe clearance and a 'gentle' heel–ground contact. Toe clearance, a factor often associated with tripping, has been reported as being about 1.0 cm (Winter, 1987). Both Winter et al. (1990) and Elble et al. (1991) failed to identify a significant difference in toe clearance between young and older adults. However, Kaneko et al. (1991) reported a reduced toe clearance height with increasing age.

WHERE TO NOW?

Reviewing the literature suggests that biomechanical analysis of locomotion of older adults has not provided convincing answers to questions related to why increasing age predisposes otherwise healthy people to falling. Would three-dimensional analysis increase the likelihood of solving this and related questions? Clearly, some questions require a three-dimensional approach. The challenge is to formulate hypotheses and experimental methods that differ from those already devised to avoid merely arriving at solutions that differ from those already existing only in the amount of data generated. Perhaps walking *per se* is too benign an activity, even for very old people, to allow partitioning of a discriminant variable set. The work of Chen et al. (1991, 1994), in which subjects are required to negotiate obstacles in the gait path, provides examples of alternative methods through which solutions may emerge. More comprehensive research efforts emphasizing within-subject relationships on mobility tasks, including ascent and descent of stairs, and transfers, such as rising from and sitting into chairs, which in many ways challenge the neuromuscular and musculoskeletal systems of older adults to a greater extent than walking, should be integrated with basic gait studies. Often, revisiting where one has previously been prior to planning the next journey is useful. In this vein, one recalls that it has been over 40 years since it was written that 'the synthesis of all the elements which simultaneously participate in locomotion, although an ideal worthy of achievement, is a task of such magnitude and difficulty, that its early attainment can not be expected' (Saunders, Inman and Eberhart, 1953).

REFERENCES

Andriacchi, T. P., Galante, J. O. and Fermier, R. W. (1982) The influence of total knee replacement design on walking and stair climbing. *J. Bone Joint Surg. (Am.)*, **64**, 1328–1335.

Bendall, M. J., Bassey, E. J. and Pearson, M. B. (1989) Factors affecting walking speed of elderly people. *Age Aging*, **18**, 327–332.

Blanke, D. J. and Hageman, P. A. (1989) Comparison of gait of young men and elderly men. *Phys. Ther.*, **69**, 144–148.

Blin, O., Ferrandez, A. M. and Serratrice, G. (1990) Quantitative analysis of gait in Parkinson patients: increased variability of stride length. *J. Neurol. Sci.*, **98**, 91–97.

Brugioni, D. J., Andriacchi, T. P. and Galante, J. O. (1990) A functional and radiographic analysis of the total condylar knee arthroplasty. *J. Arthroplasty*, **5**, 173–180.

Chen, H.-C., Ashtoon-Miller, J. A., Alexander, N. B. and Schultz, A. B. (1991) Stepping over obstacles: gait patterns of healthy young and old adults. *J. Gerontol.*, **46**, M196–M203.

Chen, H.-C., Ashtoon-Miller, J. A., Alexander, N. B. and Schultz, A. B. (1994) Age effects on strategies used to avoid obstacles. *Gait Posture*, **2**, 139–146.

Dobbs, R. J., Charlett, A., Bowes, S. G., O'Neill, C. J. A., Weller, C., Hughs, J. and Dobbs, S. M. (1993) Is this walk normal? *Age Aging*, **22**, 27–30.

Elble, R. J., Thomas, S. S., Higgins, C. and Colliver, J. (1991) Stride-dependent changes in gait of older people. *J. Neurol.*, **238**, 1–5.

Eng, J. J. and Winter, D. A. (1995) Kinetic analysis of the lower limbs during walking: what information can be gained from a three dimensional model? *J. Biomech.*, **28**, 753–758.

Ferrandez, A.-M. Pailhous, J. and Durap, M. (1990) Slowness in elderly gait. *Exp. Aging Res.*, **16**, 79–89.

Finley, F. R., Cody, K. A. and Finizie, R. V. (1969) Locomotion patterns in elderly females. *Arch. Phys. Med. Rehabil.*, **50**, 140–146.

Grabiner, M. D. and Enoka, R. M. (1995) Changes in movement capabilities with aging. *Exer. Sports Sci. Rev.*, **23**, 65–104.

Grabiner, M. D., Feuerbach, J. W. and Jahnigen, D. W. (1996) Successful recovery from a trip: control of the trunk during the initial phase following perturbation. *J. Biomech.* **29**, 735–744.

Grabiner, M. D., Koh, T. J., Lundin, T. and Jahnigen, D. W. (1993) Kinematics of recovery from a stumble. *J. Gerontol.*, **48**, M97–M102.

Hageman, P. A. and Blanke, D. J. (1986) Comparison of gait of young and elderly women. *Phys. Ther.*, **66**, 1382–1387.

Himann, J. E., Cunningham, D. A., Rechnitzer, P. A. and Patterson, D. H. (1988) Age-related changes in speed of walking. *Med. Sci. Sports Exer.*, **20**, 161–166.

Hirasaki, E., Kubo, T., Nozawa, S., Matano, S. and Matsunaga, T. (1993) Analysis of head and body movements of elderly people during locomotion. *Acta Otolaryngol (Stockh)*, **501**, 25–30.

Kaneko, M., Morimoto, Y., Kimura, M., Fuchimoto, K. and Fuchimoto, T. (1991) A kinematic analysis of walking and physical fitness testing in elderly women. *Can. J. Sports Sci.*, **16**, 223–228.

Kirtley, C., Whittle, M. W. and Jefferson, R. J. (1985) Influence of walking speed on gait parameters. *J. Biomed. Eng.*, **7**, 282–288.

Krebs, D. E., Wong, D., Jevsevar, D., Riley, P. O. and Hodge, W. A. (1992) Trunk kinematics during locomotor activities. *Phys. Ther.*, **72**, 505–514.

Larish, D. D., Martin, P. E. and Mungiole, M. (1988) Characteristic patterns of gait in the healthy old. *Ann. NY Acad. Sci.*, **515**, 18–32.

Leiper, C. A. and Craik, R. L. (1991) Relationships between physical activity and temporal-distance characteristics of walking in elderly women. *Phys. Ther.*, **71**, 791–803.

Martin, P. E. and Marsh, A. P. (1992) Step length and frequency effects on ground reaction forces during walking. *J. Biomech.*, **25**, 1237–1239.

Martin, P. E., Rothstein, D. E. and Larish, D. D. (1992) Effects of age and physical activity status on the speed-aerobic demand relationship of walking. *J. Appl. Physiol.*, **73**, 200–206.

Martin, P. E., Marsh, A. P., Burton, D. and Larish, D. D. (1993) Effect of strength improvement on postural control and preferred walking speed of elderly and young adults. In: *Conference Proceedings American Society of Biomechanics*, 17th Annual Meeting (ed. A. A. Biewener and V. K. Goel), University of Iowa, Iowa City, Iowa, pp. 63–64.

MacKinnon, C. D. and Winter, D. A. (1993) Control of whole body balance in the frontal plane during human walking. *J. Biomech.*, **26**, 633–644.

Murray, M. P., Kory, R. C. and Clarkson, B. H. (1969) Walking patterns in healthy old men. *J. Gerontol.*, **24**, 169–178.

Owings, T. M. and Grabiner, M. D. (1995) Submaximum stimulus intensity increases maximum strength in the elderly. In: *Conference Proceedings American Society of Biomechanics*, 19th Annual Meeting (ed. K. R. Williams), Stanford University, Stanford, CA, pp. 261–262.

Poulin, M. J., Vandervoort, A. A., Paterson, D. H., Kramer, J. F. and Cunningham, D. A. (1992) Eccentric and concentric torques of knee and elbow extension in young and older men. *Can. J. Sport Sci.*, **17**, 3–7.

Prince, F., Winter, D. A., Steriou, P. and Walt, S. E. (1994) Anticipatory control of upper body balance during human locomotion. *Gait Posture*, **2**, 19–25.

Saunders, J. B. D. M., Inman, V. T. and Eberhart, H. D. (1953) The major determinants in normal and pathological gait. *J. Bone Joint Surg. (Am.)*, **35**, 543–558.

Schultz, A. B. (1992) Mobility impairment in the elderly: challenges for biomechanics research. *J. Biomech.*, **5**, 519–528.

Wang, H. and Olney, S. J. (1994) Relationships between alignment, kinematic and kinetic measures of the knee of normal elderly subjects in level walking. *Clin. Biomech.*, **9**, 245–252.

Williams, K. and Bird, M. (1992) The aging mover: a preliminary report on the constraints to action. *Int. J. Aging Hum. Dev.*, **34**, 241–255.

Winter, D. A. (1987) *Biomechanics and Motor Control of Human Gait*. University of Waterloo Press, Waterloo, Ontario.

Winter, D. A. (1989) Biomechanics of normal and pathological gait: implications for understanding human locomotor control. *J. Motor Behav.*, **21**, 337–355.

Winter, D. A., Patla, A. E., Frank, J. S. and Walt, S. (1990) Biomechanical walking pattern changes in the fit and healthy elderly. *Phys. Ther.*, **70**, 340–347.

Yack, H. J. and Berger, R. C. (1993) Dynamic stability in the elderly: identifying a possible measure. *J. Gerontol.*, **4**, M225–M230.

18

The Kinematics of Treadmill Locomotion in Space

W. E. THORNTON[1], P. R. CAVANAGH[2], F. L. BUCZEK[3],
M. J. BURGESS-MILLIRON[4] AND B. L. DAVIS[5]

[1]701 Coward's Creek Road, Friendswood, Texas, USA
[2]The Center for Locomotion Studies, Pennsylvania State University,
University Park, PA, USA
[3]Motion Analysis Laboratory, Shriner's Hospital for Children, 1645 West 8th Street,
Erie, PA, USA
[4]Converse Inc., North Reading, MA, USA
and
[5]Department of Biomedical Engineering, Lerner Research Institute,
The Cleveland Clinic Foundation, Cleveland, OH, USA

INTRODUCTION

Human adaptation to microgravity includes decreased cardiorespiratory capacity, neuromuscular changes, muscle atrophy, and bone demineralization. Such adaptations may interfere with or even preclude many normal functions on return to the earth or other planets with significant gravity. Effects on landing range from transient discomfort and disability following missions of a few days, to inability to walk and significant losses of bone mineral (particularly in the lower limbs) following flights that last months (Figure 1).

It is now well accepted that exercise countermeasures must play a major role in preventing a number of adverse sequelae of prolonged exposure to microgravity (Thornton, 1990). Different forms of exercise have been employed,

Three-dimensional Analysis of Human Locomotion. Edited by P. Allard, A. Cappozzo, A. Lundberg and C. Vaughan
© 1997 John Wiley & Sons Ltd. ISBN 0 471 96949 4

Figure 1. On returning from an extended mission, a traveller may experience both temporary and chronic symptoms. (© King Features Syndicate. Reprinted with special permission of King Features Syndicate.)

ranging from walking on a slippery sheet of teflon, to tethered locomotion on a passive treadmill (Moore, 1990; Thornton and Rummel, 1977). US astronauts have not yet exercised on a motorized treadmill in 0 G, although Russian cosmonauts use a treadmill that has both passive and active modes, and the tension in the bungee tethers is nominally set at about 500 N. This device remains the centrepiece of their countermeasures for long-duration flights, including those that exceed 1 year.

In this study, we have used film from two shuttle missions to answer some initial questions concerning the kinematics of lower extremity joints and body segments during tethered treadmill running in space. The primary aim was to contrast passive treadmill running in 0 G with uphill running on a passive treadmill in 1 G. A secondary aim was to compare the kinematics of tethered running on a passive treadmill in 0 G and published data on uphill running on a motorized treadmill in 1 G (Milliron and Cavanagh, 1990).

METHODS

EXERCISE TREADMILL

The treadmill used for many of the shuttle flights, including STS-7 and STS-8 (the missions on which data that will be described in this chapter were collected), was designed with a 0.7-m running surface (Figure 2). The track on which the astronauts walked or ran consisted of a series of transverse aluminium plates linked together. Treadmill resistance was minimized by mounting the plates on precision ballbearings that travelled in a machined guide. Traction between an astronaut's feet and the track was possible because of the tethered arrangement in which four bungee cords were attached to a harness worn by the astronaut. These cords were adjustable and were connected to padded straps in the hip and

1. Track
2. Pulleys
3. Flywheel
4. Brake
5. Speed Control
6. Speedometer
7. Control
8. Generator

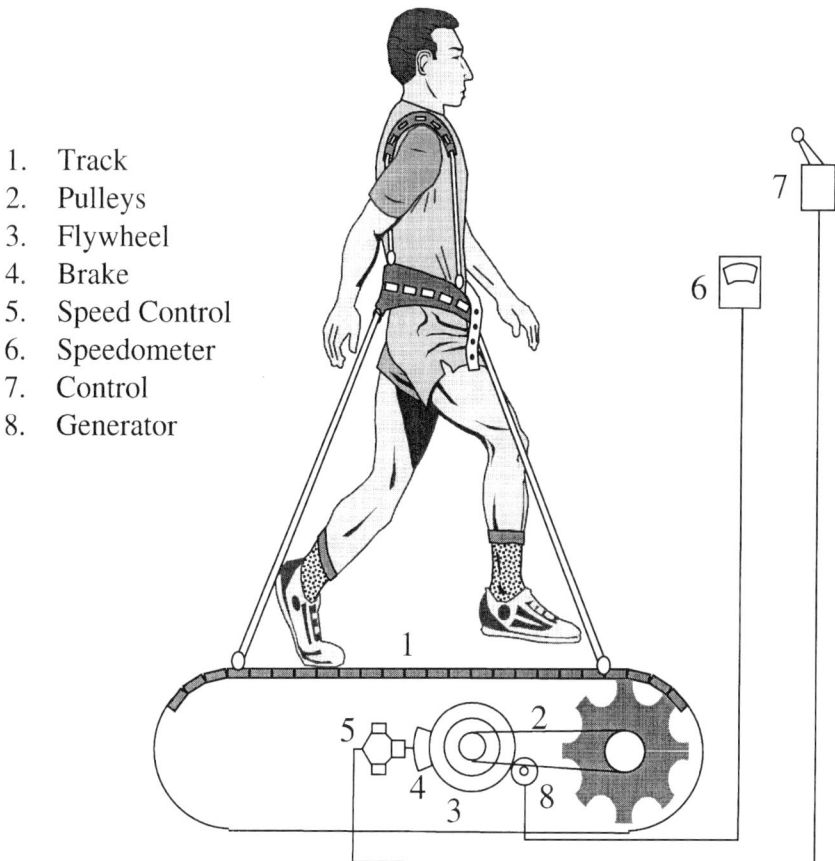

Figure 2. Passive treadmill used on NASA mission STS-7 and STS-8

shoulder regions in such a manner that the shoulder loads were approximately 35–40% of the total axial load. (Although the total tension was not measured, it was expected that loads equivalent to body weight would be generated by the tethering system.) To minimize fluctuations in the bungee tensions (ΔF) during the periods of exercise, the bungee cords were designed such that their length (L) was large compared to the oscillations (ΔL) of the upper body (i.e. since $\Delta F = K\Delta L/L$ (where K is a stiffness constant), ΔF can be minimized by increasing L).

TRIAL CONDITIONS

The primary material used in this study was 16-mm film of tethered treadmill running taken during the STS-7 and STS-8 missions (Figure 3). The camera had a nominal rate of 24 Hz (in 1 G) and a 5-mm focal length lens. Operational conditions (in flight) required that the camera be hand-held at a location approximately 1 m from the side of the treadmill. In an attempt to more fully understand the relationship between (1) 0 G and 1 G gait and (2) passive and motorized treadmill running, two other trial conditions were investigated. In the first of these, a single astronaut who was one of the four subjects for whom film in 0 G was available walked and ran in 1 G on a treadmill similar to one used on the shuttle. For these data, the same camera was used and was placed on a tripod 1.12 m from the treadmill. The orientation of this treadmill in 1 G was such that

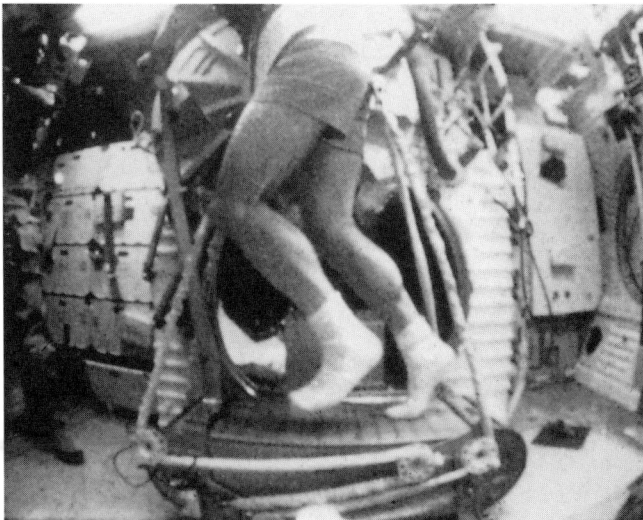

Figure 3. A single frame taken from the video recordings of astronauts exercising on the shuttle treadmill. Note the distortion caused by the camera lens. This distortion was removed by using methods described in Woltring (1980)

the subject exercised at a positive (uphill) inclination of 16.3% at speeds that were governor-controlled and set at either 1.4, 1.9 or 2.3 m/s. The second data set used for comparison was from a group of distance runners running at a fixed speed (3.4 m/s) on a motorized treadmill (Quinton) at both a level and uphill inclination of +20% (Milliron and Cavanagh, 1990).

COORDINATE DATA

All film from the shuttle treadmill (both 0 G and 1 G) was collected using a 5 mm lens, due to the restricted dimensions of the cabin. This resulted in considerable distortion in the final image. A solution to this problem was obtained by using the method described by Woltring (1980) to obtain a transformation between image space and object space.

Before the 1 G shuttle treadmill filming was conducted, a grid of known dimensions was exposed using the same camera and lens and subsequently analysed. This allowed the distortion parameters to be derived as well as the scaling factor for converting image-space coordinates into object-space coordinates. No such direct measures were available for the 0 G films, but since the camera lens was the same as for the 1 G situation, the same distortion parameters were used for both the 1 G and 0 G cases. A separate scaling factor was used for the 0 G films and was obtained using the known dimensions of the treadmill's track length, and since this was visible in all the film trials, separate scaling factors could be applied to each film sequence (since the camera position was not fixed).

Six body landmarks were used to identify segments in the films: (1) superior border of the greater trochanter, (2) lateral femoral epicondyle, (3) lateral malleolus, (4) posterior aspect of the heel, (5) head of the fifth metatarsal, and (6) most distal point of the second toe. Since no surface markers were placed on the subjects, positions of the landmarks were estimated. This was most difficult for the fourth marker in the 1 G trials, since the shoes that were worn were not as clearly demarcated as the socks or slippers that were worn in the 0 G trials. Marker 1 was also sometimes difficult to identify, due to the hip region being covered by relatively 'baggy' shorts.

All of the data obtained from the shuttle treadmill were digitally filtered with a cut-off frequency of 4.5 Hz. This frequency was the highest allowable, given the fact that the film data were collected at a rate of only 24 Hz (Winter, Sidwall and Hobson, 1974).

COORDINATE AXES

Since the primary purpose of this study was to relate 0 G treadmill locomotion to that in 1 G, a consistent axis system was required. In orbit, 'horizontal' and 'vertical' have little meaning, and for this reason a reference system relative to the treadmill bed was employed. In this treadmill axis system (TAS), the X-axis was parallel to the running surface, with the Y-axis being orthogonal to it. To allow comparisons with 0 G kinematics, the 1 G data were also transformed to this reference frame. It is emphasized that this transformation means that the 1 G data no longer reflect a gravitational reference system. In the case of the trials conducted on the shuttle treadmill in a 1 G environment at an incline of 16.3%, the gravitational force vector actually acted at an angle of about 9° to the Y-axis.

LIMB SEGMENT AND JOINT ANGLES

Due to the physical set-up of the treadmill in the space shuttle, only the left side of the astronauts was filmed. The lower extremity segment angles were considered raw data, used primarily in combination to provide desired joint angles. A segment angle was defined as the counter-clockwise angle it made with the positive X-axis, when the segment was rotated about its distal endpoint. Joint angles (hip, knee and ankle) were defined according to the convention shown in Figure 4.

The customary method for performing kinematic analyses is to make use of an initial 'standing calibration'. This involves measuring the locations of markers on a subject's lower extremity when a limb is in a neutral (upright) position. The advantage of this approach is that by relating subsequent marker positions to the initial location, it is a simple matter to determine whether the joint angle has

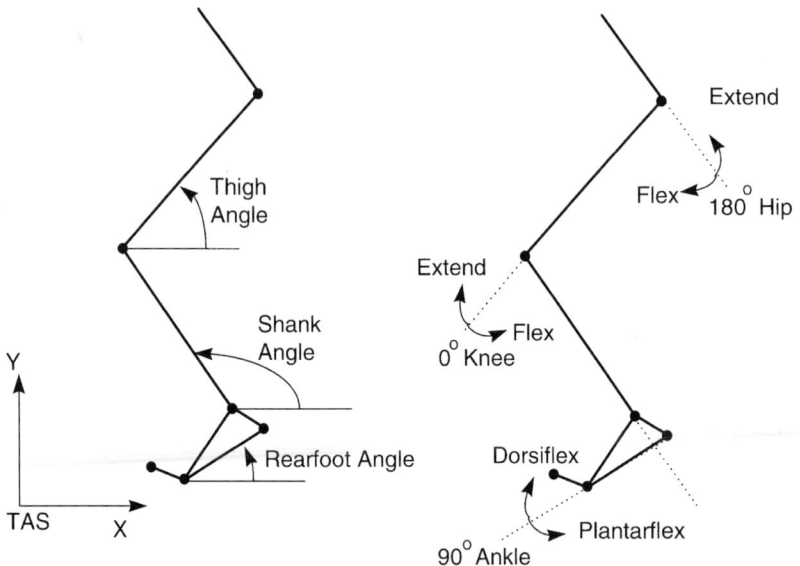

Figure 4. Conventions for determining segment and joint angles. Joint angles of 100° correspond to a flexed hip, a flexed knee and a dorsiflexed ankle

remained constant or whether flexion–extension has occurred. Without the 'standing calibration', it might appear as though the joint is flexed, whereas this may just be a result of marker location rather than anatomical factors. In the present case, since no film was available for any of the astronauts in a standing position, a point in midstance of walking was chosen that came closest to the orientation of the foot and ankle in standing. Segment lines were drawn between the 5th metatarsal head and the heel, and an angle was determined that adjusted the calculated foot segment to zero during the time that it was flat on the treadmill bed.

In a similar way, an offset angle was calculated for the trunk. If a line is drawn upwards from the hip marker, through the centre of the partially visible upper body, the angular deviation of this line from the Y-axis constitutes the trunk offset angle. This procedure was performed for each frame and it allowed the calculation of the hip angle, rather than just the orientation of the thigh segment relative to the Y-axis. For the 1 G standard treadmill data, standing calibrations were available. Correction angles were specified for the thigh, shank and rearfoot segments. The trunk offset angle was taken from the literature (Yoneda et al., 1979). The conventions used to determine joint motion (after applying the correction angles) are shown in Figure 4. All of these assumptions are typical of the compromises needed to analyse data collected under operational conditions where the rigour that can be imposed in a laboratory setting is not possible.

RESULTS

Before presenting the quantitative kinematic data, it is appropriate to present some observations based on a subjective view of the film and known biomechanical conditions. A passive treadmill must be 'driven' by the subject generating a force along the tread that must overcome the treadmill track's frictional resistance. On earth this is most easily done by elevating the treadmill, thereby producing a backward horizontal force that is a function of both the elevation angle and the subject's weight. In weightlessness, this force is a function of the subject's angle with respect to the tread and the bungee load. Also, as speed increases, so do the frictional losses and there is a concomitant increase in the degree of forward lean. Although the precise inclination of the upper body could not be determined in the film analysed in this study (due to most of the trunk being out of the view of the camera), it was apparent that the astronauts did exhibit marked forward lean which increased with speed. This style of locomotion was associated with a footstrike pattern that was characterized by an extreme forefoot strike—as opposed to the rearfoot strike seen in typical heel–toe running. It was evident that the handle on the treadmill was important for stability during gait, especially for the faster runs. The relationship between forward lean, support of the handle and footstrike pattern will be discussed later.

An initial, unexpected, result was that there was a discrepancy between the treadmill track speed as measured using the video camera and the speed (in 1 G) as set with a governor-controller. An explanation for this discrepancy could be that the camera's frame rate was higher than the nominal rate (24 frames/s). The camera's specifications stipulate that the frame rate can have an uncertainty of ±10%. If the upper limit of this range is used, then the three settings for the treadmill in 1 G (1.4, 1.9 and 2.3 m/s) are in better agreement with the speeds of 1.38, 1.86 and 2.07 m/s listed in the right-hand column in Table 1. For the remainder of this chapter it will thus be assumed that the camera had a frame rate of 26.4 frames/s rather than the nominal 24 frames/s. However, for completeness, both the nominal and 'fast camera' data are presented in Table 1.

The results describing the stride parameters for 0 G and 1 G conditions are given in Table 1. The most notable factor in the 0 G shuttle treadmill data was the low speed for both walking (0.75–1.04 m/s) and running (1.05–1.61 m/s). Even the trials that were designated as fast runs were only at speeds between 1.93 and 2.09 m/s. During locomotion on a similar treadmill in 1 G, the walking speeds were 1.38 and 1.86 m/s and the running speeds were 1.86 and 2.07 m/s. For comparison, typical distance running speeds (Cavanagh and Kram, 1990) are

Table 1. Stride parameters for locomotion on a passive treadmill in 0 G and 1 G

Activity	Stride length (m)	Stride length (stat)	Nominal speed		Speed based on a faster camera (m/s)	Stride frequency based on a faster camera (Hz)
			m/s	stat/s		
Walk (0 G)	0.83	0.47	0.69	0.39	0.75	0.90
	1.22	0.67	0.93	0.51	1.02	0.84
	1.02	0.56	0.88	0.48	0.97	0.95
	1.14	0.64	0.95	0.53	1.04	0.91
Run (0 G)	0.96	0.55	0.96	0.55	1.05	1.09
	0.98	0.52	1.06	0.57	1.16	1.18
	1.14	0.64	1.3	0.73	1.43	1.25
	1.15	0.63	1.46	0.8	1.16	1.40
	1.16	0.66	1.38	0.79	1.52	1.31
Fast run (0 G)	1.14	0.64	1.76	0.99	1.93	1.69
	1.31	0.72	1.8	0.98	1.98	1.51
	1.27	0.69	1.9	1.03	2.09	1.65
Walk (1 G)	1.15	0.63	1.25	0.68	1.38	1.20
	1.37	0.74	1.69	0.91	1.86	1.36
Run (1 G)	1.25	0.67	1.69	0.91	1.86	1.49
	1.35	0.73	1.88	1.02	2.07	1.53

The 12 rows in 0 G correspond to four astronauts walking or running at three different speeds. The trials in 1 G correspond to a single subject walking and running at two speeds between 3 and 4 m/s, while speeds of sprinting extend from 6 to 10 m/s. Thus the speeds of locomotion on the shuttle treadmill in 0 G were about 60% of those found using the same device in 1 G, while the latter were, in turn, about 60% of those typically found during overground distance running. Two factors that relate to velocity of running are stride frequency and stride length. Of these, it was found that stride length was less on the shuttle treadmill (almost certainly due to the small length of the running surface), whereas stride frequencies during running on the shuttle treadmill in 0 G (mean = 1.25 Hz) and in 1 G (mean = 1.51 Hz) were comparable to those reported for overground running in 1 G (range 1.38–1.44 Hz) reported by Cavanagh and Kram (1990).

The relationships between hip, knee and ankle motions during fast running in 0 G (mean velocity = 2 m/s) and running in 1 G (velocity = 1.97 m/s) are shown in Figures 5 to 8. The use of these kinds of angle–angle diagrams has been well described in the literature (Lamoreux, 1971). The advantage of these plots is that they are presented as a shape which is extremely sensitive to changes in an individual's style of locomotion (Milliron and Cavanagh, 1990), and they are not sensitive to errors in a camera's frame rate.

DISCUSSION

It has previously been shown that the kinematics of both running and walking gait are dependent on speed (Andriacchi, Ogle and Galante, 1972; Lamoreux, 1971; Sinning and Forsyth, 1970). In addition, it has been shown that running in simulated microgravity on a passive treadmill differs from running on a

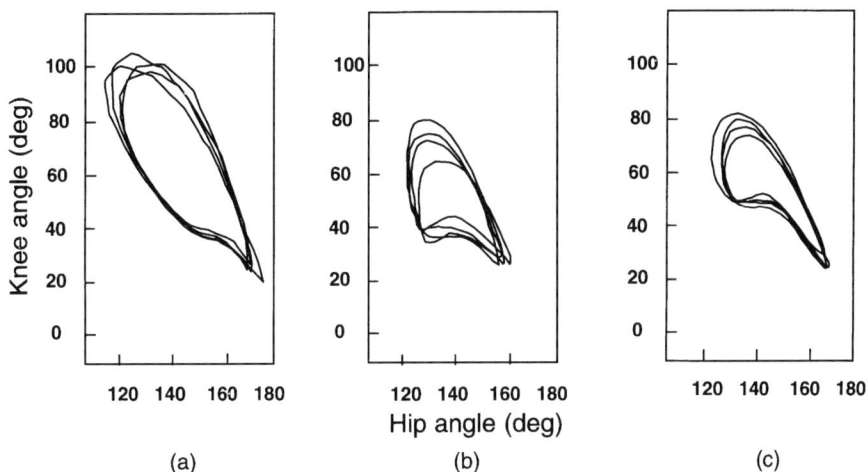

Figure 5. Hip–knee angles during fast running on shuttle treadmill in 0 G. Each panel refers to a different astronaut (a = 1.93 m/s, b = 1.98 m/s, c = 2.09 m/s)

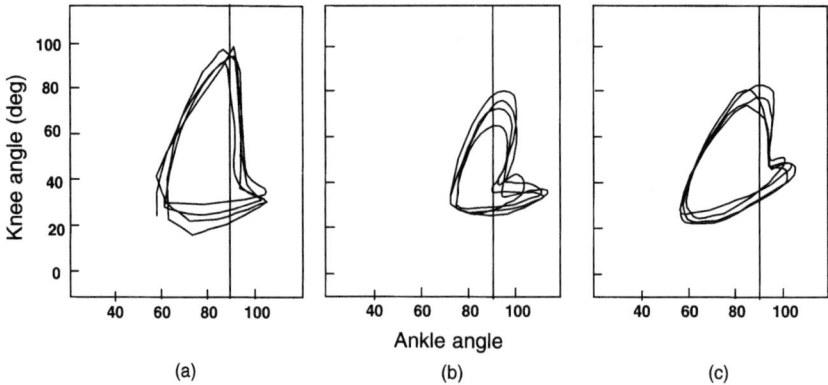

Figure 6. Ankle-knee angles during fast running on shuttle treadmill in 0 G. Each panel refers to a different astronaut (a = 1.93 m/s, b = 1.98 m/s, c = 2.09 m/s)

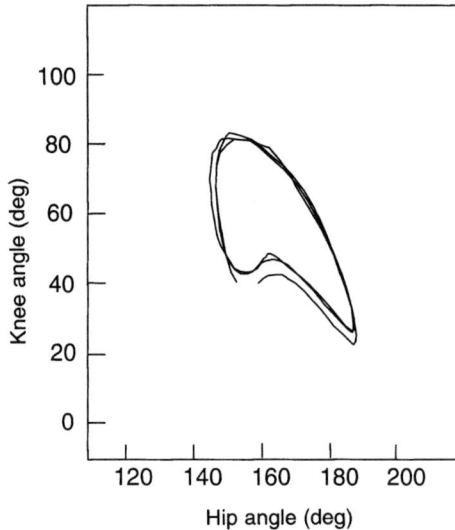

Figure 7. Hip–knee angles during running on shuttle treadmill in 1 G. (Single subject: grade = +16.3%, velocity = 1.86 m/s). Note the difference in hip angles compared to Figure 5c

motorized treadmill (Davis et al., 1996). These issues need to be taken into account when comparisons are made between gait on the shuttle treadmill in 0 G and 1 G and when trials in 0 G are compared to running on a standard motorized treadmill at various inclinations in 1 G (Milliron and Cavanagh, 1990). It should be emphasized that running in 0 G resulted in velocities that were about 60% of those achieved in 1 G. For this reason, emphasis will be placed on comparisons

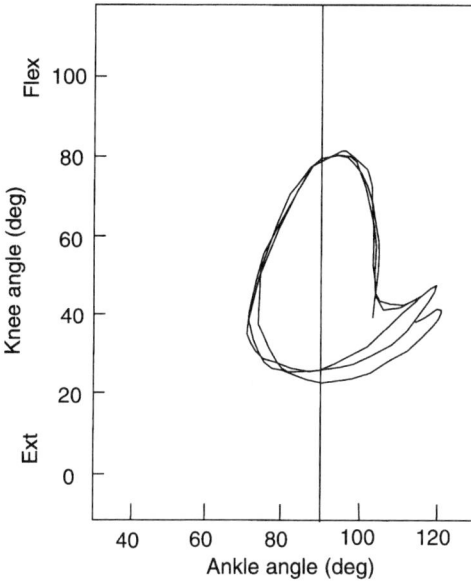

Figure 8. Ankle–knee angles during running on shuttle treadmill in 1 G. (Single subject: grade = +16.3%, velocity = 1.86 m/s)

between the fastest trials in 0 G and the slowest running trials in 1 G. The discrepancy between 0 G and typical motorized treadmill locomotion was even greater—the highest speed of 2.09 m/s in the former is markedly less than speeds reported in the literature (Milliron and Cavanagh, 1990). However, it would not have been practical to expect subjects to run on a motorized treadmill in 1 G at a velocity of about 2 m/s, since this speed would result in a very unnatural running gait. Likewise, it seems that the environmental and physical constraints that affect locomotion in 0 G preclude astronauts from running at speeds between 3 and 4 m/s.

Figure 5 shows the data of 0 G fast running. Apart from the increased knee flexion during the swing phase for one subject, there were general similarities between all the trials—for instance, it is apparent that neither the knee nor the hip joints ever became fully extended. This style of locomotion was probably a result of a variety of factors, e.g. the tension in the bungee cords that restrained the astronauts to the treadmill, the limited length of the running surface (Davis and Cavanagh, 1993), the passive nature of the treadmill, and the fact that there was a handle that offered support to the upper body. In leaning forwards, the astronauts were not only better able to grasp the handle, but, in addition, there was greater stretch in the posterior cords, which undoubtedly produced forces that made it easier to drive the treadmill track.

While horizontally directed forces may be what an astronaut focuses on as he or she exercises on a passive treadmill, what is probably of most importance to the musculoskeletal system is the degree of axial loading. While this has never been quantified during an actual NASA mission, indirect measurements using accelerometers mounted in the spacecraft structure (Dunbar, Giesecke and Thomas, 1991) suggest that treadmill exercise results in loads being applied to the lower limbs that far exceed the loads encountered during any of the other activities of daily living in microgravity. Since the accelerations that result from treadmill exercise are of concern to scientists who utilize the unique environment of 0 G for conducting their experiments (e.g. researchers who study crystal growth in weightlessness), considerable attention has recently been given to isolating treadmill vibrations from the shuttle superstructure. The emphasis in these efforts has been to design a device that transmits impact loads to the shuttle below NASA specifications and yet allows 1 G equivalent loading on the human body. The latter factor is considered important in light of work showing that impulsive loading to the lower extremity can stimulate bone formation as high as the cervical vertebra (Burr, Martin and Martin, 1983).

By selecting one trial for a single astronaut, it is possible to compare 0 and 1 G conditions at similar speeds (2.09 and 1.86 m/s respectively). Figures 5c and 7 give these results for hip–knee coordination, while Figures 6c and 8 relate to the motion at the knee and ankle. The hip–knee diagram shows that the knee joint has very similar patterns of motion, whereas the hip angle differs considerably with regard to the region where motion occurred. In 0 G the hip had a minimum flexion of 160°, while the 1 G data show a value close to 190°. The maximum flexion in the 1 G condition was about 120°, while in 0 G it was only 145°. Thus although the ranges of motion were comparable, the movement occurred in different parts of the available range. This was not surprising, since subjective analysis showed the subject to be leaning far more during the 0 G trials. With regard to the ankle–knee diagrams, it is apparent that the ankle was positioned in plantarflexion for most of the cycle (Figure 6). This was typical for most of the trials in 0 G, with only the data for one subject (Figure 6b) displaying comparable amounts of dorsiflexion and plantarflexion. The reason for this pattern was once again due to the marked forward lean which resulted in subjects running on their toes. Running on the passive treadmill in 1 G (Figure 8) produced similar kinematics to those shown in Figure 6, with the main difference being that there was more dorsiflexion of the ankle (i.e. maximum of 120°).

In the case of running on a motorized treadmill in 1 G, hip–knee and ankle–knee diagrams (Milliron and Cavanagh, 1990) allow one to ascertain that a +20% grade produces kinematics that most closely resemble the fast-run 0 G trials. The similarity in the support phase is considered most important, where, in both cases, the knee was somewhat flexed at footstrike and there was only a small amount of knee flexion after footstrike before the 'extensor thrust' phase of coordinated knee extension and hip extension began. In the case of ankle–

knee diagrams (Figures 5 and 8), the 20% uphill condition again provided the closest match.

CONCLUSIONS

Locomotion on a treadmill in 0 G will probably remain a centrepiece of NASA's exercise countermeasures programme. This form of physical activity has the potential to cause large bone and muscle forces as well as substantial metabolic loading during a period of continuous treadmill exercise. A critical concern is the provision of a treadmill which can approximate 1 G performance in space. At this point, no adequate objective measurements of in-flight treadmill kinetics or of the human response to this activity have been made.

Interpretation of the results obtained in the present study is limited by the following: (1) bungee tensions were not measured; (2) ground reaction forces were not measured in parallel with the kinematic measurements; and (3) the instrumentation used to film the astronauts could itself have been affected by microgravity. Despite these shortcomings, what is apparent is that exercise during NASA missions STS 7 and STS 8 resulted in leg motions that were similar to those found during 1 G locomotion on an inclined passive treadmill and on an active treadmill at an even steeper grade. In addition, it was apparent that the majority of the loads were transmitted through the forefoot, and one can surmise that this style of running would result in physiologically significant tensions in the calf musculature and resultant ankle compressive loading. Further speculation regarding limb loading is complicated by the fact that varying amounts of force are transmitted through (1) the treadmill handle and (2) bungee cords that act as a tether.

New generations of treadmills are being manufactured that could provide important information for planners of long-duration space missions. If these types of treadmill are flown on future missions, it will be possible to control bungee tensions more precisely, control for grade and speed, and, most importantly, provide data on the rates and magnitudes of limb loading. These data could then be incoporated into biomechanical models of the lower limb to more fully understand mechanisms of load transmission from distal to proximal structures and to optimize in-flight exercise protocols in such a way that muscle and bone loss could be reduced.

ACKNOWLEDGEMENT

Supported in part by a grant to the Penn State Center for Locomotion Studies from Krug International.

REFERENCES

Andriacchi, T. P., Ogle, J. A. and Galante, J. O. (1972) Walking speed as a basis for normal and abnormal gait measurements. *J. Biomech.*, **10**, 261–268.

Burr, D. B., Martin, R. B. and Martin, P. A. (1983) Lower extremity loads stimulate bone formation in the vertebral column: implications for osteoporosis. *Spine*, **8**(7), 6781–6786.

Cavanagh, P. R. and Kram, R. (1990) Stride length in distance running: velocity, body dimensions, and added mass effects. In: *Biomechanics of Distance Running* (ed. P. R. Cavanagh), Human Kinetics, Champaign, IL, pp. 35–63.

Davis, B. L. and Cavanagh, P. R. (1993) Simulating reduced gravity: a review of biomechanical issues pertaining to human locomotion. *Aviat. Space Environ. Med.*, **64**, 557–566.

Davis, B. L., Cavanagh, P. R., Sommer, H. J. III and Wu, G. (1996) Ground reaction forces during locomotion in simulated microgravity. *Aviat. Space Environ. Med.*, **67**(3), 235–242.

Dunbar, B. J., Giesecke, R. L. and Thomas, D. A. (1991) The microgravity environment of the space shuttle Columbia payload bay during STS 32. *NASA Technical Paper* 3141, NASA, Washington DC.

Lamoreux, L. W. (1971) Kinematic measurements in the study of human walking. *Bull. Prosthet. Res.* **10**(15), 3–84.

Milliron, M. J. and Cavanagh, P. R. (1990) Sagittal plane kinematics of the lower extremity during distance running. In: *Biomechanics of Distance Running* (ed. P. R. Cavanagh), Human Kinetics, Champaign, IL, pp. 65–105.

Moore, T. P. (1990) The history of in-flight exercise in the US manned space program. In: *Workshop on Exercise Prescription for Long-duration Space Flight* (ed. B. A. Harris Jr. and D. F. Stewart), NASA Conference Proceedings 3051, NASA, Washington DC, pp. 19–21.

Sinning, W. E. and Forsyth, H. L. (1970) Lower-limb actions while running at different velocities. *Med. Sci. Sport Exer.*, **2**(1), 28–34.

Thornton, W. E. (1990) Work exercise and spaceflight: II Modification of adaptation by exercise (Exercise prescription). In: *Workshop on Exercise Prescription for Long-duration Space Flight* (ed. B. A. Harris Jr. and D. F. Stewart), NASA Conference Proceedings 3051, NASA, Washington DC, pp. 107–115.

Thornton, W. E. and Rummel, J. (1977) Muscular deconditioning and its prevention in space flight. In: *Biomedical Results from Skylab* (ed. R. S. Johnston and L. F. Dietlein), NASA SP377, NASA, Washington DC, pp. 191–197.

Winter, D. A., Sidwall, H. and Hobson, D. (1974) Measurement and reduction of noise in kinematics of locomotion. *J. Biomech.*, **7**, 157–159.

Woltring, H. (1980) Planar control in multi-camera calibration for 3-D gait studies. *J. Biomech.*, **13**, 39–48.

Yoneda, Y., Adrian, M., Walker, F. and Dobie, D. (1979) Kinematic and kinetic analysis of sprinting and jogging. In: *Science in Athletics* (ed. J. Terauds and G. Dales), Academic Publishers, Del Mar, CA pp. 85–91.

19

Gait Data: Reporting, Archiving and Sharing

ROY B. DAVIS[1], SYLVIA ÕUNPUU[2] AND PETER A. DELUCA[2]

[1]Newington Children's Hospital, Newington, Connecticut, USA
[2]Connecticut Children's Medical Center, Hartford, Connecticut, USA

INTRODUCTION

There are no universally accepted standards or conventions for the reporting, archiving and sharing of gait analysis data. 'Gait analysis' refers to a variety of approaches that seek to qualify and quantify human locomotion. These methods range from the qualitative observations made as the subject walks down the clinic hallway or laboratory pathway to the full three-dimensional (3D) quantitative analysis of the instrumented subject based on a collection of measured values. 'Gait data' refers to a myriad of parameters that results from these evaluations that quantify and/or qualify certain aspects of locomotion. The term 'data' as used here refers not only to quantities that are measured directly, but also to the values that are computed based on these measurements. For example, distance and time measurements (data) are often combined to indicate velocity (technically 'information', but considered as 'data' in the discussion that follows). Gait reporting approaches vary depending upon the type of data collected and the motivation of the report, e.g. prospective research versus clinical interpretation. Strategies for archiving gait data are also influenced by the reasons for retaining the data, e.g. the recording of a few variables for a particular project versus the storage of all available data for comparison of gait characteristics pre- and post-treatment. The motivations for gait data sharing also dictate to some degree the approach taken. For example, it is relatively simple for a pair of individual investigators to implement a strategy for combining data sets for research

Three-dimensional Analysis of Human Locomotion. Edited by P. Allard, A. Cappozzo, A. Lundberg and C. Vaughan

purposes, but quite a challenge to develop an inter-institutional database for more general clinical and research use. The absence of standards for gait data reporting, storage and sharing means that individuals are free to adopt specific approaches that simply address the particular issue(s) at hand. The difficulty, of course, is that without a standard approach in these matters, expansion of the common gait knowledge base is retarded due to communications difficulties.

The spectrum of 'gait data' is quite broad and strategies for its reporting, archival and sharing vary. The scope of this chapter is limited to those gait data that are commonly collected, computed and provided for the interpretation of the gait-related problems of patients in the clinical setting. These gait data are provided through the quantitative measurement of the patient as she or he walks along a straight, level pathway and is monitored with respect to body segment position and orientation, ground interaction and muscle activity. Aspects of gait data reporting, storage and sharing will be presented from a perspective based on 15 years of experience in clinical gait analysis at the Connecticut Children's Medical Center (formerly the Newington Children's Hospital). The technology employed at this facility consists of an optoelectric motion measurement system (Oxford Metrics, Limited, Oxford, UK) and force platforms (Advanced Mechanical Technology, Incorporated, Newton, Massachusetts, USA). The clinical gait analysis protocol that forms the basis for this current work will be described to the degree necessary, but can be found in more detail elsewhere (Davis et al. 1991). The authors do not mean to imply that this particular approach is most proper or correct; it merely provides a vehicle for the illustration and discussion of particular points.

GAIT DATA

The spectrum of gait data associated with clinical gait analysis may be illustrated by chronologically outlining a typical gait evaluation. The entire process may be divided into a number of steps that include:

1. referral data acquisition, including medical history;
2. measurement system calibration and validation;
3. gait assesment by observation (recorded on videotape);
4. physical examination;
5. subject instrumentation and calibration;
6. motion measurement;
7. electromyographic (EMG) measurement;
8. metabolic energy consumption assessment.

Steps 5 to 8 may be repeated as needed for the evaluation of the patient while using walking aids such as lower extremity orthoses and walkers. The decision to collect motion and EMG data simultaneously (steps 6 and 7) should be based on the need to correlate precisely gait biomechanics with muscle activity and a consideration of the amount of instrumentation that may be tolerated by the sometimes small subject before gait alteration becomes an issue (because of the equipment worn).

Table 1 (A) describes the reason(s) for the referral of a patient for gait analysis and documents the demographics of the patient for administrative purposes. It is fundamentally important to determine if the specific issues that motivate a referral can in fact be addressed by gait analysis. Identifying inappropriate referrals early not only saves time and money, but also serves to educate the referring clinician with respect to the indicators for gait analysis. Relevant medical history information varies with the patient diagnosis. For the child with cerebral palsy, for example, the associated medical history data might include gestation time, birth weight, respirator use immediately following birth, birth complication(s), age at which walking began, previous surgeries and treatments, use of walking aids and bracing, and major gait problem(s), such as tripping, excessive energy consumption and joint pain. Post-gait test data (see Table 1 (B)) allow one to evaluate the timeliness, fiscal efficacy and utilization of the service provided. These issues are important in the context of quality improvement a well as the relevance of clinical gait analysis.

Table 1. Patient referral and demographic data

A.	Pre-gait test data
	Patient name
	Medical record number
	Home address
	Telephone number
	Date of referral
	Gender
	Date of birth
	Diagnosis
	Referral source, e.g. name of referring physician
	Proposed treatment(s), e.g. surgery, bracing
	Referral purpose(s)
	Walking aid(s)
	Orthoses used
	Medical history
B.	Post-gait test data
	Date of analysis
	Date of test result interpretation
	Date of gait report completion
	Number of person-hours associated with the test
	Cost of the test, i.e. amount billed
	Reimbursement for the test
	Recommended treatment(s) based on the gait analysis
	Actual treatment implemented

Protocols for measurement system calibration and validation vary. Video-based motion measurement systems can require calibration before each test period. Calibration data to be stored as part of the test record include camera positions, orientations and optical characteristics (e.g. linearization parameters), a quantitative measure of the precision of the calibration exercise, and the coefficient array(s) needed to stereometrically reconstruct three-dimensional (3D) coordinates from two-dimensional (2D) camera images. Force platform calibration information is usually provided by the manufacturer, although application-specific data related to the position and orientation of each platform at the time of the test should be stored as part of the gait record. Additional force platform data acquisition attributes such as sampling rate, signal filtering, thresh-old values, signal amplitude and phasing offsets should also be included. Plantar pressure measurement systems require routine, sometimes subject-specific, calibration procedures to relate signal amplitude to applied pressure. EMG measurement technology is generally not calibrated *per se*, although particular electrode signals can be related to physiologically relevant references such as a maximal voluntary contraction or resting muscle activation level.

For commercially available measurement systems, calibration procedures and/or data are usually specified by the equipment vendor. Routine system validation protocols that may be applied before each clinical evaluation are not commonly available. DeLuzio et al. (1993) outlined a comprehensive procedure to evaluate the accuracy and precision of a motion measurement system in a project that focused on knee dynamics. The challenge, however, is to formulate a simple protocol that can be completed before each clinical patient test so that quantitative validation results can be included as part of each gait analysis record. Validation exercises should test each measurement subsystem, some simultaneously. In the approach described by Stanhope (1994), for example, force is applied to a marker-equipped rod while simultaneously monitoring the rod displacement and force platform output. Measurement system cali-bration (and validation) should be completed before the patient arrives in the laboratory.

Immediately after arriving in the laboratory, the patient is asked to walk along the test walkway while being recorded on videotape from two perspectives simultaneously. At this time, gait abnormalities are observed and information gathered from the patient and/or family member as to the perceived gait difficulties. Currently, this qualitative information and the video recording are not stored as part of the patient's electronic (digital) record, although both are used in the gait interpretation process. Certainly, this is one area in which electronic data archiving can be utilized more effectively, particularly as more efficient data compression algorithms become available.

The patient then undergoes a physical examination to record any gait-related neuromuscular and bony abnormalities as well as to collect data specific to the gait model used in the motion data reduction process. These clinical examination data include parameters that may be measured (e.g. body weight, leg length, hip extension) or graded subjectively (e.g. knee extensor strength) (Table 2).

After the clinical examination, the patient is equipped with passive reflective markers placed on each body segment under investigation (Davis, DeLuca and Õunpuu, 1995; Cappozzo et al., 1995). Underpinning the data reduction process is the computation of an anatomically aligned, orthogonal coordinate system for each of these body segments. These body-fixed coordinate systems differ from

Table 2. Clinical examination data

Body weight
Height
Right and left leg lengths
Right and left knee diameter
Right and left ankle diameter
Distance between right and left anterior superior iliac spine (ASIS)

Hips
   Maximum flexion/strength
   Maximum extension/strength with knee at 0° of flexion
   Maximum extension/strength with knee at 90° of flexion
   Straight leg raise
   Maximum abduction/strength with hips at 0° of flexion
   Maximum abduction with hips at 90° of flexion
   Maximum adduction
   Maximum internal rotation
   Maximum external rotation
   Femoral torsion
Knees
   Maximum extension/strength with hips at 0° of flexion
   Maximum extension with hips at 90° of flexion (unilateral)
   Hamstring tightness
   Maximum flexion/strength
   Ligamentous laxity
Ankles
   Dorsiflexion with knee at 0° of flexion
   Dorsiflexion/strength with knee at 90° of flexion
   Plantarflexion/strength
Forefeet
   Inversion/strength
   Eversion/strength
Hindfeet
   Inversion
   Eversion
   Foot/thigh angle
   Bimalleolar axis
Reflexes
   Ely test
   Gastrocnemius–soleus stretch reflex
   Confusion test

coordinate systems associated with the markers on the surface of the segment (except for perhaps the pelvis) and are referred to as 'technical coordinate systems' (Cappozzo, 1984). The geometrical relationship between the anatomical and technical coordinate systems for each segment is established with a subject calibration trial, during which the subject stands quietly while markers placed on the lower extremities (including additional markers, later removed for the walking trials) are monitored by the measurement system. One may then store either the angular and linear displacements between the two coordinate systems or the position of additional markers and joint centres relative to the corresponding technical coordinate system. This subject calibration information can then be included in the data set associated with each walking trial.

The data collected during the walking trials include the trajectories of the reflective markers and force platform and EMG signals (Figure 1). In the following discussion, these data are referred to as 'collected data'. Both the 2D motion camera image data and the 3D marker coordinates stereometrically reconstructed from the camera data and often interpolated to fill trajectory gaps are included in this category.

'Intermediate values' are then computed for each frame of collected data (commonly after digital signal processing, e.g. smoothing) that include, but are not limited to:

- a technical coordinate system for each body segment;
- an anatomical coordinate system for each body segment;
- hip, knee and ankle joint centre locations;
- centre of mass location, velocity and acceleration for each body segment;
- absolute and relative rotational displacement, velocity and acceleration of each body segment.

These intermediate values are then combined to produce 'reported results' for clinical interpretation that include 'kinematics', i.e. the angular displacement of each hip, knee and ankle as well as the absolute angular orientation of the torso, pelvis and foot over the gait cycle. When force platform data acquisition is possible, the net joint moments of force and associated joint power are generally provided for interpretation as well, and are commonly referred to as 'joint kinetics'. The EMG signals collected during the walking trial are often processed in a variety of ways (e.g. digital filtering, smoothing, full-wave rectification and ensemble averaging), sometimes compared to subject-specific reference values (e.g. maximal voluntary exertion), and then provided for clinical interpretation. Stride and temporal gait variables such as walking speed, cadence, step and stride length, single and double support time, and the percentage of the gait cycle spent in stance and swing phases, are also commonly reported. Other parameters that might be included in this category of reported results are estimations of metabolic energy consumption (e.g. steady-state oxygen consumption and energy cost), mechanical energy, joint forces, and muscle lengths and moment arms.

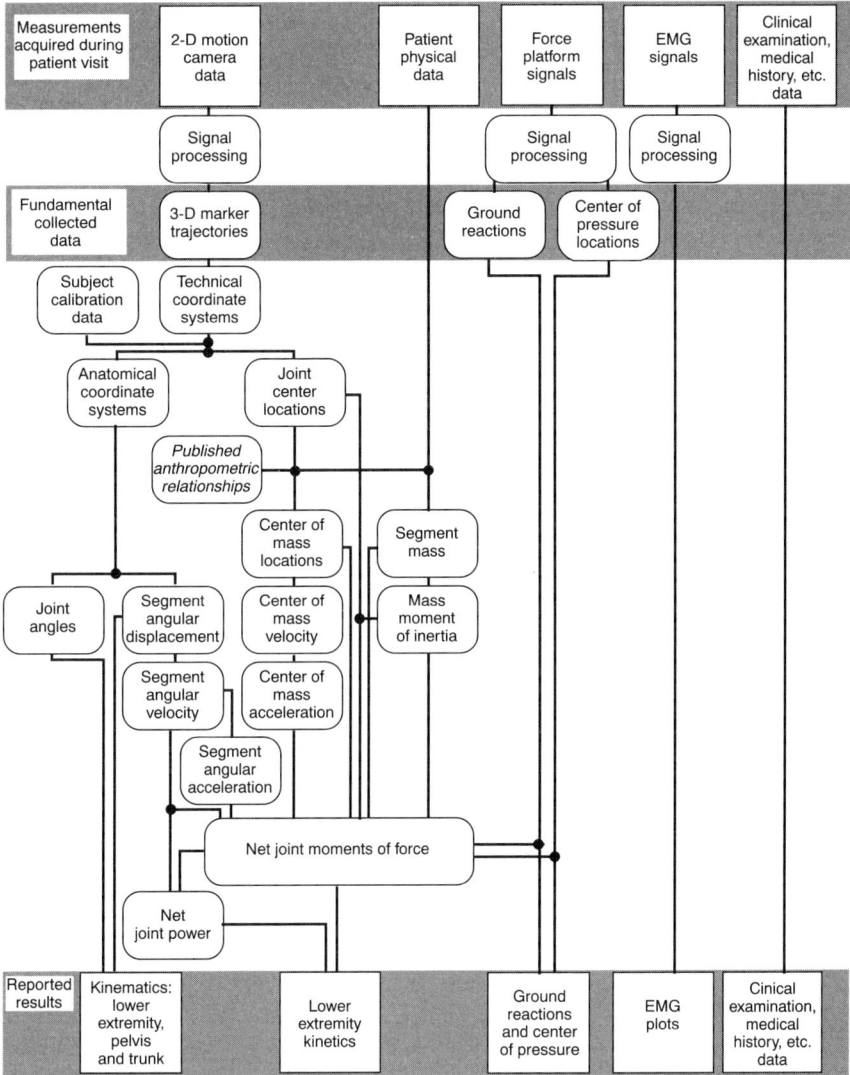

Figure 1. Gait data flow chart illustrating the measurements acquired during the patient visit, the fundamental collected data set, intermediate values, and results reported for clinical interpretation

GAIT DATA REPORTING

The fundamental objectives of the interpretation process are to identify the patient's gait attributes that are different from normal, postulate the possible causes of these abnormalities, and recommend treatment alternatives. The gait data that are made available routinely to assist in this clinical decision-making include:

- 3D angles over several gait cycles for each lower extremity joint, pelvis, and torso;
- 3D joint moments of force and power over several gait cycles for each lower extremity joint;
- Stride and temporal values such as walking speed and step length;
- Clinical examination measures;
- bi-plane video recordings;
- EMG recordings over several gait cycles of the major lower extremity muscles (or muscle groups), e.g. rectus femoris, vastus medialis, vastus lateralis, hamstrings, gastrocnemius/soleus, and tibialis anterior.

One of the fundamental problems of computerized gait analysis is the determination of effective data reporting formats that facilitate information integration and interpretation. Inconsistencies in the reporting of data have contributed to the relatively limited acceptance of gait analysis as a clinical tool. The complexity and volume of gait data further exacerbate this problem. The decisions that need to be made are: which pieces of data are relevant to the specific testing situation and how they should be presented so that all data types can be readily integrated. This section discusses some issues and ideas related to the reporting of gait data.

Before any interpretation of gait data takes place, certain background knowledge must be available to facilitate a comprehensive appreciation of the data. In the broadest sense, the interpretation team needs a well-developed understanding of normal gait mechanics that is augmented by clinical experience. With respect to a specific gait data set, the team needs to be informed of and understand the data collection and processing details specific to the laboratory where the gait data are acquired, e.g. the relationship between marker placement and resulting kinematic measures, joint and segment angle definitions, the fundamental aspects of the joint kinetic calculation, and possible sources of measurement and modelling errors. This contextual information is analogous to the 'methods section' in a published article and should be included in some way in the final gait analysis report.

To facilitate the systematic interpretation of gait analysis data, presentation formats should remain consistent between cases, i.e. organized in a repeatable manner with graphs employing established (although perhaps laboratory-specific) conventions. This enables the clinician to more readily find specific information, and consistent scaling and labelling of plots facilitates quicker comparisons between sides, conditions or persons. This encourages a pattern recognition approach by the interpreter that reduces the tendency to overinterpret the waveforms. The auto-scaling of plots impedes this process and should be avoided for these same reasons. Also, an indication of data associated with different testing conditions (e.g. barefoot versus brace testing conditions), side of the subject, and pre-treatment versus post-treatment should be readily identifiable.

Gait data can be presented as a function of time or more commonly as a percentage of the gait (or stride) cycle. Expressing data as a function of a percentage of a gait cycle facilitates the averaging of data from multiple strides for one individual and a group of similar subjects. This convention does eliminate explicit timing information that must then be obtained from temporal and stride parameters. Stance-phase (of the gait cycle) data are generally plotted before swing-phase data on each plot. The presentation of data associated with exactly one gait cycle does not facilitate the examination of the transition from swing phase to stance phase. This can be addressed by including some portion of the previous and next gait cycles in conjunction with the cycle of interest, e.g. the last 10% of the previous cycle and the first 10% of the next. Another approach that is sometimes elected is to normalize both the stance and swing phase to 60% and 40%, respectively. This method makes it difficult to quickly appreciate the stance/swing-phase ratio and symmetry and is not recommended for clinical data presentation. Data may also be reported as a function of time. This format facilitates an understanding of pelvic and torso motion, assuming that bilateral data are collected. Computing average results becomes more problematic with this approach, as does comparison of data corresponding to pre- and post-treatment or different sides.

Stride-to-stride variability for each type of data needs to be assessed before any particular stride is considered 'representative' of the patient's performance. Overlaid plots of multiple strides of kinematic or kinetic data and the plot of mean results along with the corresponding variability can both accomplish this. For EMG data, multiple (consecutive) stride data need to be plotted sequentially if the raw EMG signals are reported. It is recommended that individual walking trials be interpreted (as opposed to a patient's mean results), as this facilitates the correlation of related biomechanical events.

The identification of gait deviations is typically made by the comparison of a patient's results to those of the normal walker(s). Temporal and stride parameters are generally referenced to normal age-matched ranges. The normal reference data are generally represented in the gait literature by a mean curve in the case of joint kinematics and kinetics, and by bars for EMG. Representing the normal as a single mean curve does not allow the interpreter to appreciate the variability about the mean. Alternatively, presenting a band that represents the mean plus or minus one or two standard deviations will reveal the variability and reduce the potential for misinterpretation. That is, small deviations from normal and small differences between pre- and post-treatment are less likely to be interpreted as significant. The use of EMG bars for the normal reference does not allow the clinician to appreciate normal variability between subjects. This format also provides no indication of the possible gradual increase or reduction in amplitude of the EMG signal associated with the onset and cessation of activity. Representing normal EMG in linear envelope or ensemble average forms addresses some of these problems.

The presentation of temporal and stride data of normal ambulators is also important in the interpretation of gait data. Normal ranges should correspond to a specific patient height or leg length as opposed to age, as age is not always correlated with height. One should also be careful to appreciate differences due to gender, adolescent growth spurts, and height variations due to pathology. Normative data should be reported in conjunction with the pathological data for comparison purposes.

Segmental/joint kinematics, joint kinetics and EMG are the primary types of walking data provided in computerized gait analysis. The principal purposes of clinical gait analysis are to identify gait abnormalities and to determine the possible causes. This can be facilitated significantly by including the kinematics of the lower extremities, pelvis and trunk on one page so that within-plane interactions can be better understood. For example, severe crouch at the knee will result in the skewing of the hip motion into more flexion. This display will also assist in identifying cross-plane interactions, i.e. how motion in one plane results in deviations of motion in another. For example, crouch and internal hip rotation in stance may result in greater than normal adduction of the hips in stance.

Temporal and stride parameters are directly related to all types of data, i.e. increases in walking velocity are generally associated with elevations in EMG amplitudes, larger ranges of motion of joint kinematics, and greater peak joint kinetic magnitudes and changes in joint kinetic patterns with certain pathologies. Therefore, it is important to include temporal and stride data on all kinematic, kinetic and EMG output used to interpret changes between data collection trials or sessions. Pertinent temporal and stride parameters that should be included are: walking velocity, cadence, stride time, step and stride length, toe-off, and double support and single support times.

There are many ways of presenting EMG gait data, including the following formats: raw, full-wave rectified, linear envelope, and integrated. In any clinical interpretation of data, the raw EMG should be available for quality control, to evaluate the frequency content of the signal, and to identify the periods of muscle activity. Any additional processing impacts on one's ability to accomplish these objectives. The use of linear envelopes allows signal averaging, but smooths the signal. All the above signal formats (except integrated EMG) can be used to correlate EMG results with other types of gait data presented either as a percentage of the gait cycle or as a function of time. EMG data can also be presented in many ways with respect to magnitude. For raw and rectified EMG, magnitudes are presented in microvolts or volts. Linear envelope signals can be referenced to a known force through previous calibration trials, to the mean signal level, or to the maximum signal level in the gait cycle. The latter two approaches may distort amplitude information by either inadvertently magnifying a relatively small (amplitude) signal or attenuating a relatively large signal.

Joint kinetic data are among the most challenging data to interpret in computerized gait analysis. This is primarily because these data cannot necessarily be 'visualized' and requires some knowledge of a few mathematical and mechanical concepts. As joint kinematic and kinetic results are directly related, it is useful to arrange the kinematic, moment and power plots for a particular joint vertically in a column so that interrelationships in any specific phase of the gait cycle can be more readily appreciated. It is also useful to have these results for all lower extremity joints displayed on one page so that relationships across joints can be examined. For example, a greater than normal knee extensor moment in stance may be associated with a prolonged hip extensor moment in stance. In certain cases, pelvic and torso kinematics as well as plots of the displacement of the centre of pressure might also be useful in kinetic data interpretation.

Joint moment and power values are often referenced to subject body mass and possibly leg length to reduce intersubject variability. The shape or modulation of the curve does not change, but one must take care to note the convention before comparing specific plot magnitudes. Joint moments can also be expressed as internal or external moments. These representations both present the same biomechanical information and both are valid. One might argue that since the internal moment represents the body's response to the external load, it might therefore be more readily correlated by the clinician with the EMG data.

GAIT DATA ARCHIVING

Gait analysis data are stored to allow numerical processing after the testing period and thereby reduce the length of the patient visit. Data archiving also affords an opportunity to examine changes in the patient's gait characteristics that have occurred over time with growth or treatment, e.g. before and after surgery. By storing the data associated with multiple gait trials, one may compute mean results and quantitatively describe gait variability. Saving individual results also facilitates both prospective and retrospective clinical research by allowing the patterns associated with groups of patients to be compared and studied. The individual records collected and processed at one institution may also be combined with comparable sets of data from other institutions in multicentre research projects.

Arguably, data could be collected and processed immediately, thereby producing the 'reported results' (i.e. a hard-copy printout) without storing any of the data electronically. Practically, for the reasons cited above, however, this is not done. So, then, what gait data are to be stored? In the ideal world, one might store all of the gait data collected; this leads to redundancy with respect to both keeping multiple copies and retaining hard (paper) and soft (computer) copies of the data. If one assumes a goal of maintaining the patient archive on-line (e.g. on a computer hard disk) to facilitate prompt and less labour-intensive retrieval, then the constraints of the memory size of the secondary storage device and the seek and input/output times of the computer systems must be appreciated. For example, a typical gait test might yield approximately one megabyte of collected data (not including any digitized video images) plus another half megabyte for the intermediate and reported results. A testing volume of roughly 300 patients per year would then consume approximately one half of a gigabyte of storage space each year. One might argue that additional hardware (memory) is relatively inexpensive and becoming even more affordable as time goes on. The difficulty, however, is that as the amassed data archive grows in size, the capability of the computer system to find and retrieve particular files is reduced.

Perhaps the more appropriate approach is to ask: what portion of this gait data set needs to be stored? One approach is to store only the data collected during the clinical testing session, e.g. the motion camera images, force platform signals, and unprocessed (or raw) EMG signals. Alternatively, some or all of the intermediate values may be saved either in addition to or instead of the collected data. Finally, one might elect to archive the reported results, again in addition to, or instead of, the intermediate values and collected data.

The advantage of storing only the basic data acquired during the clinical session is that it offers the best potential to take advantage of new developments in data processing, ranging from image signal enhancement to biomechanical modelling of gait. That is, if an improved stereophotogrammetric technique emerges in the future, one may reprocess the camera images and realize a gain in the precision of the 3D marker locations. Similarly, future enhancements in filtering and smoothing strategies might improve the quality of the digital camera data and analog force platform and EMG signals. One is inclined also to expect that this data set represents the minimum size with respect to the amount of digital memory consumed, as any processing typically adds to the size of the data set. The disadvantage of this strategy is apparent in that it requires a complete reprocessing of the collected data each time they are needed in the future. The steps associated with constructing the identified 3D marker coordinate sets from the camera images require more user interaction than the remainder of the data reduction process. Hence, the approach of storing only this basic type of collected data is ill-advised, due to the repetitious labour it would require.

One might elect to save and keep on-line the most immediate products of the collected data reduction process, i.e. the 3D trajectories of the markers, ground force reactions, centre of pressure locations, and EMG values. The 2D camera data and raw force platform and EMG signals could be archived to off-line storage, such as magnetic tape, to be recovered if needed, e.g. if a marker identification error is detected. By saving only valid data, e.g. only frames that contain a 3D coordinate (real or interpolated) for all markers, additional storage efficiency is achieved. For example, in a typical walking trial with a 5-s data collection period, 6000 samples of data might be acquired from each of two (or more) force platforms. However, assume that only one of the platforms yields satisfactory data (i.e. appropriate foot placement) for 0.6 s. Then one might achieve a 94% saving in storage space by retaining only the 720 useful samples out of the 12 000 samples collected from the two platforms. In practice, this approach to data storage reduces the size of the data set associated with each clinical testing session at the Connecticut Children's Medical Center Gait Analysis Laboratory by approximately 95%, from about 1.2 megabytes to roughly 60 kilobytes, thereby increasing the on-line secondary storage capacity 10-fold. Moreover, by retaining this data at this fundamental level, one may still take advantage of improvements in gait modelling that may occur in the future. The disadvantage of this strategy is that substantial reprocessing is still required each time a data set is retrieved. The gait model data reduction process requires virtually no user interaction, however, and consequently data files may easily be recovered in a batch process. A most significant challenge relates to the fact that marker placement (and consequently a particular data set) is linked directly to a particular data reduction algorithm or gait model. In this way, the marker coordinate data set is gait model dependent. Care must be taken because the gait model used at a particular facility tends to change over time and may be quite different between institutions. This important point will be discussed further in the next section.

In addition to, or instead of, the storage of this fundamental data, one might choose to store other intermediate values and/or reported results. In general, two competing issues must be appreciated and balanced in this decision process. If intermediate values are stored after a number of data reduction steps, then these steps need not be repeated in the future when the data are retrieved. Also, if the collected data are discarded and only intermediate values are retained, then future improvements that might have been achieved in the production of those intermediate results are probably lost. For example, the advantage of saving only the reported results is that it both reduces storage requirements and obviates computational reprocessing during data retrieval in the future. This must be weighed against the relative rigidity of the approach, in that one cannot explore different gait model changes on these archived gait trials in the future. Moreover, with this strategy, one loses access to other quantities previously held as intermediate values, e.g. velocities and accelerations, that might become more important with respect to clinical gait interpretation in the future. Overall, it is probably ill-advised to discard completely all gait data except for reported results. In certain cases, however, when on-line storage space is limited, one might accept the on-line storage of certain intermediate values and reported results and off-line archiving of the collected gait data.

Certainly, there are other alternatives to the storage of particular gait variables in anticipation of gait data sharing, as outlined in the next section.

GAIT DATA SHARING

Quite often, considerations of gait data sharing focus principally on practical aspects such as file format compatibility and mode of data transfer. While these issues are important to the end goal of data transfer and sharing between facilities, they are relatively straightforward technically and can be dealt with in a variety of ways. Communications standards, improvements in technology, such as high-speed modems, and access to the Internet will lessen the relative importance of these issues in the future. This section will outline other concerns that are more difficult to deal with, specifically, what gait data are most suitable for sharing, maintenance of patients and institutional confidentiality, and database ownership.

As indicated in an earlier section, gait data are quite varied and wide-ranging. So, then, what gait data are most readily shared? Perhaps the place to begin is with the simplest scenario that involves two institutions or groups of investigators who want to combine the gait data associated with a particular group of patients to address a limited number of specific research questions. The most immediate issue technically is the compatibility of the gait data. 'Compatibility' used here refers to the equivalence of the measurements, and, to a lesser degree, the data reduction protocols used at the two laboratories. This applies to the way in which the clinical examination and patient physical data are collected, how the body segment markers are placed and aligned, the instructions that are given to the patients, how the testing environment is configured, and how the motion data are reduced. One needs to be concerned with and have an appreciation of the 'quality' of the data as well as, for example, the inter- and intra-rater reliability of the measurements made and the accuracy and precision of the several measurement systems used. An understanding of the trial-to-trial and day-to-day repeatability of the patient performance data is ultimately needed in the interpretation of research findings. For example, it is not enough to agree that the knee flexion–extension angle during gait represents movement of the shank relative to the thigh, or the angle between the long axis of the shank and the thigh. Compatibility must exist in the way in which the hip, knee and ankle centre locations are estimated and how anatomical coordinate systems for the thigh and shank and determined. This is not to say that the marker configuration must be identical between facilities, but the use of the subject calibration data at each laboratory should produce equivalent anatomical coordinate frames and joint centre locations (Figure 1). This is predicated on the assumption that each of the protocols and the precision and accuracy of the measurement systems, while different, yield comparable results. To date, it is not clear how to determine this with any significant quantitative precision. Practically, collaborators are typically selected based on less concrete (although very pertinent and germane) impressions of the 'care' that is exercised in the collection of the data as well as a mutual understanding of how data are collected and processed. The specific data shared can be limited to those quantities in which the investigators are most confident and are needed to examine a limited number of specific research hypotheses.

Unfortunately, this informal and limited approach is insufficient when applied to the creation and use of large, centralized, gait databases. The issue of collection and reduction protocol compatibility can be worked out through the development of a standard approach. It is imperative that indications of the accuracy of the protocol used be provided as part of each patient record. These 'quality assurance' data should reflect not only the capability of specific measurement equipment, but also the entire measurement system (including the abilities of the human testers).

Turning to the issue of what data should be archived in a common database, one assumes that common protocols for clinical measurements such as those listed in Table 2 might be established and these data included. Similarly, EMG signals, collected with a generally accepted electrode placement and hardware configuration, might be included in the database either in a raw form or processed in a fundamental manner, e.g. rectified and interpolated to a standard number of points per gait cycle. A more significant challenge arises when one turns to the gait kinematic and kinetic data. Should one store only certain values associated with the joint and segment dynamics, as proposed by Paul and Morris (1992)? For example, with respect to knee flexion–extension, one might consider including only the following variables: knee angle at initial contact, maximum knee angle in loading response, maximum extension (or minimum flexion) in midstance, knee angle at toe-off, maximum flexion in swing, and when in the gait cycle these maximum and minimum values were achieved. The clear advantage of this approach is that the data set size is kept relatively small; the disadvantage is that values that might be desired in the future will not be available. Cappozzo et al. (1995) suggest that the orientation and location of the anatomically referenced coordinate frames be shared, because this allows laboratories to use different marker configurations and ways of estimating the bone orientations in space. Then the joint and segmental kinematics may be reconstructed. If joint centre locations and patient body mass are added to this minimal data set, then the marker configuration and protocol flexibility is maintained and joint kinetic computation is possible. All of the other intermediate computational values may be computed as well. Data archiving at this level does, however, eliminate the possibility of the application of future improvements in the calculation of the technical frames and the use of the subject calibration data.

The technical considerations in the establishment of some type of centralized gait database are dwarfed by the administrative, social and ethical questions that the existence and use of such a database raise. Before such an effort is undertaken technologically, these fundamental questions should be answered unambiguously. Assume that several collaborators from different institutions propose to build a common gait database as part of a multicentre research project. Assume further that the central database is to be used to test 10 specific hypotheses outlined in their research proposal. The proposal contains properly informed consent and assent forms to be completed by each research subject. Assume further that these consent forms indicate the purpose of the research, the time frame of the project, the sharing of data between institutions, and the confidentiality that will be maintained with respect to the subject's medical record. Issues that should be addressed by the collaborators before proceeding further include:

- In addition to the 10 hypotheses raised in the proposal, can other questions be explored without additional permission from each research subject and each institutional review board? That is, care should be taken to construct the proposal so that the database can be used to investigate relevant clinical questions both during and after the completion of the project.
- Traditionally, the health provider institution (or individual) responsible for the collection of patient's data maintains ownership (and control) of those data. Does the institution then lose ownership of that individual's data once they have been transferred to a centralized database?
- What happens to the database after the completion of the project?
- Does any one institution 'own' the entire database at that point or is it jointly owned by all of the collaborating institutions? A reasonable approach is to provide each participating institution with a complete copy of the database at the end of the project period. This, however, does not entirely address the relevant question of database ownership and control of its future (potentially, commercial) use.
- The collaborators may or may not be able to copyright the common database because federal (USA) regulations can restrict the use of databases produced with US government support. Assuming that they can, is the copyright jointly owned by all of the participating institutions? If only one of the institutions chooses to copyright the database after completion of the project, do the other collaborators then lose access to it?
- Traditionally (and as described here), multicentre studies are structured so that the data are collected at separate sites and then the several data sets are consolidated into one large database. Do other models make more sense, such as the use of distributed database technology to link separately administered (institutional) gait databases (Robinson, 1993)?
- How is subject confidentiality to be maintained? Are data supplied to the centralized database with all subject indentification removed? Alternatively,

is the individual data set to be encrypted with a subject code? Should institutional (and perhaps clinician) anonymity be provided as well?

- What security measures (Shea, 1994; Safran et al., 1995) in the administration of the database are to be employed to adequately maintain the required level of subject confidentiality and control access to the database?

This hypothetical example outlines some of the data-sharing concerns that might arise in relation to a prospective, multicentre research project. Other cases of data sharing involve previously collected gait data that can be used for retrospective clinical research. These basic questions of patient confidentiality and database ownership and access still apply, but broader issues are raised as well. For example, assuming that technical issues such as data compatibility and uniformity of quality are satisfactorily addressed, should this type of gait database be accessible to groups other than health-care professionals involved specifically in gait analysis? Should third-party payers, patient advocacy groups and government agencies also have access to these data for outcomes and health-care economics research? Should this gait database be linked in any way to the larger public health surveillance systems (Thacker and Stroup, 1994) in the USA and abroad?

Invariably, the technical and philosophical aspects of gait data sharing will be established through the implementation of one or more large-scale multicentre research collaborations. These questions and others will be addressed according to the common laws and customs in place and accepted at that time. These issues will be challenging, with data sharing across potentially both domestic (regional) and international borders. It is hoped that the precedents (and perhaps standards) set by these efforts will lead the gait research community down the most appropriate of many pathways.

SUMMARY

Discussions of gait data reporting, storage and sharing are difficult because these terms are broadly defined. The data that are reported for the interpretation of the gait-related problems of patients in the clinical setting can vary in type and format. Different forms of the same information are useful in developing a full appreciation of an individual's gait abnormalities or deciphering a data collection artefact. Variation in reporting conventions and formats between laboratories and peer-reviewed publications impedes the communication of ideas and the expansion of our collective gait knowledge base. Practical considerations of what gait data to store are influenced by changes in computer technology and our concern about longitudinal gait–data model compatibility. Projects that seek to answer complex clinical questions and that require data from many subjects collected at a number of institutions are currently confounded by concerns about data quality and compatibility, as well as confidentiality and control of the use of the data. The ongoing challenge in these areas is to remain thoughtful as to the issues and open for discussion as progress is made toward meaningful solutions.

REFERENCES

Cappozzo, A. (1984) Gait analysis methodology. *Hum. Movem. Sci.*, **3**, 27–50.

Cappozzo, A., Catani, F., Croce, U. D. and Leardini, A. (1995) Position and orientation in space of bones during movement: anatomical frame definition and determination. *Clin. Biomech.*, **10**, 171–178.

Davis, R. B., DeLuca, P. A. and Õunpuu, S. (1995) Analysis of gait. In: *The Biomedical Engineering Handbook* (ed. J. D. Bronzino), CRC Press, Boca Raton, pp. 381–390.

Davis, R. B., Õunpuu, S., Tyburski, D. J. and Gage, J. R. (1991) A gait analysis data collection and reduction technique. *Hum. Move. Sci.*, **10**, 575–587.

DeLuzio, K. J., Wyss, U. P., Li, J. and Costigan, P. A. (1993) A procedure to validate three-dimensional motion assessment systems. *J. Biomech.*, **25**, 753–759.

Paul, J. P. and Morris, J. R. W. (1992) CAMARC II–Data exchange. What? Why? How? In: *Proceedings of the Workshop CAMARC II: Problems and Perspectives*, Rome.

Robinson, D. M. (1993) A legal examination of computerized health information. *Health Law Can.*, **14**, 40–46.

Safran, C., Rind, D., Citroen, M., Bakker, A. R., Slack, W. V. and Bleich, H. L. (1995) Protection of confidentiality in the computer-based patient record. *MD Comput.*, **12**, 187–192.

Shea, S. (1994) Security versus access: trade-offs are only part of the story. *J. Am. Med. Informat. Assoc.*, **1**, 314–315.

Stanhope, S. J. (1994) A procedure for evaluating gait analysis system performance. *Gait Posture*, **2**, 54.

Thacker, S. B. and Stroup, D. F. (1994) Future directions for comprehensive public health surveillance and health information systems in the United States. *Am. J. Epidemiol.*, **140**, 383–397.

Index